Women in Field Biology

Women are contributing to disciplines that were once the sole domain of men. Field biology has been no different. The history of women field biologists, embedded in a history largely made and recorded by men, has never been written. Compilations of biographies have been assembled, but the narrative—their story—has never been told. In part, this is because many expressed their passion for nature as writers, artists, collectors, and educators during eras when women were excluded from the male-centric world of natural history and science. The history of women field biologists is intertwined with men's changing views of female intellect and with increasing educational opportunities available to women. Given the preponderance of today's professional female ecologists, animal behaviorists, systematists, conservation biologists, wildlife biologists, restoration ecologists, and natural historians, it is time to tell this story—the challenges and hardships they faced and still face, and the prominent role they have played and increasingly play in understanding our natural world.

For a broader perspective, we profile selected European women field biologists, but our primary focus is the journey of women field biologists in North America. Each woman highlighted here followed a unique path. For some, personal wealth facilitated their work; some worked alongside their husbands. Many served as invisible assistants to men, receiving little or no recognition. Others were mavericks who carried out pioneering studies and whose published works are still read and valued today. All served as inspiration and proved to the women who would follow that women are as capable as men at studying nature in nature. Their legacy lives on today. The 75 female field biologists interviewed for this book are further testament that women have the intellect, stamina, and passion for fieldwork.

Women in Field Biology
A Journey into Nature

Martha L. Crump and Michael J. Lannoo

CRC Press
Taylor & Francis Group
Boca Raton London New York

CRC Press is an imprint of the
Taylor & Francis Group, an **informa** business

First edition published 2023
by CRC Press
6000 Broken Sound Parkway NW, Suite 300, Boca Raton, FL 33487-2742

and by CRC Press
4 Park Square, Milton Park, Abingdon, Oxon, OX14 4RN

CRC Press is an imprint of Taylor & Francis Group, LLC

© 2023 Taylor & Francis Group, LLC

Library of Congress Cataloging-in-Publication Data
Names: Crump, Martha L., author. | Lannoo, Michael J., author.
Title: Journey into nature : the story of women field biologists / Martha
L. Crump and Michael J. Lannoo.
Description: First edition. | Boca Raton, FL : CRC Press, 2023. |
Includes bibliographical references and index.
Identifiers: LCCN 2022007519 (print) | LCCN 2022007520 (ebook) |
ISBN 9781032318172 (hbk) | ISBN 9780367820350 (pbk) |
ISBN 9781003311508 (ebk)
Subjects: LCSH: Women biologists—History. | Biology—Fieldwork—History.
Classification: LCC QH305.5 .C78 2023 (print) | LCC QH305.5 (ebook) |
DDC 570.82—dc23/eng/20220603
LC record available at https://lccn.loc.gov/2022007519
LC ebook record available at https://lccn.loc.gov/2022007520

ISBN: 9781032318172 (hbk)
ISBN: 9780367820350 (pbk)
ISBN: 9781003311508 (ebk)

DOI: 10.1201/9781003311508

Typeset in Times
by codeMantra

Contents

SECTION 1 *Historical Perspective*

SECTION 2   *Current Perspectives*

SECTION 3   *Looking Toward the Future*

Preamble

A little over 150 years ago—in 1869—Myra Bradwell was the first woman to pass the Illinois bar exam. Her husband, a lawyer, taught her law because at that time no law school accepted women. The following year, she was denied admission to the Illinois bar, despite the fact that she had passed the bar exam with high honors. The Illinois Supreme Court responded that when the Illinois General Assembly gave the court the power to grant law licenses "it was not with the slightest expectation that this privilege would be extended to women."[1] Bradwell appealed to the U.S. Supreme Court, which also denied her admission to the Illinois bar. Justice Joseph P. Bradley wrote (and others concurred): "The paramount destiny and mission of women are to fulfill the noble and benign offices of wife and mother …."[2]

One hundred years ago, physician and toxicologist Dr Alice Hamilton became Harvard University's first female professor, hired as Assistant Professor in the Medical School in 1919.[3] Her position came with written limitations. She could not set foot in the Faculty Club. She could not participate in academic processions at commencement. And she was not eligible for faculty tickets at football games. If these were the written conditions, we can be sure she was not treated as an intellectual equal to the male faculty. She was never tenured and retired as an Assistant Professor in 1935.

Fifty years ago, oceanographer Elizabeth Venrick was not permitted to continue on her scientific cruise across the subarctic because the journey would be "too rough for a woman" (Chapter 6). Cetacean biologist Sue Moore was not allowed to join the crew on a fishing vessel because women "bring bad luck" (due to menstruation) to fishermen (Chapter 8). One of us (MLC) was not permitted to survey amphibians and reptiles in the Yasuní region of eastern Ecuador because it was "too dangerous" for a woman (Chapter 8). Ornithologist Mercedes Foster, ichthyologist Lynne Parenti, and others found it a challenge to be included in international fieldwork 50 years ago—women were considered a distraction (Chapter 8).

Fortunately, times have changed. We've made profound progress over the past 200 years in accepting women as intellectual equals to men, but we're not yet where we should be. From Myra Bradwell's perspective, the good news is that every year from 2016 to 2020, women have outnumbered men in law school. The bad news is that in 2020, only 37% of U.S. lawyers were women, and they make up only 24% of large law firm partners.[4] Not surprisingly, women lawyers generally are paid less than men. Women now make up about 50% of Assistant Professors in the United States. But the bad news is that the percentage decreases to 45% at the Associate Professor level and

---

[1] https://wwwnytimes.com/2021/09/09/opinion/abortion-supreme-court-religion.html; accessed 12 September 2021.
[2] https://www.supremecourt.gov/visiting/exhibitions/LadyLawyers/section1.aspx; accessed 12 September 2021.
[3] https://faculty.harvard.edu/dr-alice-hamilton; accessed 10 July 2019.
[4] https://msmagazine.com/2021/06/03/women-lawyers-stop-attrition-workers-quit-law; accessed 12 September 2021.

36% at the Full Professor level.[5] While 48% of academic fathers achieve tenure, only 27% of academic mothers do so. Male full professors at research-intensive schools earn an average of $10,000 more per year than their female Full Professor colleagues.[6]

Field biology follows the pattern of these other intellectual endeavors, in that women fought to be accepted in what was largely a "man's world." For centuries, women's place has been assumed to be in the home. The history of women field biologists, embedded in a history largely made and recorded by men, has never been written. Compilations of biographies have been assembled, encyclopedia-like,[7] but the narrative—their story—has never been told. In part, this is because many expressed their passion for nature as writers, artists, collectors, and educators during eras when women were excluded from the male-centric world of natural history and science. Given the preponderance of today's professional female ecologists, animal behaviorists, systematists, conservation biologists, wildlife biologists, restoration ecologists, and natural historians, it is time to tell this story.

Field biology, simply defined, is studying nature in nature. Such study ranges from a backyard tea-time break to observe hummingbirds slurping nectar from trumpet flowers (about the extent of what it was thought women could/should be doing two centuries ago) to mammalogist George Schaller's grand Karakorum and Himalayan expeditions to investigate the behaviors of snow leopards[8] (the sort of activity women today are doing, and doing as well as men).

Field biologists tend to be a special type of person. Biologist and naturalist E. O. Wilson observed they have a lot more "gee whiz" or "sense of wonder" than other kinds of scientists.[9] Field biology reflects the lifestyle currently being sold by outdoor clothing companies. In contrast to the images in these slick advertisements, however, field biology is mostly a lot of hard work. After physically demanding days of collecting data, field biologists stay up late skinning birds, pinning insects, or writing meticulous field notes in the dim light of a headlamp or maybe a full moon. Such hard work was once assumed, even by many women themselves, to be too strenuous for females to undertake. Field biology involves self-sacrifice. There is a cost—travel, sleep, comfort, health, perhaps relationships and motherhood. For these committed biologists, however, fieldwork is both the means and the end to a life well-lived and thoroughly tested.

The history of Western women field biologists is intertwined with men's changing views of female intellect and with increasing educational opportunities available to women. For a broader perspective, we profile selected European women field biologists, but our primary focus is the journey of women field biologists in North America. On both continents, before women were permitted to receive a formal higher education, they gained knowledge of natural history from their fathers, husbands, or

---

[5] https://www.chronicle.com/article/why-we-need-more-women-full-professors; accessed 12 September 2021.

[6] https://www.chronicle.com/article/why-we-need-more-women-full-professors; accessed 12 September 2021.

[7] For example, Bonta (1991), Gates (1998), Gates and Shteir (1997), Norwood (1993), Ogilvie (1986), Rossiter (1982), and Schiebinger (1989).

[8] Schaller (1980), Lannoo (2018).

[9] Royte (2008).

other men, or they taught themselves. Some followed gender-role expectations and expressed their passion for nature as writers, artists, and educators. Others defied convention and explored wilderness areas and collected specimens. Once women could enroll in institutions of higher learning, they could take biology courses. And once they could do that, they could become biologists.

Think of the individual stories of the women highlighted here as threads in a woven tapestry. Each woman followed a unique path. For some, personal wealth facilitated their work. Some worked alongside their husbands. Some neither married nor had children and pursued their passions unencumbered by family obligations. Many served as invisible assistants to men, receiving little or no recognition. Others were mavericks who carried out pioneering studies and whose published works are still read and valued today. The individual threads represented by these women's stories form the overall story of women in field biology that we tell here. Each woman served as inspiration and proved to the women who would follow that women are as capable as men at studying nature in nature. The 75 female field biologists interviewed for this book are further testament that women have the intellect, stamina, and passion for fieldwork.

REFERENCES

Bonta, M. M. 1991. *Women in the field: America's pioneering women naturalists.* College Station: Texas A&M University Press.

Gates, B. T. 1998. *Kindred nature: Victorian and Edwardian women embrace the living world.* Chicago: University of Chicago Press.

Gates, B. T., and A. B. Shteir, eds. 1997. *Natural eloquence: women reinscribe science.* Madison, WI: University of Wisconsin Press.

Lannoo, M. J. 2018. *This land is your land: the story of field biology in America.* Chicago: University of Chicago Press.

Norwood, V. 1993. *Made from this Earth: American women and nature.* Raleigh: University of North Carolina Press.

Ogilvie, M. B. 1986. *Women in science: antiquity through the nineteenth century. A biographical dictionary with annotated bibliography.* Cambridge, MA: The MIT Press.

Rossiter, M. W. 1982. *Women scientists in America: struggles and strategies to 1940.* Baltimore, MD: John Hopkins University Press.

Royte, E. 2008. Night moves. *New York Times Book Review*, June 22.

Schaller, G. B. 1980. *Stones of silence: journeys in the Himalaya.* Budapest: André Deutsch.

Schiebinger, L. 1989. *The mind has no sex? Women in the origins of modern science.* Cambridge, MA: Harvard University Press.

Acknowledgments

We thank Chuck Crumly for his enthusiasm and support for a book on the untold history of women in field biology. MLC gratefully acknowledges support of a book grant from the Alfred P. Sloan Foundation during research and manuscript preparation.

MLC thanks Judy Brodie, Karen McKree, and Al Savitzky for advice and perspective as this project came together; MJL thanks Bill Souder, Genevieve Arlie, Scott Canfield, Américo Iglesias-Lopez, Mike Urban, Pete Lannoo, and Sue Lannoo for the same, and thanks Bridget Olson for her perspectives on her career accomplishments and challenges.

Many individuals provided input concerning which women to highlight in our story, especially for women to interview for Chapters 7 and 8. For numerous suggestions, MLC gives special thanks to Dee Boersma, Jane Brockmann, Robert Espinoza, Lee Fitzgerald, Karen Lips, Lucinda McDade, Erin Muths, Al Savitzky, Jessica Ware, and Sidney Woodruff. MJL thanks Paul Dayton, Mark Edlund, and Caitlin Nagorka for suggestions. We are deeply indebted to all the women we interviewed. Their individual stories of pathways and perspectives help to tell the story of women in field biology today. The interviews were carried out under IRB (International Review Board) protocols: MCL = #11377; MJL = #2008249105.

We thank John Anderson, Caitlin Eschmann, Bethany Krebs, Karen McKree, Al Savitzky, and Rochelle Stiles for reviews of part or all of the manuscript. We thank Sue Lannoo for proofreading the submitted draft.

We dedicate this effort to the memories of Monique Halloy, Mary Henry, Rochelle Renken, Meg Stewart, and Cathy Toft, first-class women field biologists and long-time friends and colleagues who died way too young.

Authors

Martha L. Crump is Adjunct Professor in the Biology Departments at Utah State University and Northern Arizona University. She has extensive field experience working with amphibians in Latin America, with fieldwork concentrated mainly in Costa Rica, Brazil, Ecuador, Argentina, and Chile. Crump received the Distinguished Herpetologist Award from The Herpetologists' League (1997) and the Henry S. Fitch Award for Excellence in Herpetology from the American Society of Ichthyologists and Herpetologists (2020). She is Past-President of the Society for the Study of Amphibians and Reptiles.

Crump is the author of over 70 scientific papers, one of six authors of *Herpetology* (college-level textbook), coauthor of *Extinction in Our Times: Global Amphibian Decline*, and author of five popular scientific books, most recently *A Year with Nature: An Almanac* (University of Chicago Press, 2018). *In Search of the Golden Frog* (University of Chicago Press, 2000) is Crump's travel/adventure/memoir story of fieldwork in Central and South America. She is also author of five children's books, including the award-winning *The Mystery of Darwin's Frog*.

Michael J. Lannoo is Professor of Anatomy and Cell Biology at Indiana University and an affiliate of the Illinois Natural History Survey at the University of Illinois and Purdue University. He has considerable tropical and polar field experience in addition to his primary research emphasis on temperate systems. In 2001, Lannoo received the Parker/Gentry Award for Excellence and Innovation in Conservation Biology through The Field Museum of Natural History, Chicago, IL. This award honors "an outstanding individual, team or organization whose efforts are distinctive and courageous and have had a significant impact on preserving the world's natural heritage, and whose actions and approaches can serve as a model to others" (see http://www.parkergentry.fieldmuseum.org/2001).

Lannoo is the author/editor of over 100 scientific papers and 7 popular scientific books, including *Leopold's Shack and Ricketts's Lab: The Emergence of Environmentalism* and most recently *This Land is Your Land: The Story of Field Biology in America* (University of Chicago Press, 2018).

Section 1

Historical Perspective

1 Introduction

The ancient Greek philosopher Aristotle (384–322 BCE) played a key role in developing a way of understanding the natural world by focusing on knowledge gained mainly through direct observation rather than belief.[1] He largely wrote about what he observed rather than what he was told. His *Historia Animalium*, the first systematic and comprehensive study of animals, became the primary source of zoological knowledge for 2000 years. Aristotle's sometimes keen insights into nature were, however, offset by his misguided view of women. He wrote that women are immature, imperfect, and deficient, and he asserted that women's place is in the home, controlled by their husbands. While he conceded that both men and women could be courageous, he wrote that the quality of their courage was different. Men's courage is "in commanding." Women's courage is "in obeying."[2]

The history of women in science reflects Aristotle's views. For centuries, women were deemed intellectually inferior to men and incapable of both understanding and doing science. While the story of women's struggles in science is relatively well known,[3] the history of women in one of the most physically challenging and male-dominated areas of science—field biology—has not been well documented. Much has changed for the better over the centuries. When Ellen Swallow, an environmental chemist, was an undergraduate at Vassar College in the late 1860s, she wrote that her chaperone, Miss Lyman, was shocked to see ladies in Canada working out of doors.[4] Today, though, most of us can conjure the image of a young Jane Goodall interacting with chimpanzees at her Gombe Stream study site (Figure 1.1). Imagine this scene being repeated tens of thousands of times, with women examining geologic formations, plants, insects, fishes, amphibians, reptiles, game and nongame birds, mammals, and Indigenous cultures, in streams, lakes, prairies, wetlands, forests, deserts, and tropical islands, and a more general picture emerges. It took humanity only 200,000 years of social evolution for this vision to become close to a reality.

ORIGINS: EUROPE

During the Early Middle Ages of Western Europe, from the collapse of Roman civilization in the fifth century to the early tenth century, women, especially married and noblewomen, enjoyed some degree of respect due to the chivalric system and the cult of the Virgin Mary.[5] Educational opportunities for women were mostly centered in monasteries and convents, where women could control their own destinies rather

---

[1] Anderson (2013).
[2] Ogilvie (1986).
[3] Bonta (1991), Gates (1998), Gates and Shteir (1997), Norwood (1993), Ogilvie (1986), Rossiter (1982), Schiebinger (1989).
[4] Hunt (1958), p. 32.
[5] Ogilvie (1986), p. 9.

DOI: 10.1201/9781003311508-2

FIGURE 1.1 Jane Goodall interacting with a young chimpanzee at her Gombe Stream study site in Tanzania. (Photo used with permission of the Jane Goodall Institute. © the Jane Goodall Institute/By Hugo van Lawick.)

than live as the legal property of their fathers or husbands. These early institutions also served as health clinics.[6] Women tended medicinal gardens and administered to the sick herbs, purges, bloodletting, and other therapies passed down through oral tradition.

The High Middle Ages, lasting from the tenth to the thirteenth century, were generally more peaceful and secure times that allowed people—mostly men—to focus on new ideas. Women were not permitted to attend classes at the newly established universities. Thus, independent and intellectually curious women during this era continued to enter convents, where they could exercise their creativity and become educated.[7] One of these was Hildegard von Bingen (1098–1179 CE), a natural historian who established St. Rupert's, a Benedictine abbey on the Rhine River, in Germany.[8] von Bingen is considered by many the foremost natural historian of her time; her reputation was such that she advised bishops, popes, and kings.[9] She wrote at least 14 books on religious philosophy, medicine, and natural science. Her greatest work, *Physica,* continues to be a practical handbook detailing the curative powers of nearly 1,000 plants and animals. von Bingen's natural history writings were based primarily on her own observations. She was a woman ahead of her times.

The Late Middle Ages, from about 1250 to 1450, initially brought famines, the plague, wars, and social unrest. By the middle of the fifteenth century, not only had the human condition improved, but also the printing press had been invented. Mass-produced books provided access to written information that earlier had been accessible only to men through academic associations. Noble and wealthy women were permitted to attend university classes in Italy but generally not elsewhere in Western

[6] Minkowski (1992), p. 289.
[7] Ogilvie (1986), p. 9.
[8] Lipscomb (1995), p. 324.
[9] Halsall (2018).

Europe, where the presence of women was thought to "disrupt serious intellectual endeavor."[10]

Toward the end of the Late Middle Ages, male scholars and artists concentrated less on religious thinking and more on understanding the nature of people, the world, and people's place in the world. This new outlook favored critical thinking over superstition and emphasized individual rights. A few intellectuals even dared to ask whether these individual rights should apply to women and whether perhaps women should have access to better education and more control over their own lives. Most of society, however, did not accept these ideas, and women typically adhered to some subset of acceptable roles.

The European view of nature changed after the "discovery" of the New World began ongoing contact between Europe and these continents. By the sixteenth century, a new paradigm had developed: plants and animals could be studied for their own sake, not as an appendage of medicine. It became acceptable and exciting to reveal nature's secrets. Universities established botanical gardens, and private gardens became popular. A publishing explosion yielded herbal and other types of natural history books, including the *Aberdeen Bestiary* (1542) and Conrad Gessner's five-volume *Historiae animalium* (1551–1558 and 1587), today considered the beginning of "modern" zoology. Collecting natural objects became fashionable as a social activity and intellectual pursuit.[11] The European elite supported networks of collectors who provided them with unusual, rare, and exotic objects. These private collections, ranging from fine art to natural wonders, were called "cabinets of curiosities." The cabinets often were entire rooms, essentially miniature museums. Margaret Cavendish Bentinck, Duchess of Portland (1715–1785), was a renowned collector of natural objects. As a child, Bentinck collected seashells and other treasures. Her collecting instinct continued throughout her life, but she also funded natural history expeditions to bring back exotic curiosities.

The Renaissance, with its emergent philosophies, discoveries, and writings, further changed how humans viewed nature. In 1660, the Royal Society was founded in London. Its motto was *Nullius in verba*—"Take nobody's word for it."[12] Five years later, the society established the *Philosophical Transactions of the Royal Society*, the world's oldest continuously published scientific journal. The period saw the emergence of modern inquiry based on the scientific method of observation, experimentation, empiricism, and inductive reasoning. It became fashionable for women to stay informed about scientific discoveries, though generally at a superficial level. Some independent women played a professional role in science, but because they were excluded from male organizations and activities, their achievements often went unrecognized. Men continued to assert that women were intellectually inferior to them.

During the Age of Reason/Age of Enlightenment, from the late seventeenth century through the eighteenth century, women were still perceived as a distraction to male scholarly pursuits. In fact, they were deemed so dangerous to academic life

---

[10] Schiebinger (1989), p. 151.
[11] Simmons (2016), p. 59.
[12] Anderson (2013), p. 52.

that until the late nineteenth century, faculty at the University of Oxford and the University of Cambridge were required to be celibate.[13] Paris was the exception. During the seventeenth and eighteenth centuries, well-educated and socially prominent women offered their private sitting rooms for informal gatherings called *salons*, where men and women discussed science. Parisian women gained access to current scientific knowledge through hosting *salons*, although a major purpose of the gatherings was to identify and support talented young men.[14]

Enlightenment philosopher Jean Jacques Rousseau (1712–1778) strongly influenced intellectual thought during the eighteenth century. He wrote of women's duties:

> To please, to be useful to us, to make us love and esteem them, to educate us when young, and take care of us when grown up, to advise, to console us, to render our lives easy and agreeable; these are the duties of women at all times, and what they should be taught in their infancy.[15]

Rousseau emphasized the parallels between physical and mental strength and believed that "participation in science required a certain strength that women simply lack."[16]

Rousseau, however, believed that botany was the exception because its study was neither "complicated" nor "difficult;" it required only patience.[17] He wrote:

> I am convinced that at all times of life, the study of nature lessens the taste for frivolous amusements, prevents the tumult of the passions, and provides the mind with profitable nourishment by filling it with an object worthy of contemplation.[18]

In essence, nature study would keep women out of trouble. Botany was considered acceptable for women also because it was associated with herbal healing and gardening, respectable hobbies for women. There was a caveat, however: women should not exceed the level of amateur in their botanical studies. Not all women agreed.

ORIGINS: NORTH AMERICA

Before women could become scientists, they needed access to education. As will be discussed in Chapter 2, it took a long time for society to believe that women should be educated. By the late nineteenth century, women professionals in the eastern United States tended to become school teachers, while women who had trained west of the Appalachians found opportunities to become researchers. With their expanded curricula, which included the new science of ecology, Land Grant colleges, especially in the Midwest but also at Cornell (the only Ivy Land Grant), created opportunities for women field biologists, as did the University of Chicago and the California Academy of Sciences.

---

[13] Schiebinger (1989), p. 151.
[14] Schiebinger (1989), pp. 31–32.
[15] Ogilvie (1986), p. 13.
[16] Schiebinger (1989), p. 236.
[17] Schiebinger (1989), p. 243.
[18] Schiebinger (1989), p. 242.

Ecology forms the framework of much of field biology, so it is informative to examine the emergence of women in this field. In her treatise on the history of women ecologists, Langenheim[19] lists five women who received degrees in ecology between 1902 and 1916. Three had trained in the Midwest and two in the East. During the period bookmarked by the First and the Second World Wars, American midwestern and western institutions continued to graduate many women field biologists. Of the nine women Langenheim lists as receiving degrees in ecology between 1917 and 1945, six had trained in the Midwest, two had trained along the East Coast, and one in England. After the Second World War, educational opportunities for women expanded, and traditional curricula in the East began to include disciplines incorporating field data. Harvard, Yale, Duke, Florida, Colorado, and a number of other schools hosted significant ecological programs and graduated their first women PhDs, while the midwestern and western schools continued to contribute. Between 1949 and 1971, Langenheim lists 28 women who received degrees in ecology—13 from schools west of the Appalachians, 14 from eastern schools, and 1 in England.

After the Civil Rights Acts were passed in the 1960s, many of the problems faced by women were finally acknowledged, and women were afforded more opportunities. Of the 13 women Langenheim lists as receiving degrees between 1972 and 1975, 10 came from west of the Appalachians and 3 from Ivy League schools. The percentage of prominent women ecologists who were married and had children increased in the decade after the Second World War. As women became more accepted as ecologists, they found they could be both scientists and mothers.

While the overall trend is for an ever-increasing acceptance of women as professional equals, many of the issues women historically faced in field biology remain today. Among the problems that persist are microaggressions, sexual harassments, glass ceilings, and unequal pay.[20] These problems vary by institutional type and institution. Academic institutions and some NGOs are among the best at providing equal rights and opportunities; federal agencies tend to be middle of the road; and state agencies are variable but when bad, can be among the worst. In certain institutional types and regions, the old boy network remains strong and embarrassingly resilient to change.

REFERENCES

Anderson, J. G. T. 2013. *Deep things out of darkness: a history of natural history.* Berkeley, CA: University of California Press.

Bonta, M. M. 1991. *Women in the field: America's pioneering women naturalists.* College Station, TX: Texas A&M University Press.

Gates, B. T. 1998. *Kindred Nature: Victorian and Edwardian women embrace the living world.* Chicago: University of Chicago Press.

Gates, B. T., and A. B. Shteir, eds. 1997. *Natural eloquence: women reinscribe science.* Madison: University of Wisconsin Press.

---

[19] Langenheim (1996).
[20] Nicholson et al. (2008), Jones and Solomon (2019).

Halsall, P., ed. The life and works of Hildegard von Bingen (1098–1179). Internet History Sourcebooks Project. Fordham University. https://sourcebooks.fordham.edu/med/hildegarde.asp; accessed 22 January 2018.

Hunt, C. L. 1958. *The life of Ellen Richards, 1842–1911*. Anniversary Edition. Washington, D.C.: American Home Economics Association.

Jones, M. S., and J. Solomon. 2019. Challenges and supports for women conservation leaders. *Conservation Science and Practice*; doi: 10.1111/csp2.36; accessed 17 November 2021.

Langenheim, J. H. 1996. Early history and progress of women ecologists: emphasis upon research contributions. *Annual Review of Ecology and Systematics* 27:1–53.

Lipscomb, D. 1995. Women in systematics. *Annual Review of Ecology & Systematics* 26: 323–341.

Minkowski, W. L. 1992. Women healers of the Middle Ages: selected aspects of their history. *American Journal of Public Health* 82:288–295.

Nicholson, K. L., P. R. Krausman, and J. A. Merkle. 2008. Hypatia and the Leopold standard: women in the wildlife profession 1937–2006. *Wildlife Biology in Practice* 4:57–72.

Norwood, V. 1993. *Made from this Earth: American women and nature*. Raleigh, NC: University of North Carolina Press.

Ogilvie, M. B. 1986. *Women in science: antiquity through the nineteenth century. A biographical dictionary with annotated bibliography*. Cambridge, MA: The MIT Press.

Rossiter, M. W. 1982. *Women scientists in America: struggles and strategies to 1940*. Baltimore: John Hopkins University Press.

Schiebinger, L. 1989. *The mind has no sex? Women in the origins of modern science*. Cambridge, MA: Harvard University Press.

Simmons, J. E. 2016. *Museums: a history*. New York: Rowman & Littlefield.

2 Pre-1880 (Late Age of Discovery)

The Age of Discovery carried forward the thinking of the Age of Enlightenment by emphasizing reasoning and individualism over tradition. It also brought profound changes in the biological sciences and the way we think about nature. In 1833, English polymath William Whewell coined the word science, which replaced the term natural philosophy. Charles Darwin (pushed by Alfred Russel Wallace's similar thinking) published his theory of evolution by natural selection in 1859, and Gregor Mendel proposed his laws of inheritance in 1866. The germ theory of disease was finally accepted, and Louis Pasteur invented pasteurization. But change had not yet come regarding society's perception of women, who were still considered a distraction to male scholarly pursuits. Gradually, secondary education became more available for women, and in some countries women were permitted to attend colleges and universities.

As with most factual histories, the story of women field biologists does not lay out "straight, tense and inevitable ... as if life had been made and not happened."[1] Instead, it proceeded in fits and starts; there were dead ends and times when progress in one discipline raced ahead of advancements in all others. Some disciplines, such as botany, followed Rousseau's opinion and were heavily represented early, while others such as anthropology and archeology had to wait until the twentieth century. In this noise, however, there can be detected enough of a signal to create a framework of understanding. During the Age of Discovery, women interested in field biology were scattered and isolated both geographically and temporally. Eventually, they would become collegial in a process that was at first gradual and then accelerated following the formation of women's schools and colleges. As they were recruited as faculty, women began intellectually influencing women.

EUROPE

Here, we highlight 16 pioneering field biologists representing eight countries. Maria Sibylla Merian, Catharina Helena Dörrien, and Amalie Dietrich were German; Jeanne Baret and Jeanne Villepreux-Power were French; Elisabeth Christina von Linné was Swedish; Marie-Anne Libert was Belgian; Anna Atkins, Elizabeth Andrew Warren, Anna Worsley Russell, Marianne North, Mary Anning, and Eleanor Ann Ormerod were British; Ellen Hutchins was Irish; Elisabetta Fiorini Massanti was Italian; and Olga Fedchenko was Russian.

The scientific interests of these women included plants, invertebrates, and fossils. Recall that Rousseau (1712–1778) felt that women could study botany, because it

---

[1] Maclean (1976), p. 127.

DOI: 10.1201/9781003311508-3

required little physical strength or mental fortitude. Indeed, by the late 1700s, botany had become a popular focus of study for women, particularly those belonging to the higher social classes. Dörrien, Baret, Linné, Atkins, Warren, Hutchins, Massanti, Russell, Fedchenko, Dietrich, and, in a sense, Libert were botanists. Merian and Ormerod were entomologists; Villepreux studied marine invertebrates; and Anning collected fossils. Merian, Dörrien, Atkins, Russell, and North were also illustrators. Merian, Dörrien, and Linné were inspired by their fathers and Baret by her lover. Most of these women worked near home; a few traveled widely, often alone. Merian sailed to Suriname; Dietrich explored Australia; North explored much of the Southern Hemisphere; and Baret circumnavigated the globe.

Maria Sibylla Merian (1647–1717) was born in Frankfurt am Main, Germany.[2] Her father owned a publishing house that printed high-quality books detailing natural history and other subjects. He died when she was three. A year later, her mother married the painter and art dealer Jacob Marrell, who encouraged Merian to paint. Because the painters' guild forbade women to use oils, she learned how to paint with watercolors and to engrave. At a young age, Merian became fascinated with insects. At 13, she began raising silkworms on mulberry leaves and painted, in fine detail, stages of their life cycle. She recorded silkworm metamorphosis at a time when it was thought that insects arose through spontaneous generation from rotting organic matter.

In 1665, 18-year-old Merian married her stepfather's apprentice, Johann Graff. Five years later, the couple and their 2-year-old daughter moved to Nuremberg. There, Merian continued to paint despite a local painting guild that banned participation by women. Germany was in the throes of witch hysteria, with thousands of suspected witches already having been executed. Because many insects were considered poisonous and spawn of the Devil, persons associated with insects were suspected of being witches. Popular belief held that butterflies could be witches in disguise. Undeterred, Merian reared caterpillars in her kitchen and continued her investigations. In 1679, she published the first of her two caterpillar, or *Raupen*, volumes (English translation: *The Wondrous Transformation of Caterpillars and Their Remarkable Diet of Flowers*). The first volume contained 50 copperplates, all engraved by Merian, depicting the lepidopteran life cycle, with caterpillars associated with their host plants. In 1683, she published the second volume of *Raupen*, containing another 50 plates.

Not long afterward, Merian left her husband and moved with her elderly mother and two daughters to an experimental Labadist colony in Holland.[3] At that time, Labadists were traveling overseas to convert Indigenous peoples to Christianity and establish new colonies. Those who visited Suriname brought back specimens of bizarrely shaped and outrageously colored butterflies, moths, and beetles. Merian's interest was piqued, and she began planning an eventual trip to South America.

After her mother died in 1690, Merian and her daughters moved to Amsterdam. The Netherlands provided a far more permissive culture for women than did her native Germany and even accepted women as professional artists. After 8 years of

---

[2] This profile is based in large part on Todd (2007) and Pieters and Winthagen (1999).
[3] Labadists were members of a Protestant religious movement that lasted from the 1660s to the 1730s.

working as a scientific illustrator in Amsterdam, she was financially independent. In 1699, Merian sold her paintings and unnecessary belongings, wrote her will, and boarded a ship for the 2-month voyage to Suriname with her 21-year-old daughter, Dorothea.[4] She financed her travels herself and was the first European explorer to travel to the New World not financed by a patron or government; she could pursue her objectives in her own way, accountable only to herself. Her biographer, Kim Todd, wrote:

> In this environment, her painting shifted, absorbing the techniques she'd learned in Amsterdam, as well as the vibrant colors, the eeriness that got under the skin. … The pictures almost pour over the sides of the page, and show their subjects glimpsed from strange angles: a view from beneath a banana branch, the underground life of a cassava root. Unlike her European plants and caterpillars, centered and balanced, these portraits show asymmetry … Even the chips in Merian's study book expanded—the extravagance of wings resulting in an extravagance of parchment.[5]

The fieldwork was challenging. Many of these tropical butterflies and moths were harder to catch than European species. The rain forest seemed impenetrable with its dense tangle of vines and understory. Many insects were too high in the trees for her to reach. Eventually she used a ladder, but that would carry her only so far toward the canopy. Thus, tracking the life cycle of some species proved impossible. To deal with the staggering abundance of insect life, Merian consulted the knowledge and enlisted help from local Indigenous peoples and enslaved Africans, who became her liaisons with the rainforest world.

Fewer than 2 years into her stay in Suriname, Merian began to feel feverish, weak, and shaky. In the summer of 1701, she packed her notes, sketches, paintings, and specimens of plants and animals, including butterflies, cocoons, beetles, and pickled snakes, turtles, iguanas, geckos, and hummingbirds. She and Dorothea boarded a ship to Amsterdam.

Back home, Merian recovered and began her book. To help make ends meet, she sold some of her preserved specimens to collectors. In 1705, she published *Metamorphosis Insectorum Surinamensium*, her magnum opus, consisting of 60 copperplate engravings in folio depicting changes in color and form during the life cycles of Surinamese butterflies, moths, beetles, bees, and flies as well as engravings of a few spiders, amphibians, and reptiles. The book brought her world-fame among art collectors and naturalists alike. *Metamorphosis* is widely considered to be one of the most beautiful natural history books ever published.[6]

In 1717, Merian died from a stroke. Shortly before her death, Russian Czar Peter the Great, an avid collector and owner of a massive cabinet of curiosity, purchased many of Merian's original drawings. This collection, now belonging to the Academy of Sciences in St. Petersburg, is the best collection of Merian's original drawings in the world.[7] Over the next century, her style became the standard for insect illustration. The next influential book to illustrate life in the New World was Mark Catesby's

---

[4] Lewis-Jones and Herbert (2016).
[5] Todd (2007).
[6] Pieters and Windhagen (1999).
[7] Etheridge (2011).

Natural History of Carolina, Florida and the Bahama Islands, published in 1729–1747. Much about this book echoes *Metamorphosis*, from the size and layout to the ecological composition of his paintings.[8] Merian's influence of portraying segments of biological communities extended to other naturalist painters, including William Bartram, John James Audubon, Prideaux John Selby, and John Gould.[9] Maria Sibylla Merian, a female natural historian of the late seventeenth/early eighteenth century, had shaped a new way of viewing and portraying nature. In 1867, the Artis Library was built in Amsterdam to promote the knowledge of natural history. Merian's name appears on the façade of the building, nestled among 35 famous men of science, including Aristotle, Pliny, and Linnaeus.[10]

Catharina Helena Dörrien (1717–1795) was born into a scholarly family in Hildesheim, Germany. She was educated at home by her father, where she studied geography, history, religion, Latin, and science. Her parents both died before she was 20, and by age 30 she had moved to Dillenburg to work as a private teacher for her friend's children. Her friend's husband, Anton von Erath, encouraged her botanical interests, and she began collecting, studying, and illustrating the flora of the region. Dörrien was one of the first German botanists to use Linnaeus' classification and nomenclature system. She published a much-praised catalog of the plants of Orange-Nassau principality; as a result, she became the most celebrated female German naturalist of her time. Her research led to her election to the Botanical Society of Florence, Berlin Society of Friends of Nature Research, Regensburg Botanical Society, and the Berlin Society.[11] Dörrien challenged the way German women were being educated, with its emphasis on practical household skills. She argued that in addition to maintaining a household, women should receive a well-rounded education, including history and science—an education that would fulfill them intellectually.

Jeanne Baret (1740–1807) was born the daughter of poor farm workers in the Loire valley of France.[12] She was an "herb woman," schooled in the largely oral tradition of medicinal plants. The internationally renowned botanist Philibert Commerson lived near her village. The two met in the early 1760s; she knew the curative power of plants and he wished to learn more. In 1764, 26-year-old Baret and Commerson moved to a fashionable apartment in Paris, near the city's botanical garden the Jardin de Roi. In 1764, Baret gave birth to a son. Commerson wanted nothing to do with the boy; the newborn went to a foster mother and died a few months later.

In 1766, Louis Antoine de Bougainville received permission from King Louis XV to circumnavigate the globe. His expedition was the first to have a professional naturalist on board. Commerson volunteered for the position. Because women were not allowed on ships, Baret could not accompany him, but Commerson could bring an assistant "equally capable of pressing specimens and dressing his master."[13] Commerson and Baret colluded to have her go along as his assistant, disguised as a

---

[8] Etheridge (2011).
[9] Etheridge (2011).
[10] Etheridge and Pieters (2015).
[11] Maroske and May (2018).
[12] Ridley (2010).
[13] Ridley (2010).

FIGURE 2.1   Jean (Jeanne Baret). Art by Bronwyn McIvor, drawn after an imagined portrait of Baret, 1817, artist unknown. (Used with permission of Bronwyn McIvor.)

man (Figure 2.1). Two ships would make the expedition. *La Boudeuse* was a frigate, and the larger and faster vessel. The *Étoile*, a storeship, would carry goods ranging from food to bartering trinkets.

Within a few days of the ships' launch in late 1766, rumors circulated. Commerson's assistant was behaving strangely. The sailors relieved themselves at the "heads." Baret never did. When confronted, Baret claimed to be a eunuch, playing to the fear of eighteenth-century sailors of being captured by pirates and sold into slavery in the Ottoman Empire, to be circumcised or castrated. When the expedition neared the equator, the equatorial virgins stripped naked for the ritualized "Crossing of the Line." Baret endured the immersion fully clothed.

In June 1767, Commerson and Baret collected their first plant specimens, around Rio de Janeiro. Commerson's leg had become ulcerated, so he found a comfortable place to rest while Baret collected. One of her finds was a showy red vine, later named by Commerson in honor of the expedition commander: *Bougainvillea*. Baret thought the scarlet bracts and seeds might cure blood poisoning. She collected boughs, made poultices for Commerson's leg, and saved a handful of seeds to take back to France for cultivation.

Throughout the expedition, Baret worked harder than she imagined a male assistant would work so as not to arouse suspicion. At sea, she dried botanical specimens and inspected them for signs of mold and insect infestation. Nonetheless, the crew became increasingly suspicious. In Papua New Guinea, they decided it was time to determine the assistant's gender. On 11 July 1768, while washing her clothes with the other servants, she was "inspected" at gunpoint, then gang-raped. Commerson feigned surprise to learn that Baret was a woman. Bougainville remained silent about the rape. Two months later, Baret knew she was pregnant.

In early November, the expedition landed at Port Louis on Île de France (now Mauritius), their first French-controlled land in over 22 months at sea and home to the foremost French botanical garden outside Paris. Commerson and Baret stayed in Mauritius to assist with botanical work. Baret was given her own room in the servants' quarters, her first privacy in nearly 2 years. In April 1769, 9 months after the rape, Commerson and Baret temporarily left Port Louis for the settlement of Flacq. There a coffee planter and his wife adopted Baret's baby.

Commerson died in early 1773. His will stipulated that Baret would receive 600 livres and could live in his apartment for 1 year while she organized his natural history specimens to be sent to the Royal Collection. While working as a barmaid in Mauritius, Baret met a soldier. They married, returned to France, and Baret received her inheritance. Later, the Ministry of Marine acknowledged Baret's bravery and exemplary behavior during the Bougainville expedition and awarded her a yearly pension of 200 livres.

The collections of Baret and Commerson, buried in boxes in government storehouses, survived the bloodiest phase of the French Revolution when libraries were burned and museums were razed. A national library and national natural history museum were later established for the enjoyment of *all* French citizens. Today, parts of their collection from the Bougainville expedition are housed in this museum, built on the grounds of the former Jardin du Roi. Jeanne Baret died in obscurity at the age of 67.[14] She had botanized on the Strait of Magellan, in rain forests, and in habitats in between and was the first woman to circumnavigate the globe. She had provided France with valuable natural history specimens, yet she lived at a time when being a female botanist simply wasn't acknowledged. Today she is celebrated as a woman who pursued her passion, undeterred by tragic hardships.

Elisabeth Christina von Linné (1743–1782), the daughter of Carolus Linnaeus, was the first female Swedish botanist. She was likely tutored at home, along with her brothers and her father's students. At age 19, Linné noticed flashes of light emanating from orange nasturtium flowers in the family's garden at twilight.[15] She published a paper with the Royal Swedish Academy of Sciences describing this phenomenon, which is now named after her: the Elizabeth Linnaeus Phenomenon.[16] She died at the age of 39. After her death, Erasmus Darwin (grandfather of Charles Darwin) referred to Linné's paper in one of his publications. Later, the English Romantic poets William Wordsworth and Samuel Taylor Coleridge, both of whom were interested in

[14] Ridley (2010).

[15] Vargues (2013); https://www.nybg.org/blogs/science-talk/2013/12/flashes-in-the-twilight/; accessed 16 March 2018.

[16] Scientists initially believed that "flashing flowers" were an electrical phenomenon or phosphorescence. Today it is thought that quirks in our retinal circuitry cause us to perceive twinkling colors at twilight in the absence of actual plant illumination.

FIGURE 2.2 Marie-Anne Libert; image in the public domain.

botany, read Darwin's paper and alluded to the phenomenon of flashing flowers in their writings.

The plant pathologist Marie-Anne Libert (1782–1865) was born in Malmedy, Belgium. Her father recognized her intellectual potential and facilitated her education. Libert was first educated by Sépulcrine nuns, and then attended a girls' boarding school in Prüm (now part of Germany).[17] After returning home, Libert became fascinated with nature. She collected, identified, and cataloged local plants and taught herself Latin, which she used in describing more than 200 new taxa. Her talents went beyond description. When late season blight first attacked potatoes in northern Europe in the mid-1840s, it was Libert who identified the culprit as a fungus. She described the pathogen and named it *Botrytis vastatrix*. Her identification of this pathogenic fungus contributed to the founding of plant pathology.[18] She later described other pathogens, including the fungus *Fusarium coeruleum* that causes dry rot in potatoes. Libert was elected an associate member of the Linnaean Society of Paris in 1820. Emperor Friedrich-Wilhelm III awarded her a gold medal of merit. In 1862, at the age of 80, she became the first woman to join the Belgian Royal Society of Botany (Figure 2.2).

---

[17] Maroske and May (2018).
[18] Maroske and May (2018).

Anna Atkins (1799–1871), born in Kent, England, was a botanist and one of the earliest female photographers. Growing up, Atkins was surrounded by science, influenced by her zoologist/chemist/mineralogist father who was affiliated with the Royal Society and the British Museum. In her early 20s, Atkins illustrated her father's translation of Lamarck's *Genera of Shells*. Later, she found she could better capture the details of marine algae and other botanical specimens using cyanotypes.[19] Using this process, she photographed seaweeds seemingly floating over blue backgrounds, transforming algae into art. Atkins published the world's first natural history book with photographic images, *Photographs of British Algae: Cyanotype Impressions* (1844).[20] The book contains hundreds of algal species native to Great Britain. Later she collaborated with her close friend Anne Dixon on two additional books using cyanotypes. Atkins gave her specimens to the British Museum in London. She was elected a member of the Botanical Society of London in 1839 (Figure 2.3).[21]

Ireland's first female botanist, Ellen Hutchins (1785–1815), was born in Ballylickey, County Cork. She made detailed watercolor paintings and meticulously prepared specimens, recording over 400 species of vascular plants, about 200 species of algae, 200 liverworts and mosses, and 200 lichens. Although as an illustrator she never published under her own name, she was admired by leading botanists as a major contributor to the field.[22] Her specimens, many of which are now housed in the Natural History Museum in London, are still highly valued. More than 200 of her drawings reside in the archives of the Royal Botanic Gardens, Kew. Other specimens and drawings are housed at Trinity College, Dublin; the Linnaean Society, London; and the New York Botanical Garden. She died just before her 30th birthday. Each year Ireland celebrates Ellen Hutchins Festival in Bantry Bay, West Cork. The event features botanical workshops, exhibitions of botanical art, nature walks, and talks as part of National Heritage Week.

The year after Ellen Hutchins was born in Ireland, Elizabeth Andrew Warren (1786–1864), botanist and marine algologist, was born in Truro, Cornwall. She was fascinated by the seaweeds and other plants she found while exploring the shoreline and spent most of her life in the village of Flushing, collecting along the southern coast of Cornwall. British women at that time had no opportunity for higher education. To make up for that, she corresponded with other botanists and worked closely with several scientific societies, including the Royal Horticultural Society of Cornwall. The Royal Horticultural Society of Cornwall sponsored annual competitions for the best and rarest botanical specimens. Warren won most of the prizes. When she was put in charge of organizing efforts to record and collect indigenous Cornish plants, she contributed nearly 75% of the specimens. Warren corresponded with fellow botanist William Hooker, learned much from him, and sent him numerous specimens for Kew Gardens over the years. The Royal Cornwall Museum in

---

[19] Cyanotyping was an early printing process that uses chemicals and sunlight to produce cyan-blue prints.

[20] Gates (1998).

[21] Cyanotypes of British Algae by Anna Atkins (1843). The Public Domain Review; https://public-domainreview.org/collections/cyanotypes-of-british-algae-by-anna-atkins-1843; accessed 17 March 2018.

[22] www.ellenhutchins.com/ellen-hutchins; accessed 15 March 2018.

FIGURE 2.3   Anna Atkins; image in the public domain.

Truro houses Warren's plant collections. Museum visitors can see her specimens, most of which retain their pigment and are in excellent condition today.

Countess Elisabetta Fiorini Massanti (1799–1879) was born into an aristocratic family in Terracina, in the Papal States (territories in the Italian Peninsula under the rule of the pope from 756 to 1870). The family moved to Rome, where, under her father's supervision, she studied literature, music, history, geography, Latin, English, French, and German.[23] She studied botany with several naturalists. Massanti published her first work at age 24—descriptions of 30 species of angiosperms, including their locations and phenology. Her most significant contributions dealt with ferns, mosses, fungi, lichens, and algae. Her last work was *Florula del Colosseo* (*Flora of the Colosseum*), a survey of plant diversity at the Colosseum that revealed a substantial loss of species since a survey undertaken two decades earlier. She mourned the loss and wrote that the archeological restoration of the Colosseum was destroying nature's contribution to the ancient amphitheater. Massanti's sentiments regarding the destruction of the Colosseum's flora contrasted with those of many people in the late 1800s who felt that humans had the "right" to conquer nature. Her career made her the most celebrated Italian female botanist of her time. She was a member of five scientific academies and was elected a member of the Pontifical Academy of the New Lincei.

---

[23] Logan (1999).

FIGURE 2.4   Marianne North; image in the public domain.

Described as "perhaps the ablest and most outstanding woman field botanist of her time,"[24] Anna Worsley Russell (1807–1876) was born in Bristol, England. The Worsleys, one of the leading Unitarian families in Bristol, were active in the local intellectual scene. Several members of her family encouraged her interest in natural history, first with insects and later with plants. In 1839, she published *Catalogue of Plants, Found in the Neighbourhood of Newbury*, which included the first record of over 60 species in the Berkshire area. She soon became a member of the Botanical Society of London and was active in specimen exchanges. Well known and respected in Britain, Russell regularly exchanged information with other botanists. Her field-work took her to many areas from southwest England to southern Scotland. She left more than 730 drawings of fungi to the British Museum.

English artist Marianne North (1830–1890) inherited a fortune when her father died. She used her wealth to travel the world and paint rare and exotic plants. Friends, including Charles Darwin, Joseph Hooker, and Francis Galton, supported her passion and wrote letters of introduction for her.[25] She visited Africa, Australia, New Zealand, Borneo, Japan, Java, Ceylon, India, Brazil, the Seychelles, Chile, the United States, and Jamaica. Defying convention, she usually traveled alone. North was ambitious and sought recognition, compensating for what she lacked in formal education with wit and bravado (Figure 2.4).[26]

When Sir Joseph Hooker, Director of Kew Gardens, recommended that North be awarded the Order of the Cross of Malta for her botanical contributions, he discovered the only women eligible were royalty and wives of high-ranking colonial officials.[27] Undeterred, North suggested the idea of a Marianne North Gallery at Kew Gardens and offered to pay for its design and construction. Hooker accepted. The

[24] Creese (2000), p. 32.
[25] Lewis-Jones and Herbert (2016).
[26] Gates (1998).
[27] Gates (1998).

gallery opened in 1882 and is still open; visitors can see 832 of her paintings—most of her life's work—lining the walls. Her paintings stand out because of their vivid colors, scientific accuracy, and natural settings.[28] North's travel writings and auto-biography helped to popularize science. With her painter's eye, she saw what many others did not, and she vividly described in words what she saw.

Olga Alexandrovna Fedchenko (1845–1921), born in Moscow, began collecting minerals, shells, insects, and birds' eggs and assembled an herbarium as a child.[29] At 16 years, she was illustrating natural history, translating biology texts from English and German into Russian, and corresponding with foreign naturalists.[30] At 22, she married naturalist and geologist Alexei Fedchenko. The following year, the Moscow Society for Natural Science invited Alexei to join a team for scientific exploration of recently conquered Russian Turkestan. Fedchenko joined as an unpaid member to study the plants. In preparation, the Fedchenkos studied natural history collections in Italy, France, and Sweden. During the expedition, she collected over 1,500 botanical specimens. After her husband died in a climbing accident in 1873, the Moscow Society for Natural Science asked Fedchenko to continue publishing the couple's natural history studies. Once that was completed, she continued to study plants. She undertook her last fieldwork, in Turkestan, at age 70. She was recognized as the leading authority on the flora from Turkestan and was widely known abroad as well as in Russia.[31] In 1906, Fedchenko became the second female corresponding member of the Russian Academy of Sciences (Figure 2.5).

Botany was a field that attracted many nineteenth-century women naturalists, but it wasn't the only one. Some women seem to have been born collectors and continued this passion throughout their lives. One of them was Mary Anning, fossil collector extraordinaire.

Mary Anning's (1799–1847) career was both serendipitous and—once she had success—stereotypical of nineteenth-century women field biologists. She was born one of ten children (only she and a brother survived to adulthood) in the seaside resort town of Lyme Regis in southwest England.[32] Her father was a carpenter, and whether by disposition or economic necessity, he walked the nearby cliffs hunting ammonites and other fossils that he sold as curios to the well-off who visited the spa village on holiday.[33] Anning inherited her father's adventuresome spirit.[34] She adored him and learned the curio trade. After he died, Anning, then only 11 years, took up his occupation, reasoning that it was the only option to keep the family financially solvent.[35]

---

[28] Gates (1998).
[29] Creese and Creese (2015), p. 71.
[30] Creese and Creese (2015), p. 72.
[31] Creese and Creese (2015), p. 74.
[32] Emling (2009), p. 14; Lyme Regis is situated on the southern shore of England, to the west of large erodible limestone cliffs called Black Ven, a series of depositions that date to the Jurassic Period (200–145 million years ago [mya]), in the Mesozoic Era [The Age of Reptiles]). Black Ven contains, among its many strata, a 100-foot thick, fossil-rich layer near the sea called Blue Lias ("blue flat stone"), which dates from the Lower Jurassic, 200–195 mya. Storms blowing from the southwest produce waves that crash into and erode these cliffs, exposing fossils buried near the ever-eroding surface.
[33] Emling (2009), p. 7.
[34] Emling (2009), pp. 19–27.
[35] Emling (2009), p. 1.

FIGURE 2.5   Olga Fedchenko; image in the public domain.

But tea table curios were not the only fossils buried in the limestone and shale-layered formation known as Blue Lias. The remains of early vertebrates, especially extinct marine forms—some of them very large—were also present. Anning's familiarity with the cliffs, combined with her fearlessness, dexterity in climbing them, and talent for spotting promising surface anomalies,[36] made her uniquely suited to find these fossils.

Anning made her first important scientific find in May 1811, when she was only 12 years old. Over the course of a year, she exhumed the specimen, which had fins and was about 17-feet long.[37] A local benefactor bought the fossil and donated it to the "London Museum."[38] Five years later, Charles Konig, assistant curator at the British Museum, noting that the specimen looked partly like a fish and partly like a lizard, combined the Greek words for these two animals and came up with the name *Ichthyosaurus*.[39] In 1822, the species was fully described and given the specific epithet *platyodon*. In 1821, she discovered another smaller species *Ichthyosaurus vulgaris*.[40]

---

[36] The same keen search image that would serve Mary Leakey so well over a century later.

[37] Emling (2009), p. 34.

[38] Actually, William Bullock's Museum of Natural Curiosities.

[39] Now *Temnodontosaurus*.

[40] Emling (2009), p. 76.

There were other significant finds as well. Following a December storm in 1823, Anning excavated the complete skeleton of a 9-foot-long reptile with an extraordinarily elongated neck, which became known as *Plesiosaurus*. Richard Grenville, the Duke of Buckingham, purchased the skeleton for £110—at the time the highest price ever paid for a fossil.[41] In late 1828, Anning found a fossil fewer than 4-feet long with hollow bones, claws, and wings, scales, numerous vertebrae, and a long tail ending in a diamond shape. The head was missing. It was a pterosaur, the first unearthed outside of Germany. From Anning's specimen, and two others with skulls, the species was designated *Dimorphodon macronyx*.[42] Toward the end of 1829, Anning discovered an ancient cartilaginous ray that appeared ancestral to both modern sharks and rays, designated *Squaloraja polyspondyla*. In 1834, the Swiss anatomist Louis Agassiz visited Lyme Regis and spent time on the cliffs with Anning and her collecting partner Elizabeth Philpot. The women had collected 34 species of fossil fish. Impressed with their acumen, Agassiz later named the fossil fishes *Acrodus anningiae* and *Belenostomus anningiae* for Anning, and *Eugnathus philpotae* for her colleague.[43]

Anning made scientific discoveries beyond fish and large ancient reptiles. In late 1826, in collaboration with Buckland, a geology lecturer at University of Oxford, she began describing and studying coprolites (fossilized dung).[44] That same year, she cracked open an elongated belemnite fossil and discovered a dried-up ink sac. Her partner Elizabeth Philpot added water and reconstituted the ink, which was then used by Lyme Regis artists to create drawings of the sorts of creatures that had produced the ink.[45]

Anning was a life-long learner and kept a journal, which ran to four volumes.[46] She taught herself vertebrate anatomy and practiced scientific illustration. As she escorted scientific luminaries around her collection sites, she discovered her anatomical knowledge on par with theirs.[47] Her financial situation alternated between adequate and insolvent. To fill the gaps, she opened a curio shop, "Anning's Fossil Depot,"[48] to sell invertebrate specimens to tourists. Anning succumbed to breast cancer in 1847. Her gender and social class had prevented her from publishing her own findings; understandably, she resented the fact that men described the fossils she discovered. While she was never a household name during her lifetime, in 2010,

---

[41] Emling (2009), p. 83.
[42] Emling (2009), p. 116; lymeregismuseum.co.uk; accessed 8 July 2020.
[43] Emling (2009), pp. 168–169.
[44] Emling (2009), pp 104–105.
[45] Emling (2009), p. 109.
[46] Only the last one survives.
[47] Emling (2009), p. 59; Emling (2009), p. 83: In 1824, Lady Harriet Silvester, visited Anning and wrote: the extraordinary thing about this young woman is that she has made herself so thoroughly acquainted with [her] science that the moment she finds any bones she knows to what tribe they belong. She fixes the bones on a frame with cement and then makes drawings and has them engraved. ... It is certainly a wonderful instance of divine favor—that this poor, ignorant girl should be so blessed, for by reading and application she has arrived to that degree of knowledge as to the habit of writing and talking with professors and other clever men on the subject, and they all acknowledge that she understands more of the science than anyone else in the kingdom.
[48] Emling (2009), p. 98.

the Royal Society of London included Mary Anning on the list of ten British women most influential in British science history.

Another collector was Amalie Dietrich (1821–1891), born in Saxony, Germany, the daughter of a purse-maker. In 1846, she married Wilhelm Dietrich, a botanist who acquired and sold natural history specimens. Wilhelm taught Dietrich the Latin names for plants and techniques for their preparation. In 1861, she had finally had enough of his dominance and faithlessness; she left him and moved to Hamburg with her 14-year-old daughter Charitas. Dietrich landed a job making botanical specimens. The following year, shipping magnate and natural historian Johan Caesar VI Godeffroy offered her a 10-year contract to collect natural history specimens in Australia for the Godeffroy Museum. She left Charitas with friends and headed off on an adventure.

Fearless, Dietrich often spent months alone in the Australian bush, with only her pet cat and collecting gear.[49] She pressed plants, preserved insects, spiders, and other small animals in spirits, and collected corals, shells, and other treasures, which she shipped back to Hamburg. Dietrich was "tough as an old boot, both physically and emotionally."[50] She was "brilliant, contrary, obstinate and not the easiest person to get along with."[51] She published nothing in her own name but was respected throughout Europe for her stamina, bravery, and collections that provided research material for many scientists. Shortly before her death in 1891, she wished to attend a male-only anthropological conference in Berlin. She begged the doorman to let her listen from the back. He spoke to an official, who recognized her name. Not only was she allowed in, she was escorted through the hall and introduced to the committee.[52]

Some women focused on marine invertebrates. Jeanne Villepreux-Power (1794–1871) was born in the village of Juillac in southern France, the oldest daughter of a shoemaker. At the age of 18, she walked more than 400 km to Paris to find work and became a successful society dressmaker.[53] She met and married James Power, a wealthy English merchant. From 1818 to 1843, they lived in Sicily, where Villepreux-Power taught herself about the island's natural surroundings and collected fossils, minerals, butterflies, and seashells. In 1842, she published *Guida per la Sicilia*, an inventory of the island's natural environment.

Villepreux-Power was especially interested in cephalopods. She had access to live animals, but to study them closely, in 1832, she invented three types of aquaria.[54] One was a glass aquarium in which she could place animals for observation—today's classic aquarium; two other types were submersibles. Villepreux-Power focused her studies on the paper nautilus, *Argonauta argo*, a pelagic octopus. For the next 10 years, she tackled questions like how the nautilus acquired its paper-thin shell, and the function of the membranes of two unsuckered arms. She reared young and found

[49] Sydney (1976); https://trove.nla.gov.au/newspaper/article/51603723; accessed 21 February 2018.
[50] Sydney (1976); https://trove.nla.gov.au/newspaper/article/51603723; accessed 21 February 2018.
[51] Sydney (1976); https://trove.nla.gov.au/newspaper/article/51603723; accessed 21 February 2018.
[52] Ogilvie (1986).
[53] www.malacsoc.org.uk/malacological_bulletin/BULL34/JEANNE.htm; accessed 16 March 2018.
[54] www.malacsoc.org.uk/malacological_bulletin/BULL34/JEANNE.htm; accessed 16 March 2018.

they produce their own shells and that the membranes of the unsuckered arms secrete a substance that repairs broken shells. These had been mysteries since the writings of Aristotle and Pliny the Elder. Villepreux-Power gained prominence throughout Europe for this research.[55]

Villepreux-Power's other passion was preserving nature. She laid the foundations of aquaculture in Sicily by advocating that overfished rivers could be repopulated by raising young fish in cages until they were large enough for reintroduction. Villepreux-Power was the first woman member of the Catania Accademia, a correspondent member of the London Zoological Society, and a member of 16 other scientific academies. In 1997, a large crater on the surface of Venus, discovered by the Magellan probe, was named after her: Villepreux-Power.

By the mid-1800s, natural history societies had become commonplace. With them came increasing interest in entomology and publication of field guides. This was also a time of transition in the study of insects, moving from amateur to more professional natural history. Insect study and collecting became popular with Victorian women, some of whom, such as Eleanor Ann Ormerod, made significant contributions.

Eleanor Ann Ormerod (1828–1901) was born on an 800-acre estate in Gloucestershire, England. As a child, she was fascinated with insects. By age 24, Ormerod was studying entomology and collecting insects. While helping to manage the estate grounds, she recognized the need for increased knowledge about insect pests.[56] In 1868, she responded to a call in the *Gardeners' Chronological and Agricultural Gazette* for the collection of insects both harmful and beneficial to British agriculture and horticulture.[57] Within a decade, she was editing annual reports concerning injurious insects. She became an expert on the botfly, Hessian fly, uses of pesticides, and the ecology of insectivorous birds and their prey. In 1878, she was elected to the Entomological Society of London. A few years later, she was consulting formally for the Royal Agricultural Society of England and informally with agriculturists across the country. In the early 1880s, she wrote *Guide to the Methods of Insect Life* and *Text-Book of Agricultural Entomology*. In 1900, she was the first woman to be awarded an honorary doctor of law degree from the University of Edinburgh (Figure 2.6).

Ormerod's inheritance gave her independence. To establish herself, in the beginning of her career she took no remuneration for her work.[58] Even later, she generally refused payment, believing that doing so kept her work free of external influence. In 1882, after she had lectured on injurious insects, an accomplished botanist publicly praised Ormerod for doing so much without men's help. Ormerod responded that in fact she *had* relied on the generosity and support of men to carry out her work.[59] The journal *Nature* heralded her as "A Lady Entomologist," a reflection of the Victorian era.

[55] www.malacsoc.org.uk/malacological_bulletin/BULL34/JEANNE.htm; accessed 16 March 2018.
[56] Kampmeier (2011).
[57] Clark (1992).
[58] Gates (1998).
[59] Clark (1992).

FIGURE 2.6   Eleanor Ann Ormerod; image in the public domain.

NORTH AMERICA

Early Eurasian migrants to North America and their Indigenous descendants relied on the native flora and fauna for survival, using their surrounding rich biodiversity for medicines, food, clothing, tools, and spiritual and sacred connections to Earth. They knew the flora and fauna intimately. In contrast, the plants and animals were largely unknown to the Europeans who "discovered" the New World in the late fifteenth century. Tales of exotic animals and plants filtered back to Europe, which enticed adventurous naturalists to visit the New World to record and collect these biological wonders, from snakes that rattle to hummingbirds, which were thought to be crosses between bees and birds. Eventually, the New World produced its own natural historians. John Bartram (1699–1777), botanist, horticulturist, explorer, and Pennsylvania Quaker farmer, is considered America's first native-born naturalist.

The early colonial naturalists in the New World first focused on documenting, describing, and cataloging the landscapes and species they encountered. By the mid-1800s, the U.S. government sponsored mapping expeditions to explore the ever western-moving frontier. Scientific discovery became an integral part of many of these mapping expeditions, in the form of intensive biological surveys conducted by naturalists. Three elements—accessibility to abundant flora and fauna, a culture that valued nature and the study of that nature, and institutional rewards for discovery—account for the popularity of natural history surveys during this period.[60] Naturalists collected fossils, plants, and animals and sent them east to the Smithsonian Institution.

By the latter third of the 1800s, many more areas had become accessible by road and railroad, yet were still relatively wild and unsettled by Europeans. The West was a naturalist's paradise. Survey work continued well into the early twentieth

---

[60] Kohler (2006), pp. 15–16.

century, with collections sent to the Smithsonian Institution, to civic museums (e.g., the California Academy of Sciences, San Francisco; the American Museum of Natural History, New York; the Field Museum, Chicago; and the Academy of Natural Sciences, Philadelphia), and to academic institutions (e.g., the Museum of Comparative Zoology, Harvard; the Museum of Vertebrate Zoology, Berkeley; and the University of Michigan Museum of Natural History, Ann Arbor).

Attitudes toward nature were changing. Wilderness became less threatening as more people embraced the new culture of outdoor recreation, including camping, walking and hiking, mountain climbing, sport hunting, fishing, bird-watching, and amateur naturalizing.[61] These outdoor experiences provided a source of intellectual and emotional enrichment and encouraged a view that made scientific interest in nature acceptable.[62] The increased bond with nature also fostered interest in reading nature essays. Often written by women, the classic nature essay was a short account of an author's experience with some aspect of the natural world. Easterners, curious about the unusual creatures being collected in the West, were eager to see the wonders firsthand, which led to animals being stuffed and displayed in museum dioramas or captured alive for display in zoos. Naturalists also collected seeds and plant cuttings, which they sent to botanical gardens in the East and to Europe.[63]

By the late 1800s, vast areas of the country had been cleared for homesteads, agriculture, and ranching. These changes brought increasing concern regarding disappearing species, from passenger pigeons to bison. A new breed of naturalists had been born. The emphasis on natural history studies expanded from collecting and cataloging to promoting conservation and studying life histories and behavior of animals in their natural environments.

Opportunities for men versus women naturalists differed greatly during the early years. Men were employed in survey and collecting work by the government, museums, and the U.S. Biological Survey. Most women naturalists were writers, artists, or educators. There were exceptions, however, for example, women who collected specimens for scientific study and display. The accomplishments of early women naturalists often went unrecognized, either because as women they were not taken seriously or because their contributions were attributed to the men with whom they worked. Many of these women are just now being recognized for their contributions.

To understand the challenges confronted by these women, we need to consider the educational and political climate of their times. Not surprisingly, education in colonial America reflected European traditions. Based on the English model, when New England children were educated, they often began at "dame schools"—groups of children educated by a woman in her own home (basically kindergarten). Girls from wealthy families might be educated by a governess or attend a convent school, but even then, the subjects generally were limited to reading and writing. Middle-class girls were often left out, as usually only boys were educated. Neither boys nor girls from lower-class families received education. In the South, boys were often educated by private tutors; girls were sometimes allowed to sit in on the lessons.

---

[61] Kohler (2006), p. xii.
[62] Kohler (2006), pp. 47, 56.
[63] Moring (2002), p. 3.

Pennsylvania saw innovative experiments in women's education during the 1700s. What would become the Bethlehem Female Seminary, the first boarding school for young girls in America, was founded in Germantown, Pennsylvania, in 1742. Three years later, the school was relocated to Bethlehem, Pennsylvania.[64] Girls were taught reading, writing, arithmetic, geography, history, grammar, music, and German— plus sewing and needlework. The school's philosophy was that when you educate women, you educate families. Arguments for reform in women's education intensified after the Revolutionary War (1775–1783). In 1787, the Young Ladies Academy was founded in Philadelphia, Pennsylvania. Subjects included mathematics, chemistry, and natural science. The school served as a model for many other women's academies and seminaries that opened in the late 1700s and early 1800s. Many of these focused on preparing young women to become teachers. In 1783, Washington College, in Chestertown, Maryland, appointed its first female teachers—the first of any American college or university.

In 1821, Emma Willard established the Troy Female Seminary in Troy, New York—the first endowed American institution to provide a college-level education for women comparable to that available to men. In 1826, the first public high schools for girls were opened, in Boston and New York. Established in 1833, Oberlin Collegiate Institute (renamed Oberlin College in 1850) in Oberlin, Ohio, was the first U.S. coeducational liberal arts college; the first year's class consisted of 15 women and 29 men. Women were admitted to the baccalaureate program in 1837, and that same year African American women were admitted to the college (African American men had been admitted since 1835). In 1836, the Georgia Female College (renamed Wesleyan College in 1917) was founded in Macon, Georgia, and is the world's oldest college chartered to grant degrees to women. The first women's college west of the Rocky Mountains was the Young Ladies Seminary, established in Benicia, California, in 1852. Two years later, Mary Atkins, the first woman to graduate from Oberlin College (in 1845), became principal of the Young Ladies Seminary. In 1855, the University of Iowa became the first U.S. coeducational public or state university. Once the Civil War broke out, so many men were fighting that spaces opened up for women in colleges and universities willing to admit women.

Women demanded more than the right to education. They wanted to be treated as equal to men. In 1769, the colonies had adopted the British law declaring that women could not own property in their own name or keep their own earnings. It wasn't until 1900 that every state had reversed this law, even though the Industrial Revolution (late 1700s to early 1900s) opened up opportunities for women to work outside the home. In 1920, the Nineteenth Amendment was ratified, finally giving women the right to vote—but they still were shut out from most positions that would allow them to effect political change for women. Not until 1963 did Congress pass the Equal Pay Act, assuring (in principle) equitable wages for the same work regardless of color, religion, or gender of the worker. Many early women natural historians were active in suffrage and abolition movements as well as other social issues such as women's educational opportunities.

---

[64] In 1913, it became the Moravian Seminary and College for Women and is now the coeducational Moravian College.

Following European tradition, many eighteenth-century American women chose their natural history focus based on gender-role expectations. Women's assumed role was to nurture. Thus, tending gardens became a popular hobby. Some women wrote about horticulture or local flora, and some collected plants. Nature study, especially focused on plants and birds, became an increasingly popular hobby for women during the nineteenth century. Many of these early natural historians expressed their passion and educated the public about nature as writers, educators, and lecturers; some argued for conservation. Some independent adventurers, ignoring accepted gender codes, made expeditions and collected plants and animals. Others helped to establish botanical gardens, natural history museums, and women's colleges.

Nine women are highlighted here: Martha Daniell Logan, Jane Colden, Almira Hart Lincoln Phelps, Susan Fenimore Cooper, Graceanna Lewis, Elizabeth (Cary) Agassiz, Mary Treat, Martha (Dartt) Maxwell, and Catherine Furbish. They were all born in the East. About half (Logan, Colden, Phelps, and Furbish) studied botany; three (Cooper, Lewis, and Treat) were broad-based natural historians; Agassiz was interested in ichthyology and facilitated science education for women; Maxwell was a taxidermist. Only Maxwell traveled west.

Martha Daniell Logan (1704–1779) was born into a prominent family in St. Thomas Parish, South Carolina. Her father had arrived in South Carolina from Barbados in 1679. He owned 48,000 acres of land, had a nursery business,[65] and became a prominent merchant before becoming Lieutenant Governor of South Carolina. She learned how to read and write, and cultivate plants.[66] The year after her father died, the 14-year-old married George Logan, Jr. The young couple moved to the plantation she inherited from her father, where she became an authority on southern gardening and sold roots, cuttings, and seeds at her nursery. Logan anonymously published a column entitled "Gardener's Kalender" in the *South Carolina Gazette* in which she shared traditional lore as well as factual knowledge.[67]

Logan's writings were the first American works on horticulture. In 1760, John Bartram, considered the father of American botany, visited Logan's garden. He wrote to Peter Collinson, a botanical merchant contact in London: "I received a lovely parcel (of seeds and cuttings) in the Spring from Mistress Logan … Her garden is her delight and she has a fine one."[68] Bartram then sent Logan's seeds and plants to Carolus Linnaeus and to the Kew Gardens in London, continuing the introduction and early cultivation of many New World plants in Europe.[69]

Recognized as America's first female botanist, Jane Colden (1724–1766) was born in New York City. Her father, a Scottish immigrant, was a physician. When she was four, the family moved to a 3,000-acre wilderness estate in Orange County, New York. She was educated at home by her mother. Her father, an amateur botanist, corresponded with many well-known American and European botanists, including

---

[65] www.scencyclopedia.org/sce/entries/logan-martha-daniell/; accessed 21 June 2018.
[66] www.scencyclopedia.org/sce/entries/logan-martha-daniell/; accessed 21 June 2018.
[67] Shearer and Shearer (1996), p. 266.
[68] Shearer and Shearer (1996), p. 265.
[69] Philippon (2001), p. 16.

Linnaeus, and sent them plant specimens. Impressed with his daughter's curiosity with the natural world, the elder Colden taught her Linnaeus's new system of classification.

By 1757, Colden had compiled a catalog of over 300 local plants from the lower Hudson River Valley.[70] Her manuscript included ink impressions of leaves, sketches of living plants, and notes on natural history and medicinal use. Through her father, Colden met and corresponded with leading naturalists.[71] Like Logan, Colden collected and exchanged seeds and plants with other naturalists in America and Europe. At 37, she married a physician and discontinued her botanical work. She died of complications from childbirth 4 years later. While she lived, she received no formal recognition for her botanical accomplishments. Then, during the Revolutionary War (1775–1783), Colden's botanical notes and drawings were discovered and sent to England; they have been housed in the British Museum ever since.[72] In the 1990s, the Jane Colden Native Plant Sanctuary was established in New Windsor, New York.[73]

Without access to higher education, North American women wanting to study nature learned from their fathers or other male mentors, or they taught themselves. Some progressive educators, however, fought to improve higher education for women.

Educator Almira Hart Lincoln Phelps (1793–1884) was born in Berlin, Connecticut, the youngest of 17 children. She was educated at home and encouraged to think independently.[74] At age 16, she taught for a year at the Berlin Academy, the beginning of a long teaching career dedicated to educating young women. The following year, in 1810, she moved to Middlebury, Vermont, to live and study with one of her sisters. Her sister's nephew, who boarded at the home, attended the local all-male Middlebury College. He shared with Phelps his classwork on mathematics and philosophy, and she realized the striking differences between male and female educational opportunities. She spent the rest of her life trying to correct this disparity.

After her husband died in 1823, Phelps taught at the Troy Female Seminary in New York. While there, she became especially interested in botany. Noting a lack of introductory textbooks, in 1829 she wrote and published her first and most famous textbook: *Familiar Lectures on Botany*. Filled with stories, poetry, folklore, practical skills as well as scientific knowledge, the book was an alternative to the standard dry, technical approach used at men's schools. The book was widely used in female seminaries and schools. It inspired women to get out of the parlor and into their gardens, fields, and woodlands to explore their surroundings so that they could impart a love and understanding of the natural world to their children. Over the next 43 years, her textbook underwent numerous printings and sold more than 275,000 copies.[75] Not long after the first printing of her botany book, Phelps published her second book, *Lectures to Young Ladies*, in which she outlined her views that women needed a

---

[70] https://www.womenhistoryblog.com/2020/07/jane-colden-botanist-new-york.html; accessed 3 June 2018.

[71] Ogilvie (1986), p. 61.

[72] Norwood (1993), p. 59.

[73] https://www.womenhistoryblog.com/2020/07/jane-colden-botanist-new-york.html; accessed 3 June 2018.

[74] Almira Phelps. Educator and Author of Science Textbooks. www.womenhistoryblog.com/2013/08/almira-phelps.html; accessed 1 June 2018.

[75] Norwood (1993), p. 19.

well-rounded education, because, as mothers and teachers they would be educating the next generation of men.

Phelps remarried in 1831 and stepped back from teaching to raise her husband's six children and another two. In her spare time, she wrote additional textbooks, including *Dictionary of Chemistry*, *Geology for Beginners*, *Chemistry for Beginners*, *Botany for Beginners*, and *Lectures on Natural Philosophy*. Phelps's textbooks became the standard in schools across the United States and Canada.[76] In 1859, at age 66, Phelps was elected a member of the American Association for the Advancement of Science.

Phelps was a paradox. After the Civil War, she fought for equal educational opportunities for women, yet joined the Women's Anti-Suffrage Association and wrote articles strongly opposing women's right to vote. Perhaps her lasting legacy is that many of her students became teachers who taught science to their female students. These teachers then spread Phelps's ideas from the Northeast into the rest of the country, "carrying with them the seeds of a new way for the educated woman to spend her time—in pursuit of a better knowledge of her natural environment."[77]

Susan Fenimore Cooper (1813–1894), daughter of novelist James Fenimore Cooper,[78] has been praised as "one of the first American environmentalists and the first American woman to write essays on nature."[79] She claimed that her maternal grandfather inspired her love of nature.[80] She spent her teenage years in Europe,[81] returning to her hometown of Cooperstown (founded by her paternal grandfather), New York, when she was 23. For 2 years she recorded her seasonal observations of plants and animals made while roaming the gardens, fields, and woods. Woven throughout her descriptions of the seasonal changes in landscape, plants, and animals, she detailed changes in her own domestic activities, from spring through winter. In her journal, Cooper speculated about bird migration and noted declines in passenger pigeons, fishes, and game birds; she observed that beavers and otters were rare and that she no longer saw deer and bear.

The journal formed the basis for her book *Rural Hours* (1850), which she authored anonymously ("By a Lady"). The book became a bestseller. She lamented that most Americans knew little about their native animals and plants and wrote: "Had works of this kind been as common in America as they are in England, the volume now in the reader's hands would not have been printed."[82] She wrote that native forests should be preserved for pragmatic, esthetic, moral, and religious reasons, arguing that Euro-Americans could no longer exhaust natural resources and move west.[83] Her words came at a time when few worried about environmental degradation. *Rural Hours* was republished ten times between 1850 and 1998.

---

[76] Almira Phelps. Educator and Author of Science Textbooks. www.womenhistoryblog.com/2013/08/almira-phelps.html; accessed 1 June 2018.

[77] Norwood (1993), p. 9.

[78] Whose novels include *The Last of the Mohicans* (1826), *The Prairie* (1827), and *The Deerslayer* (1841).

[79] www.womenhistoryblog.com/2013/12/susan-fenimore-cooper.html; accessed 20 May 2018.

[80] Bonta (1995), p. 2.

[81] www.womenhistoryblog.com/2013/12/susan-fenimore-cooper.html; accessed 20 May 2018.

[82] Norwood (1993), p. 25.

[83] Norwood (1993), p. 34.

In contrast to her progressive thinking on the environment, Cooper remained conservative about women's roles in society. While she supported higher education for women and believed that women should get equal pay for equal work, like Almira Phelps, 20 years her senior, Cooper did not believe that women should have the right to vote. Throughout *Rural Hours*, she promoted the idea that women should use nature as a springboard for moral education and religious meditation[84] and that the close observation of flowers reflected proper female virtues of modesty, constancy, sisterhood as well as nurturing roles and family responsibilities. Her acceptance by society no doubt reflected the fact that she stayed within accepted gender expectations. At the same time, her observational nature writing paved the way for later American women natural history writers.

Graceanna Lewis (1821–1912) was born to Quaker farmers in Chester County, southeastern Pennsylvania. Her father died when she was three, leaving her mother to nurture her growing curiosity in nature. Her underlying interest in nature was rooted in her Quaker beliefs emphasizing that nature study allowed one "better to comprehend its Author."[85] She spent over 50 years teaching, lecturing, and writing about natural history. As with many other nineteenth-century women natural historians, Lewis was a social activist. Before the Civil War, she was involved in abolitionist causes and housed runaway slaves as part of the Underground Railroad to Canada. Later she focused on temperance issues and women's rights and was an activist in the women's suffrage movement. In keeping with her Quaker beliefs, she was a pacifist. After the Civil War, now in her 40s, Lewis turned her natural history interests to birds.[86] In 1868, she published *Natural History of Birds*. Two years later, she was granted membership in the Academy of Natural Sciences in Philadelphia. Lewis supported herself by lecturing, teaching, and writing.[87] She entered her 50 watercolors of Pennsylvania tree leaves in the 1893 Chicago Columbian Exposition and earned a bronze medal. Throughout her life, she remained convinced that "a reasonable acquaintance with the objects of nature may be a fitting preparation for our advancement in eternity."[88] After her death, she was praised as "the only woman in Pennsylvania who has done any original work in natural science."[89]

Elizabeth (Cary) Agassiz (1822–1907) was born in Boston, Massachusetts, as the second of seven children. Her parents' families, both from England, had come to America during the 1600s. As a child, she had frail health and was educated at home. In 1846, she met the Swiss-born natural historian/biologist/geologist Louis Agassiz. The following year, Agassiz joined the faculty of Harvard University, and the couple married in 1850. In 1856, she organized a school for girls in her Cambridge home. Her husband and other Harvard professors taught courses at the school. Louis Agassiz founded the Harvard Museum of Comparative Zoology in 1859 and served as its first director. Elizabeth Agassiz worked closely with her husband, helping with

---

[84] Norwood (1993), p. 30.
[85] Bonta (1991), p. 18.
[86] Norwood (1993), p. 78.
[87] Bonta (1995), p. 10.
[88] Bonta (1991), p. 29.
[89] www.womenhistoryblog.com/2014/12/graceanna-lewis.html; accessed 22 June 2018.

his scientific research. She closed her school in 1863 and then helped to organize and manage the Thayer Expedition (1865–1866), a 15-month study of the distribution of Brazilian freshwater fishes. She accompanied her husband on the expedition and kept a detailed journal as the self-appointed clerk.[90]

A few years later, Agassiz helped to organize the Hassler Expedition (1871–1872), a deep-sea dredging trip along the Atlantic and Pacific coasts of the Americas.[91] Again, she accompanied her husband and recorded detailed notes. In 1873, she helped her husband plan and administer the coeducational Anderson School of Natural History on Penikese Island in Buzzard's Bay, Massachusetts, the first seaside marine laboratory in the country. Her husband died in December, following the opening of the school in July 1873. Although the school lasted only 2 years, it served as a model for other summer schools that taught natural history to teachers.

Agassiz's first natural history book, written under her husband's direction, was *Actaea, a First Lesson in Natural History* (1859). In collaboration with her stepson Alexander, she revised that book as *Seaside Studies in Natural History* (1865), considered a well-written textbook and field guide on marine zoology.[92] In 1868, she coauthored *A Journal in Brazil* with her husband. After he died, she edited and published a two-volume biography of her husband, *Louis Agassiz: His Life and Correspondence* (1885).

In her later years, Agassiz turned her interests to helping other women acquire an education in science. She was one of the seven founders of the Society for the Collegiate Instruction of Women, nicknamed the "Harvard Annex," established in 1879 after an unsuccessful, prolonged struggle to convince Harvard to accept women. The annex was a private program in which women could take college courses unofficially, taught by willing Harvard professors. Courses included Greek, Latin, and other languages, history, philosophy, natural history, physics, among others. In 1894, the Harvard Annex became Radcliffe College. Elizabeth Agassiz served as Radcliffe's first president (Figure 2.7). (Radcliffe College and Harvard University finally merged in 1999, establishing the Radcliffe Institute for Advanced Study at Harvard.)

Mary Treat (1830–1923) was born in Trumansburg, New York. Although nature had always fascinated her, it wasn't until 1868, after she and her husband moved to Vineland, New Jersey, that she focused her attention on pest insects. Vineland was a new community, built with the idea of making it a center of fruit culture. Insect pests created problems, however, and Treat decided to study them. Over the next decade, she published ten articles in *American Entomologist* on agricultural pests, describing their life histories in detail.[93]

Treat also studied and wrote about non-pest insects, birds, wildflowers, and other plants. The Pine Barrens, located just 5 miles east of Vineland, is characterized by sandy, acidic soil low in nutrients, home to carnivorous plants. Treat was fascinated by them and became an expert. She was particularly interested in *how* the plants

---

[90] Ogilvie (1986), p. 24.
[91] Ogilvie (1986), p. 24.
[92] Ogilvie (1986), p. 24.
[93] Bonta (1991), p. 43.

FIGURE 2.7   Elizabeth Cary Agassiz; image in the public domain.

function. One such plant was bladderwort. After many hours of observing bladders (utricles) capturing microscopic animals (zooplankton) under a microscope, she discovered a depression at the entrance of the utricle that serves as a lure. Just beyond there is a valve which, when touched, springs back and entraps the animal.[94]

Treat frequently contacted scientists who could guide her research. One of these was botanist Asa Gray, who put her in touch with leading naturalists, including Charles Darwin and Sir Joseph Hooker, Director of Kew Gardens in London. Knowing that Darwin was interested in carnivorous plants, Treat wrote to him and shared her observations about the capture mechanism in bladderworts. He agreed that she was correct and that he had been wrong about thinking that insects used their heads as wedges to crawl into the plants. He later wrote in his book *Insectivorous Plants*: "Mrs. Treat of New Jersey has been more successful than any other observer in understanding how bladderworts capture insects."[95] Treat later studied sundews, butterworts, Venus flytraps, and pitcher plants.[96]

Treat published over 70 articles in popular and scientific journals, some of which formed the basis for several books (Figure 2.8). In her most popular book, *Home Studies*

---

[94] Bonta (1991), p. 44.
[95] Bonta (1991), p. 43.
[96] Bonta (1991), p. 44.

FIGURE 2.8   Mary Treat; image in the public domain.

in Nature (1880), she reiterated the themes of Susan Fenimore Cooper 30 years earlier—that one need not travel far from home to enjoy the pleasures of nature and that all of nature is a household, each organism playing a cooperative and harmonious role.[97]

The "Colorado Huntress" Martha (Dartt) Maxwell (1831–1881) was born in Dartt's Settlement, Tioga County Pennsylvania, into a family of women who had for generations displayed "prodigious energy, tenacity of purpose, and a fierce independence of judgment."[98] She was drawn to the woods and the animals that lived there by her grandmother.[99] Her father died of scarlet fever, and after her mother remarried, the family moved to Baraboo, Wisconsin, where Maxwell and her family set down roots. One day when she was 13 and her stepsister Mary was a toddler, Maxwell spotted a rattlesnake in the cabin next to Mary. She grabbed her stepfather's gun and killed the snake. As she grew up, she became accomplished with firearms.[100]

Maxwell briefly attended Oberlin College,[101] where she became increasingly interested in natural history. Back in Baraboo,[102] in 1853 she met James Maxwell, widower, prominent businessman, and father of six. The following year, at age 22, she married him. In 1858, their daughter Mabel was born, and 2 years later they headed for the gold fields of Pikes Peak, Colorado.

The couple lost money. Martha Maxwell ran a boarding house and invested the money in land and mining claims. In early 1861, she bought a ranch claim with a one-room cabin on the Platte River. The couple stayed long enough to comply with the settlement requirements. Then, when they left the cabin for a few days, an

---

[97]   Norwood (1993), p. 42.
[98]   Benson (1986), p. 1.
[99]   Bonta (1991), p. 31.
[100]   Bonta (1991), p. 32.
[101]   Benson (1986), p. 16.
[102]   Benson (1986), p. 34.

itinerant German taxidermist jumped their claim. During the subsequent confrontation, Maxwell noticed his mounted birds and mammals and decided "I wish to learn how to preserve birds & other animal curiosities in this country."[103]

Maxwell returned to Baraboo, where Professor Edward F. Hobart of the Baraboo Collegiate Institute had decided to assemble a collection of mounted birds and mammals. Knowing of her interest, he approached Maxwell with a job and she agreed. She modified existing techniques by using the animal's skinned body to make a plaster manikin, which she then covered with its preserved hide. She positioned her specimens on backgrounds replicating the animals' natural environments to promote a more realistic impression,[104] and she kept captive wild animals in her home so she could study their behaviors.

When Maxwell returned to Colorado in 1868, she wanted her taxidermy work to expose people to the beauty and wonder of Colorado wildlife so that they might value and protect it. She camped out by herself in pursuit of game, accompanied by her water spaniel. She wrote during one of these trips,

> Clothes damp, boots hard and stiff, frost a quarter of an inch thick on everything outside the tent, and no hope of warmth and breakfast, till, from under sheltering rocks and logs, enough fuel can be gathered for a fire.[105]

Maxwell's "hunting costume," as she called it, consisted of a medium-length dress over bloomers, jacket, wide-brimmed hat, and long, hobnailed boots.[106] She also called it "a gymnastic suit of neutral tint and firm texture," reminiscent of garments being worn by women in the dress reform movement (Figure 2.9).[107]

Maxwell first displayed her specimens in the fall of 1868, at the Colorado Agricultural Society. It featured the essence of the classic Maxwell presentation—many attractive and well-crafted mounted specimens arranged in a semblance of a natural and lifelike setting.[108] The judges gave her the highest award possible, 50 dollars and a diploma.[109]

Seeking more scientific knowledge to inform her displays, in early 1869 she wrote a letter to the Smithsonian Institution in Washington, D.C., which began a correspondence with Assistant Secretary Spencer Baird. He referred her to the best sources for information and offered her a 20% discount on Smithsonian books. In turn, Baird asked her for help in building up the Smithsonian's collections.[110]

In 1874, Maxwell established the Rocky Mountain Museum on the main street of Boulder, Colorado, to display her collection. She described it as a scientific institution, "a kind of academy of science, perhaps an adjunct to the State university." She hoped that the museum would be both educational and entertaining[111] and be of

---

[103] Benson (1986), p. 70.
[104] Benson (1986), p. 70.
[105] Benson (1986), p. 81.
[106] Moring (2002), p. 171.
[107] Benson (1986), pp. 92–93.
[108] Benson (1986), p. 82.
[109] Benson (1986), p. 82.
[110] Benson (1986), pp. 103–104.
[111] Benson (1986), p. 97.

FIGURE 2.9   Martha Maxwell; image in the public domain.

interest to the young, awakening in them a love for a nature that was wholesome and refining.

The Colorado legislature invited Maxwell to display her collection as part of the Colorado pavilion for the 1876 Centennial in Philadelphia.[112] As always with Maxwell's work, visitors were intrigued by the natural, lifelike displays; most had never seen anything like it.[113] Visitors were eager to learn about her and her adventures, and about the West. Major articles about Maxwell and her work appeared in fashionable magazines.[114]

Following the Centennial, Maxwell, by then in her mid-forties, continued to struggle with her finances.[115] She went to Boston, and in early 1878 she enrolled in classes in Ellen Swallow's Women's Laboratory of the Massachusetts Institute of Technology (MIT). To make ends meet, she gave taxidermy lessons.[116] After battling health issues, she died at age 49.[117] Maxwell was one of many museum workers who died young in those days, perhaps poisoned from arsenic used to prepare the hides.[118]

[112] Benson (1986), p. 115.

[113] Benson (1986), p. 133.

[114] Benson (1986), p. 134.

[115] Benson (1986), p. 150.

[116] Benson (1986), p. 154.

[117] Benson (1986), p. 191.

[118] https://legionofhonor.famsf.org/poisons-part-ii-arsenical-world-taxidermy; accessed 20 June 2020.

Maxwell saw her taxidermy work as "an effort to prove women's capabilities in both art and science."[119] Reflecting this belief, she had placed a sign at the front of her Centennial diorama: "Woman's Work." Today a remnant of her Centennial exhibition remains in the Smithsonian's Arts and Industries Building. It is still labeled "Woman's Work."

Catherine Furbish (1834–1931) was born in Exeter, Maine, and soon after, her family moved to Brunswick, home of Bowdoin College. Her father raised tomatoes[120] and taught her plant names. In large part because of the presence of Bowdoin, Brunswick was a liberal, open-minded town with an active cultural life that Furbish absorbed.[121] During the Civil War, she made bandages for battleground hospitals but wished her health would improve so she might do more[122] (she was weak from a vague illness diagnosed as neuralgia[123]). By the end of the Civil War, when Furbish was 30, she began taking painting lessons and resigned herself to the life of an unmarried woman.[124]

The spring of 1870 was abnormally mild, and Furbish spent a great deal of time outdoors. She began drawing and painting flowering plants using watercolors[125]; she rendered the conifers in black on white, which she hoped would resemble engravings.[126] When depicting hardwood trees, she painted the flowers and the fruit, but often left the foliage uncolored.[127] Her illustrations often resembled herbarium sheets, and she took great care to get the plant morphology correct. In the field, she made careful drawings of stems, leaves, flowers, and roots. She used magnification and the camera lucida[128] to aid accuracy.[129] Her references included Asa Gray's *Manual of Botany of the Northern United States*,[130] and she developed correspondence with Gray and his successors.[131] Her biographers wrote,

> In 1870 Furbish produced an outpouring of scientific and artistic work unequaled at any part of her life. As if by some miracle, the painter of Maine's flora suddenly appeared, at the height of her powers, fully equipped, fully mature.[132]

After Furbish's parents died, she moved to a small house that had a second story bathed in sunlight, and it was here she located her easel.[133] She had decided to depict

---

[119] Norwood (1993), p. 214.
[120] Unless otherwise indicated, this account was modified from Graham and Graham (1995).
[121] Graham and Graham (1995), p. 8.
[122] Graham and Graham (1995), p. 18.
[123] Graham and Graham (1995), p. 18.
[124] Graham and Graham (1995), p. 21.
[125] Graham and Graham (1995), p. 73.
[126] Graham and Graham (1995), pp. 66–67.
[127] Graham and Graham (1995), p. 68.
[128] A technique where the image viewed in a microscope was projected onto a piece of paper for important structures to be accurately traced as a basis for further illustration.
[129] Graham and Graham (1995), p. 38.
[130] Graham and Graham (1995), p. 75.
[131] Graham and Graham (1995), p. 75.
[132] Graham and Graham (1995), p. 37.
[133] Graham and Graham (1995), p. 61.

FIGURE 2.10 Catherine Furbish; image in the public domain.

all the flowering plants found growing in Maine.[134] Her biographers describe her house as "a repository for the collection of thousands of pressed plants she obtained throughout the state, and a temporary storage place for her already considerable body of botanical watercolors" (Figure 2.10).[135]

Her lifestyle alternated between traveling, often to collect plants, then returning to Brunswick to organize and paint. She often spent the warm months in search of specimens, boarding with families and being guided by farmers and timbermen. Because she depicted all life history stages of her plants, it was often necessary to revisit places; for example, once to observe flowers, and then again to observe seeds.[136]

Furbish made her chief discovery in 1877, in Aroostook County, which borders Canada. Along the bank of the St. John River, she discovered a new plant species, later named *Pedicularis furbishiae*—Furbish's Lousewort.[137] It was known only from the margins of this river and was never common. The plant was collected again in 1946, but in 1975 the Smithsonian Institution's Report of Endangered and Threatened Plant Species of the United States listed it as "probably extinct." In 1976, Charles Richards, a botanist with the University of Maine, surveyed the St. John's River for endangered plants as a prerequisite for permits to be issued for the construction of

---

[134] Graham and Graham (1995), p. 61.
[135] Graham and Graham (1995), p. 61.
[136] Graham and Graham (1995), p. 68.
[137] Graham and Graham (1995), pp. 75–81.

the Dickey-Lincoln hydroelectric power plant. Much to everyone's surprise, he discovered populations both upstream and downstream of the proposed dam site. The species was petitioned for Endangered Species status, and Furbish's Lousewort (not without a great deal of controversy) helped to halt construction of the dam.[138]

It took Furbish 40 years to complete the *Flora of Maine*. She was 75 in the summer of 1908, when she presented the book to Bowdoin College. It consisted of 1,326 sheets assembled in 14 volumes, each 20 inches by 16½ inches, bound in leather. She also gave the college two volumes containing mushroom paintings and her miscellaneous work.[139] Furbish sent her collection of around 4,000 herbarium sheets to the Gray Herbarium at Harvard.[140] Her work was done, but she lived another 22 years, dying of heart failure at age 97.

The accomplishments of the women profiled here are impressive, especially in light of the various forms of adversity they faced. They had no common origin and some did not derive from any scholarly traditions or schools. They did however, as individuals, show what women interested in field biology could do, both intellectually and in terms of physical strength and stamina. After an honest consideration of their undertakings, men could no longer say that women were not up to the demands required by field biology or, for that matter, any form of science. And with that realization, the barriers to women began lowering.

REFERENCES

Benson, M. 1986. *Martha Maxwell: Rocky Mountain naturalist*. Lincoln, NE: University of Nebraska Press.

Bonta, M. M. 1991. *Women in the field: America's pioneering women naturalists*. College Station, TX: Texas A & M University Press.

Bonta, M. M. 1995. *American women afield: writings by pioneering women naturalists*. College Station, TX: Texas A&M University Press.

Clark, J. F. M. 1992. Eleanor Ormerod (1828–1901) as an economic entomologist: 'pioneer of purity even more than of Paris Green'. *British Journal for the History of Science* 25:431–452.

Creese, M. R. S. 2000. *Ladies in the laboratory? American and British women in Science, 1800–1900: A Survey of Their Contributions to Research*. Lanham, MD: Scarecrow Press.

Creese, M. R. S. and T. M. Creese. 2015. *Ladies in the laboratory IV: Imperial Russia's women in science, 1800–1900*. New York: Rowman & Littlefield.

Emling, S. 2009. *The fossil hunter: dinosaurs, evolution, and the woman whose discoveries changed the world*. New York: Palgrave Macmillan.

Etheridge, K. 2011. Maria Sibylla Merian: The first ecologist? In *Women and science: figures and representations—17th Century to Present*, eds. V. Molinari and D. Andreolle, pp. 31–50. Newcastle upon Tyne: Cambridge Scholars Publishing.

Etheridge, K. and F. F. J. M. Pieters. 2015. Maria Sibylla Merian (1647–1717): pioneering naturalist, artist, and inspiration for Catesby. In *The curious Mr. Catesby: a "truly ingenious" naturalist explores new worlds*, eds. E. C. Nelson and D. Elliot, pp. 39–56. Athens, GA: University of Georgia Press.

138 Graham and Graham (1995), pp. x–xi.
139 Graham and Graham (1995), pp. 137–142.
140 Graham and Graham (1995), p. 14.

Gates, B. T. 1998. *Kindred nature: Victorian and Edwardian women embrace the living world.* Chicago: University of Chicago Press.

Graham, A. and F. Graham Jr. 1995. *Kate Furbish and the flora of Maine.* Gardiner, ME: Tillbury House Publishers.

Kampmeier, G. E. 2011. Great women in entomology. *American Entomologist* 57:202–204.

Kohler, R. E. 2006. *All creatures: naturalists, collectors, and biodiversity, 1850–1950.* Princeton, NJ: Princeton University Press.

Lewis-Jones, H. and K. Herbert. 2016. *Explorers' sketchbooks: the art of discovery & adventure.* San Francisco, CA: Chronicle Books.

Logan, G. B. 1999. Italian women in science from the Renaissance to the nineteenth century. Dissertation. Ottawa: University of Ottawa.

Maclean, N. 1976. USFS 1919: The ranger, the cook, and a hole in the sky. In *A river runs through it and other stories*, pp. 125–217. Chicago: University of Chicago Press.

Maroske, S. and T. W. May. 2018. Naming names: the first women taxonomists in mycology. *Studies in Mycology* 89:63–84.

Moring, J. 2002. *Early American naturalists: exploring the American West 1804–1900.* New York: Cooper Square Press.

Norwood, V. 1993. *Made from this Earth: American women and nature.* Raleigh, NC: University of North Carolina Press.

Ogilvie, M. B. 1986. *Women in science: antiquity through the nineteenth century. A biographical dictionary with annotated bibliography.* Cambridge, MA: The MIT Press.

Philippon, D. J. 2001. Gender, genus, and genre: women, science, and nature writing in early America. In *Such news of the land: U.S. women nature writers*, eds. T. S. Edwards and E. A. De Wolfe, pp. 9–26. Hanover: University Press of New England.

Pieters, F. F. J. M. and D. Winthagen. 1999. Maria Sibylla Merian, naturalist and artist (1647–1717): a commemoration on the occasion of the 350th anniversary of her birth. *Archives of Natural History* 26:1–18.

Ridley, G. 2010. *The discovery of Jeanne Baret: a story of science, the high seas, and the first woman to circumnavigate the globe.* New York: Crown Publishers.

Shearer, B. F. and B. S. Shearer, eds. 1996. *Notable women in the life sciences: a biographical dictionary.* Westport, CT: Greenwood Press.

Sydney, M. 1976. At home with Margaret Sydney. *The Australian Women's Weekly*, 18 February 1976.

Todd, K. 2007. *Chrysalis: Maria Sibylla Merian and the secrets of metamorphosis.* New York: Harcourt, Inc.

Vargues, L. 2013. Flashes in the twilight. New York Botanical Garden. Science Talk. 30 December 2013.

3 1880 to 1916 (Gilded Age)

The years from 1880 to the beginning of the First World War saw an increasing number of women engaged in nature study and field biology. Women wrote, taught, and collected specimens of plants, animals, and fossils; many conducted research in fields ranging from botany to entomology, herpetology, and ornithology. Increasingly, women focused on the need to protect nature. In England, Beatrix Potter exemplified both what women could accomplish in science and the resistance to such accomplishments by a traditional scientific establishment. In North America, the influence of women's colleges began to be felt within the scientific community. Ellen Swallow Richards, the first female student and, later, faculty member at Massachusetts Institute of Technology (MIT), was the intellectual and energetic equal of any scientist of her time. Elizabeth Britton, Mary Agnes Chase, Mary Dickerson, and Edith Patch also had successful careers at eastern institutions. At the same time, women such as Katherine Brandegee, Alice Eastwood, Florence Merriam Bailey, Annie Alexander, and Louise Kellogg were taking advantage of opportunities out West, particularly in California.

Most of us associate Beatrix Potter (1866–1943) with the personages of Peter Rabbit, Mrs. Tiggy-Winkle, and Jemima Puddle-Duck. Less well known is that she was an accomplished naturalist, mycologist, and conservationist. Helen Beatrix Potter was born in Kensington, London. As a child, her life was filled with art, literature, science, fantasy, and travel.[1] She was inspired and enriched by nature, beginning as a child with visits to the Scottish countryside and later in the Lake District of northwest England. Potter exhibited an early talent for drawing and painting and was encouraged by her parents, both of whom sketched. Her earliest sketchbook, at age nine, contains watercolors of caterpillars, each accompanied with descriptive notes and other observations. Another early sketchbook contains drawings of ice-skating rabbits, decked out in jackets, hats, and scarves. By the late 1880s, in her early 20s, her scientific and artistic interests turned to fungi.

In 1892, during a summer holiday in Perthshire, Scotland, Potter met the respected naturalist and amateur mycologist Charles McIntosh. She showed him her drawings of fungi. The two discussed the biology and classification of fungi, and McIntosh shared techniques for drawing fungi under the microscope. They agreed that he would send her fresh specimens, and she would draw them and send back her illustrations. After that meeting, Potter spent hours in the Natural History Museum in London learning fungal taxonomy.

In 1895, Potter collected *Boletus granulatus*, a fungus that was "very slimy & had yellowish milky drops on the spores and stem."[2] She began asking questions: Do spores survive the winter, and if so, in what form do they reappear; how do fungi reproduce; and are there differences across species? Few mycologists had studied

---

[1] Lear (2007); much of the information here has been extracted from this book.
[2] Lear (2007), p. 102.

DOI: 10.1201/9781003311508-4

fungal reproduction. Potter speculated that spores germinate, and she pursued this line of research.

She germinated spores from over 50 species, kept detailed records, and wrote a manuscript "On the Germination of the Spores of Agaricineae" for presentation at the Linnaean Society, at that time a men's-only club. She arranged for George Massee, chief mycologist at the Royal Botanic Gardens at Kew, to submit the paper. It was read on 1 April 1897 but not selected to be published.[3] Like other women of her time who attempted to gain recognition for their scientific research from the Linnaean Society, Potter was not taken seriously. The paper was withdrawn a week later. She continued to experiment and draw germinating spores, but soon after her discouraging experience she quit trying to achieve academic recognition.

On 1 April 1997, 100 years after Potter's paper was read, the Linnaean Society offered an official apology. The executive secretary publicly acknowledged that she had been "treated scurvily" by the Linnaean Society.[4] The society admitted that 100 years earlier it had displayed sexism in its handling of her research and was antagonistic toward her as an amateur and as a woman.

Much as Potter loved exploring the woods and painting fungi, she also loved fantasy and writing tales about animals. *The Tale of Peter Rabbit*, published in 1902, was the first, but there were many more books to come. In 1905, Potter used a small inheritance from an aunt and royalties from her children's books to purchase Hill Top Farm in Near Sawrey. She subsequently purchased additional farms, totaling over 4,000 acres bordering Hill Top, to preserve the landscape and culture. Today her property, much of the Lake District National Park, is England's largest national park and has been designated a World Heritage Site.

Potter's contributions as a naturalist are threefold. First, millions of children have bonded with nature through her 23 children's books. *The Tale of Peter Rabbit* has been translated into 36 languages, and more than 45 million copies have been sold, making it one of the world's all-time best-selling books. Second, she is celebrated as an amateur mycologist. Her illustrations of fungi are so scientifically accurate that still today, mycologists consult them to identify species of mushrooms. Third, she gave England a lasting treasure when she bequeathed the land she loved to the National Trust and challenged all of us to think about land preservation.

In North America, first women's, then coeducational colleges began opening up in the East. In the West, women attending coeducational colleges gained not only increased professional opportunities but also the ability to use their degrees in meaningful ways. Here, we highlight the stories of nine of these women. The first, Ellen Swallow, was a naturalist but professionally an environmental chemist. We include her here because of her profound influence on women's education, her strong belief in women's ability to become scientists, and the fact that she served as a strong role model for young women scientists.

Ellen Henrietta Swallow Richards (1842–1911) was born in Dunstable, Massachusetts. Her father taught her history, science, and logic and was a major influence in her life.

---

[3] Lear (2007), pp. 123–124.
[4] Lear (2007), p. 482.

Her ability to learn quickly and retain knowledge, traits exhibited from childhood, would later earn her the nickname "Ellencyclopedia." She was in her element in the outdoors.[5] She would lie on the ground and watch insects and other invertebrates, and she collected rocks and fossils. It wasn't long before she had learned the names of most of the plant and animal species on the farm.[6] When she was 16, Swallow's family moved to Westford, where she attended Westford Academy. She especially enjoyed foreign languages, mathematics, and science. When Swallow applied to Vassar, her entrance exam scores were so high she was placed in the third year.

Astronomer Maria "Comet" Mitchell, Vassar's first female instructor, reinforced Swallow's love of learning. Mitchell's advice to her students was "Study as if you were going to live forever; live as if you were going to die tomorrow," and "Do not falter because you are women."[7] She also said, "Let no one suppose that any woman in all the ages has had a fair chance in science," and "We especially need imagination in science. It is not all mathematics, nor all logic but is somewhat beauty and poetry."[8]

While at Vassar, Swallow wrote,

> … my aim is … to make myself a true woman, one worthy of the name, and one whose aim is to do all of the good she can in the world and not be one of the delicate little dolls or silly fools who make up the bulk of American women, slaves to society and fashion.[9]
>
> The only trouble [here] is they won't let us study hard enough. … They are so afraid we shall break down. … Can girls get a college degree without injuring their health?[10]

After graduating, Swallow sought a chemical company that would employ women. None would. A company executive suggested she apply to Boston Tech (today's MIT) to further her studies. In late 1870, she became the first woman accepted to an American institution of science and technology. During much of that first semester she was on her own. "I was not then allowed to attend any classes."[11] Whereas the male students worked with lab partners, her assignments were given to her privately, and she worked alone. After a few months, she was allowed limited interactions with both the young men and the instructors who were beginning to like and depend upon her. Noted chemist William Ripley Nichols, who had argued against her admission, declared "When we are in doubt about anything we always go to Miss Swallow."[12]

Massachusetts had recently established America's first state Board of Health. In Boston, industrial pollution was everywhere. There were open sewers, and the streets were littered with garbage and manure. Cholera and typhoid fever were rampant. Sixty percent of the Irish-born children in Boston died before adulthood.[13] Swallow joined others in addressing these health issues by working on a study of Boston's water quality.

---

[5] Swallow (2014), p. 7.
[6] Swallow (2014), p. 7.
[7] Swallow (2014), p. 25.
[8] Swallow (2014), p. 27.
[9] Swallow (2014), p. 30.
[10] Swallow (2014), p. 31.
[11] Swallow (2014), p. 41.
[12] Swallow (2014), p. 42.
[13] Swallow (2014), p. 45.

A year after she graduated, she started The Woman's Laboratory at MIT. It was the first laboratory of its kind in the world devoted to teaching science to women, offering classes in chemistry, mineralogy, and natural science. After the first year, Swallow wrote, "Our students have proved that the most severe training does not make women repulsive and does not unfit them for housewifely duties," and "The capability of women to carry through a severe course of scientific education without injury to body or mind is now established."[14] The Women's Lab was open 7 years and hosted more than 500 students, including the naturalist/taxidermist Martha Maxwell (profiled in Chapter 2). In 1878, Swallow was elected Fellow of the American Association for the Advancement of Science. (Just 3 years earlier, her former teacher, astronomer Maria "Comet" Mitchell, had been elected a Fellow of the American Association for the Advancement of Science.) In 1883, the Laboratory of Sanitary Chemistry was built, open to any student qualified to study at MIT. Swallow taught there, and with this position became MIT's first paid female instructor, although she was never permitted to pursue a PhD.[15]

Swallow's public health career focused on improving people's lives through public and private health initiatives. In 1890, she helped to open the first New England Kitchen, designed to provide inexpensive, nutritious food for those in need. She set up a demonstration kitchen at the 1893 World's Fair in Chicago. Later, she helped establish a successful lunch program for high school students in Boston.[16] With these activities, she helped develop the discipline of home economics.[17] Her work helped to pass food and drug act legislation in Massachusetts and Connecticut, and the building of the nation's first modern sewage treatment plant—the Lawrence Experiment Station in Lowell, Massachusetts. Swallow created the term "Human Ecology,"[18] wrote 15 books, including the popular *Air, Water, and Food*, and coauthored 3 others. In 1910, Smith College awarded her an honorary Doctor of Science degree. A year later she died, and the Ellen Richards Research Prize, an international award for women in science, was created. Of her, it has been said "Her whole life has been an example of what a woman of determination [and extraordinary talent] can do" (Figure 3.1).[19]

Botany remained one of the most "acceptable" fields of science for women. In a paper written in 1982, Emanuel Rudolph, an historian of botany, identified 1,185 women as having an active interest in botany during the 1800s.[20] The vast majority of the women were active in botany during the last three decades of the century (only 68 women, or 6% were active between 1800 and 1870). Much of this trend is likely explained by the greater access to education for women after the Civil War. Sixty-seven percent of the 1,185 women were located in the Northeast; 28% identified as being married. About 10% of the women received a bachelor's degree (119 individuals), 3% a master's degree (36 individuals), and 1% a PhD (17 individuals);

---

[14] Swallow (2014), p. 69.
[15] Swallow (2014), p. 70.
[16] Swallow (2014), p. 111.
[17] Swallow (2014), p. 113.
[18] Swanson (2013).
[19] Swallow (2014), p. 131.
[20] Rudolph (1982).

FIGURE 3.1 Ellen Swallow Richards (1842–1911). Image in the public domain.

an additional 17 women had medical or dental training. About 23% of the women (332 individuals) were plant collectors. Indeed, an examination of collections in major U.S. herbaria from the nineteenth century reveals that a significant number of specimens were collected by women. About 15% of the women in the sample had a profession (80% of them were teachers). Rudolph reported that although few of these women became professional botanists, they contributed significantly to the development of botany as a profession in three ways: (1) carrying out scientific research, (2) making plant collections, and (3) actively supporting botanical clubs and societies. Unfortunately, their contributions are often overlooked. Following are the stories of four of these women who focused their passion for nature on plants during the period from 1880 to America's entry into the First World War.

Elizabeth Gertrude Knight Britton (1858–1934) was born in New York City. She spent her early childhood near Matanzas in northern Cuba, where her father operated a sugar plantation and furniture factory. She returned to New York City and graduated from Normal College (established in 1870 as a women's college; later renamed Hunter College), where she specialized in botany. Britton began her career by teaching at Normal College as an assistant in natural sciences. In 1879, she joined the Torrey Botanical Club, the first botanical society in America (founded in 1867). She was especially interested in mosses and became one of the first American bryologists (scientists who study mosses, liverworts, and hornworts).

In 1885, she married Nathaniel Lord Britton, then an assistant in geology at Columbia College (now Columbia University) and resigned her position at Normal College. More interested in botany than geology, her husband became instructor of botany at Columbia 2 years after they married. In the summer of 1888, the couple spent 2 months in England. He worked on classifying Bolivian botanical collections at Kew Gardens in London, while she worked on mosses at the Linnaean Society—on the upstairs floor, since women were not admitted to the main floor. They both loved Kew Gardens, with its herbarium, library, and its mission to grow as many plants as possible from all over the world. Britton asked her husband why there couldn't be something comparable in New York City. A seed was planted, and they took the idea back to the Torrey Botanical Club. By June 1895, $250,000 had been raised and 250 acres of Bronx Park had been set aside for the New York Botanical Garden. Additional funds had been pledged for buildings, and a stellar board of managers had been elected: Cornelius Vanderbilt president, Andrew Carnegie vice president, J. P. Morgan treasurer, and Nathaniel Lord Britton secretary. In 1896, Britton's husband resigned from Columbia College to become Director of the New York Botanical Garden.[21] She assisted, but also continued her own botanical work, collecting in the Dismal Swamp in Virginia, the Adirondacks, and the Appalachians. She deposited her specimens in the Garden's herbarium and wrote about 350 popular and scientific articles, mostly on mosses.[22]

Soon after the New York Botanical Garden opened in 1902; Britton helped organize the Wild Flower Preservation Society of America and became its driving force until the mid-1920s. Her work resulted in legislation protecting native plants. She also continued her research with mosses. In 1912, she was officially named honorary curator of mosses at the New York Botanical Garden (a mere formality since she had been curating them all along). The couple made 23 trips together to the West Indies, collecting plants for the New York Botanical Garden. They had no children. In Nathaniel Britton's obituary in the *Journal of the New York Botanical Garden*, Marshall Howe wrote: "In a way that was very real for them, the New York Botanical Garden was their child."[23] The couple had made the garden one of the best botanical and horticultural gardens in the world. It is still one of the best (Figure 3.2).

Katherine (Layne) Brandegee (1844–1920) was born in Tennessee, but her father's wanderlust took the family west, first to Utah, then Nevada, and when Brandegee was nine, to a farm near Folsom, California. In her early twenties, she married a hard-drinking constable and taught school to pay his debts. After he died,[24] Brandegee, then 30, enrolled as a medical student at the University of California, Berkeley. Three years later, she had her medical degree but found that few people wanted to be treated by a female doctor.

Brandegee became interested in medicinal plants, then all plants. She went to San Francisco where she worked at the California Academy of Sciences, at that time the center of botanical research in the western United States.[25] In 1883, when the curator

[21] Bonta (1991), p. 127.
[22] Lipscomb (1995), p. 338.
[23] Bonta (1991), p. 131.
[24] Bonta (1991), p. 86.
[25] Bonta (1991), p. 86.

FIGURE 3.2   Elizabeth Gertrude Knight Britton (1858–1934). Image in the public domain.

of botany at the academy retired, she was offered the position. This offer came at a time when most women employed by herbaria and botanical gardens were illustrators. Since the academy's founding in 1853, it had followed an enlightened policy, which was scoffed at in the East: The academy approved of and invited women to work in every department of natural history.[26] Brandegee quickly organized the herbarium's collections and began conducting collecting trips. In 1884, she launched the *Bulletin of the California Academy of Sciences* and became its editor.

In 1889, she married civil engineer Townshend Stith Brandegee, one of the most renowned botanical collectors in the West. They made an unusual couple: her husband, a small, quiet intellectual from Yale University, and Brandegee, a large, robust, outspoken Californian. Shortly before their marriage, Townshend inherited $40,000, and they were financially independent for the rest of their lives. The couple honeymooned while walking and botanizing 500 miles from San Diego to San Francisco.[27]

In 1892, Brandegee brought her protégée, Alice Eastwood, to the academy as joint curator, giving Eastwood her entire salary of $75/month. By 1895, Eastwood became the sole curator, allowing the Brandegees to retire. They moved to San Diego, built their own herbarium and library, planted a botanical garden on a mesa above the

---

[26] Bonta (1991), p. 86.
[27] Bonta (1991), p. 88.

city, and spent time doing what both loved best—collecting plants. Brandegee had to be exceedingly careful and had some collecting trips cut short, because she was a diabetic before the days of insulin treatment. On 18 April 1906, the San Francisco earthquake struck the coast of northern California. The herbarium at the California Academy of Sciences was destroyed. Soon afterward, the Brandegees donated their entire collection of 100,000 plants and their library to Berkeley. They moved to Berkeley, where they worked in the University Herbarium until their deaths.

Brandegee left an enduring legacy to botany in several ways. She labored to make the standards of western botanists comparable to those of eastern workers. Her collections helped to determine range distributions of many plant species. Questioning the legitimacy of many newly named species, she sought out intermediate forms and showed that in some cases the "new species" were merely subspecies of existing forms. Significantly, Brandegee paved the way for other western women botanists to work as field collectors and as paid professionals. One of these women was Brandegee's protégée, Alice Eastwood.

Born in Toronto, Canada, Alice Eastwood (1859–1953) spent her early years living on the grounds of the Toronto Asylum for the Insane, where her father was superintendent.[28] Eastwood's Uncle William, an experimental horticulturist, encouraged her interest in plants, giving her books and teaching her Latin names. Eastwood was educated at a convent outside of Toronto where a French gardener-priest further encouraged her botanical interests. After moving to Denver, she graduated valedictorian from East Denver High School, and then taught there.

Eastwood spent her summers in the High Rockies collecting plants for her herbarium. She traveled alone, on foot or horseback, hauling her heavy wooden plant presses. She designed her own collecting outfits consisting of buttoned denim shirts with bustles made from heavy cotton nightgowns.[29] In the field, she often encountered rough miners, cowboys, and ranchers but claimed she never experienced "the slightest discourtesy."[30] Eastwood felt that carrying a gun showed fear, so she traveled unarmed. She quickly earned herself a reputation as Colorado's most accomplished botanist.[31] Her collections formed the nucleus of the herbarium at the University of Colorado, Boulder, the only formal collection of plants in the state at that time.[32] English naturalist Alfred Russel Wallace visited Denver in 1887, intending to explore 14,000-foot Grays Peak. Eastwood volunteered to be his guide. Together for 3 days, the world-famous naturalist in his 60s and the amateur woman botanist in her 20s shared what Eastwood called "a glorious adventure."[33]

In 1890, Eastwood visited the California Academy of Sciences in San Francisco, where she met Katherine Brandegee. Brandegee invited Eastwood to write for their magazine *Zoe* and help organize the academy's herbarium. Two years later, Eastwood accepted Brandegee's offer of the botanical curatorship position and remained curator until 1949. The only time during that span when she wasn't working at the academy

[28] Lipscomb (1995), p. 336.
[29] Shearer and Shearer (1996), p. 103.
[30] Bonta (1991), p. 94.
[31] Bonta (1995), p. 84.
[32] Ogilvie (1986), p. 80.
[33] Bonta (1995), p. 95.

FIGURE 3.3   Alice Eastwood (1859–1953). Image in the public domain.

was for 6 years after the 1906 earthquake and subsequent fire destroyed the academy. Eastwood became an instant heroine when she dashed back into the building and saved over 1,200 irreplaceable type specimens.[34] With foresight, she had placed all the types together where she could find them in case of emergency.[35] After the fire, the rescued specimens were all that remained of the academy's botanical collections. Eastwood began recollecting, using her own meager resources. Finally, she received the call she had been waiting for: The California Academy of Sciences had decided to reestablish the institution in Golden Gate Park, and they wanted her to return and rebuild the herbarium.[36]

Eastwood became known as the Grand Old Lady of the Academy (Figure 3.3). She is said to have been a curmudgeon, as, for example, she refused to accept honorary degrees. She had taught herself and had never received a college degree, so no thank you to an honorary one. One of her assistants said: "Her impatience could be as violent as her kindness and generosity were great, and the force and bite of

---

[34] Type specimens are the single specimen (or small group) that are identified in the original description of the species and are irrevocably associated with the species' name. They are thus immensely valuable in serving as a reference when defining that species, especially in comparison with other, similar, species.

[35] Bonta (1995), p. 84.

[36] Bonta (1995), p. 99.

that impatience were dreaded by all who ever encountered it."[37] When she retired in 1949, at age 90, Eastwood was invited to serve as honorary president of the Eighth International Botanical Congress in Stockholm, Sweden. She presided while seated in a chair used by the renowned Swedish botanist and taxonomist Carolus Linnaeus. Eastwood published 2 books, over 300 articles, and promoted public awareness of the importance and need to protect native plants. She never married. The herbarium, which she thought of as her child, was her life; she added 340,000 specimens to it.[38]

Mary Agnes Merrill Chase (1869–1963) was born in Iroquois County, Illinois. Her father, an Irish railroad blacksmith, died when she was two. She and her four siblings were raised by her mother and grandmother in Chicago. Her family had so little money she worked rather than attend high school. At 19, she married an editor, William Chase, but became a widow the following year when he died of tuberculosis.

Chase's interest in plants began as a hobby after her 13-year-old nephew persuaded her to take him into the field to collect plants. Later, in her spare time, she began a study of the flora of northern Illinois. By chance, one day while collecting in a swamp, she met the bryologist Reverend Ellsworth Jerome Hill. He needed an illustrator, and Chase offered to help. She turned out to be skilled. Hill could not pay her, but instead taught her botany and microscopy. He put her in contact with the botany curator at the Field Museum of Natural History in Chicago, which led to Chase illustrating two publications. In 1903, she became the illustrator for the USDA's Bureau of Plant Industry in Washington, D.C.

On her own time, she worked in the USDA grass herbarium and published her first scientific paper in 1906. The following year, she was appointed scientific assistant for Albert Spear Hitchcock in systematic agrostology (study of grasses). Chase became an international authority on grasses and is now referred to as America's "Dean of Agrostologists." She attributed her love of grasses to a childhood experience. Unable to find flowers, she picked a bouquet of grasses for her grandmother. Her grandmother told her that grasses did not have flowers. Chase told her grandmother to look more closely and she would see tiny flowers in each spike.[39] Chase, of course, was right.

Chase collected widely in the United States and donated all her specimens to the USDA herbarium. In 1913, she collected grasses in Puerto Rico, her first foreign expedition. She visited museums in Europe and claimed, "Botanizing in herbaria does not afford the same pleasure as does botanizing in the field but is not without its thrills of discovery."[40] Chase wrote *First Book of Grasses* to teach the nonprofessional how to identify grasses and to appreciate their worth and beauty. In 1924, she collected grasses in Brazil and returned with 20,000 botanical specimens, including 500 grasses. After she returned, she was promoted to associate botanist. Four years later she returned to Brazil and collected with Ynés Mexía (profiled in Chapter 4). The two women climbed the eastern side of the Pontão Crystal peak, a place in which no botanists had ever collected. Chase called the trip the "hardest physical feat of my

[37] Bonta (1995), p. 102.
[38] Bonta (1995), p. 86.
[39] Norwood (1993), p. 83.
[40] Bonta (1991), p. 136.

FIGURE 3.4   Mary Agnes Chase (1869–1963). Image in the public domain.

life,"[41] during which she collected over 4,500 grass specimens. In 1935, she moved into Hitchcock's position at the Bureau, where she remained until her mandatory retirement at age 70. Chase continued her work for another 24 years, without pay (Figure 3.4).

Chase was another socially active natural historian who supported progressive reform movements. She was a suffragist, jailed in 1918 and again in 1919 while demonstrating for women's right to vote.[42] She was also a pacifist and contributed part of her salary to Quaker causes. She was a confirmed prohibitionist.[43] She stood up for what she believed and encouraged women to believe in their own abilities.

Like her friend on the West Coast, Alice Eastwood, Chase was proud of her self-education. Unlike Eastwood, however, she accepted an honorary degree. In 1958, at age 89, the University of Illinois awarded her an honorary Doctor of Science degree—her first college degree. Additionally, she was made an Honorary Fellow of the Smithsonian Institution and a fellow of the Linnaean Society. She stopped working 5 months before her death. She died at age 95, the same day she entered a nursing home, respected by fellow scientists throughout the world. Barely 5 feet tall, she had become a botanical giant with only a grammar school education.

One of America's most renowned and influential early ornithologists, Florence Merriam Bailey (1863–1948) was born in Locust Grove, New York, the youngest of four children. Her father, a successful businessman, retired shortly before her birth to manage the family farm. Her mother had a degree from Rutgers College. Bailey was sickly as a child, and perhaps because of this, her parents encouraged outdoor activities. When she was nine, she accompanied her father and brother Hart (then 17 years old) on a rugged 2-month camping and collecting trip to Florida.[44]

Bailey attended a private preparatory school in Utica and from 1882 to 1886 studied at Smith College. She enrolled as a "special student" at Smith because she

---

[41] As quoted in Bonta (1991), p. 141.
[42] Norwood (1993), p. 83.
[43] Bonta (1991), p. 134.
[44] Bonta (1991), p. 186.

lacked the educational background to be admitted in the degree program.[45] This stigma affected her most of her life, making her feel inadequate, especially when she compared herself to her prominent and influential brother Hart and her father.[46] (Hart developed and then lead the Bureau of Biological Survey, which eventually became the U.S. Fish and Wildlife Service.) At Smith, she honed her interest in birds and eventually in bird conservation.[47] She kept records on bird migration and wrote articles for *Audubon Magazine* about the lives of birds. Shortly after George Bird Grinnell started the first Audubon Society in 1886, Bailey and a friend founded the Smith College Audubon Society.[48] She organized the society's first bird walks and recruited the famed naturalist John Burroughs to lead them. Later, Bailey and a friend conducted their own birding trips. Within 3 months, nearly 100 young women (one-third of the student body) had become members.[49] In 1885, a year before her graduation, she became the first woman associate member of the American Ornithologists' Union (AOU).

After graduating, she returned home and cared for her parents, did social work, and when healthy continued her bird observations and writing. She penned a series for Grinnell's *Audubon Magazine* entitled "Hints to Audubon Workers: Fifty Common Birds and How to Know Them." Her narratives often relied on John Burroughs' descriptions of bird habitats and behaviors. She quoted Emerson and Thoreau as well as contemporary naturalists, many of whom, such as John Burroughs, Henry Henshaw, Ernest Seton Thompson, George Bird Grinnell, and David Starr Jordan, she knew through her brother. This series demonstrated the broad knowledge that would come to characterize her work, including articles she would publish in *The Auk*, *St. Nicholas*, *Our Animal Friends*, *American Agriculturist*, and *Observer*.

As a young woman, Bailey struggled to maintain her health (she may have contracted tuberculosis from her mother). Her brother was concerned about both of them and took the family to California, where he hoped the curative benefits of "western air" would restore their health. Indeed, they all felt better after spending the spring at the uncle's ranch in Twin Oaks. That is, until Bailey developed typhoid fever.

Despite her ill health, Bailey published her first book, *Birds Through an Opera Glass*, in 1889 when she was 26. She used the field technique developed by the avid bird-watcher Olive Thorne Miller (and perfected by Margaret Morse Nice, profiled in Chapter 4), where she would:

> steal in through the bushes in her leaf-colored gown, open her camp-stool cautiously at the foot of a tree whose dark trunk would help conceal her, pull down a branch before her and, with note-book ready, carefully raise her opera-glass and focus it upon the nest she wanted to study. And there she would sit in silence, stoically defying tormenting gnats and mosquitoes, patiently waiting and watching to see what might befall.[50]

---

[45] Moring (2002), p. 220.
[46] Moring (2002), p. 220.
[47] Shushkewich (2012), p. 95.
[48] Shushkewich (2012), p. 97.
[49] Shushkewich (2012), p. 99.
[50] Kofalk (1989), p. 53.

Birds Through an Opera Glass was favorably reviewed in *The Auk*. Ornithologist (and cofounder of the AOU) William Brewster wrote:

> As an observer, Miss Merriam is unmistakably keen, discriminating, and accurate; as a writer, always simple and true, at times highly vigorous and original. Her attractive little book may be cordially recommended to all who wish to study our familiar birds, either with or without an opera glass.[51]

During the winter of 1892–1893 she was still sick, coughing constantly. The following spring, her mother died, and Bailey went to Utah to restore her health. She described her experiences in her second book, *My Summer in a Mormon Village*, published in 1894. While in Utah, her health improved, and she began once again keeping field notes. She spent the winter of 1893–1894 on the West Coast, associated with Stanford. While there, she met Alice Eastwood, and they became lifelong friends. She also learned to ride horses astride. This experience led to her third book, *A-Birding on a Bronco*, published in 1896, and illustrated by a young Louis Agassiz Fuertes.

Bailey moved to Washington, D.C., where she lived with her brother Hart, head of the U.S. Biological Survey. She continued to write, helped found the Audubon Society of the District of Columbia, and taught their first bird classes to teachers and the public. In the late 1890s, she began working with the Committee on the Protection of North American Birds, formed to combat the millinery driven trade in birds. She combined her extensive field notes with various magazine articles she had published into *Birds of Village and Field* (1898), a book for beginners. The following year, at age 36, she married Vernon Bailey, chief naturalist of the U.S. Biological Survey. A grandniece wrote, "Vernon was perfect for her, a very simple gentleman. She fussed over details he wouldn't have. He was outgoing and everyone adored him."[52]

In 1900, the Baileys went to the Southwest,[53] where they surveyed the wildlife of the Mexican Border, New Mexico, and California. Her husband liked to travel light and live hard; rations were simple. She kept up, and by the time they had finished their first summer alone together in the field, there was no doubt this lifestyle suited them. Their interests, while overlapping, took separate avenues: He focused on mammals, she on birds.

The Baileys were gracious and charming. Naturalist and conservationist Margaret Murie (profiled in Chapter 5) recalled that when new survey biologists brought their young wives to Washington, the women often felt overwhelmed. The Baileys hosted a dinner party with other survey members and their spouses to make the women more comfortable. Her influence on these wives was more than casual; by this time, Florence Bailey was a role model for women interested in pursuing a career (Figure 3.5).

In 1902, Bailey published *Handbook of Birds of the Western United States*. It was 600 pages long and had about as many illustrations, again done by Fuertes. Ornithologist Frank Chapman considered it "the most complete text-book of regional

---

[51] Kofalk (1989), p. 51.

[52] Kofalk (1989), p. 161.

[53] Bonta (1995), p. 97.

FIGURE 3.5   Florence Merriam Bailey (1863–1948). Image in the public domain.

ornithology which has ever been published."[54] Zoologist and conservationist William
Temple Hornaday relied on it. Bailey's biographer noted that besides bringing the
outdoors into her pages, with this book Bailey showed once and for all that following
birds into the woods was not the "eccentricity of conduct" for a woman that one early
ornithologist had pronounced it.[55]

Bailey's biographer also noted that independent women such as Bailey did not
suffer in the West; they could be exactly who they were. When men or women kept
to their own business, they were generally accepted, and Bailey became an object of
curiosity. As she walked on one mountain trail, an old prospector came up behind
her. His only comment, noted in her journal, was on her field glasses: "There's a lady
carries her eyes in her fist."[56]

After the *Birds of New Mexico* project was shelved because of the death of the
original author, Dr Wells W. Cooke, in 1916, Bailey was appointed special assistant
and tasked to finish it. *Birds of New Mexico* was finally published in 1928 by the
New Mexico Department of Game and Fish, with drawings/photographs by Ernest
Thompson Seton, Robert Ridgway, and Olaus Murie. Most of the original plates were
done by Major Allan Brooks, a Canadian bird artist.

---

[54] Kofalk (1989), p. 104.
[55] Kofalk (1989), p. 106.
[56] Kofalk (1989), p. 107.

In 1929, Bailey was elected a Fellow of the AOU, the first women so designated. Later that year, the AOU awarded her its Brewster Medal for the most important book on birds of the Western Hemisphere. In 1933, the University of New Mexico awarded her an honorary LLD degree. In the summer of 1933, Vernon Bailey retired from the U.S. Biological Survey after 46 years of service. The couple continued their travels and collaboration through the 1930s. In 1939, she published her last book, *Among the Birds in the Grand Canyon Country.*

Annie Montague Alexander (1867–1950) was born in Hawaii, the daughter of a businessman who, with his partner, controlled much of the islands' shipping and sugarcane industries.[57] Father and daughter shared a love of adventure. When Alexander was 15, her family moved from Maui to Oakland, California. After high school, she studied at the Lasell Seminary for Young Women in Auburndale, Massachusetts. In 1889, the family toured Europe. Alexander stayed behind in Paris and studied art at the Sorbonne. After an eye condition caused her to abandon art study, she returned to Oakland.

In 1899, Alexander and her friend Martha Beckwith explored southern Oregon, botanizing and bird-watching. Beckwith had found paleontology fascinating and encouraged Alexander to audit paleontology classes at Berkeley. Alexander attended a class taught by John C. Merriam and was hooked. She loved collecting fossils and had an uncanny ability for knowing just where to dig. She financed some of Merriam's fieldtrips, paying for expenses not covered by the department's meager budget, such as transporting heavy fossils back to Berkeley.[58]

In 1901, Merriam helped Alexander organize her own expedition to the Fossil Lake region of south-central Oregon. Fossils at the site included woolly mammoths, giant ground sloths, miniature horses, and camels.[59] Merriam insisted that it was inappropriate for a woman to travel alone in the company of men. Thus, a second woman was invited on the expedition. Alexander returned with over 100 bones, ranging from rodents and birds to camel and elephant remains, and many *Hipparion* (ancestor of modern horses) fossils. In 1902, she financed an expedition to Shasta County, California, where she found three ichthyosaur skeletons. Merriam described and named a new species of ichthyosaur after her: *Shastasaurus alexandrae.*[60] The following year, Alexander sponsored a second expedition to Shasta County. At their first campsite, she unearthed what she thought was another *Shastasaurus*. It was even better. Merriam described it as a new genus of ichthyosaur and named it *Thalattosaurus alexandrae.*[61]

In 1904, Alexander and her father shared a hunting safari in British East Africa (modern Kenya). They trekked more than 800 miles, accompanied by porters and beaters.[62] Both father and daughter hunted large mammals, and she photographed plants and wildlife. During their visit to Victoria Falls on the Zambesi River, a falling

---

[57] Anderson (2013), p. 236.

[58] Stein (2001), p. 24.

[59] Stein (2001), p. 24.

[60] Stein (2001), p. 31.

[61] Stein (2001), p. 33.

[62] Anderson (2013), p. 237.

3-foot boulder crushed her father's left leg and side. He died the next morning.[63] Alexander, devastated, buried her beloved father in a small cemetery in Livingstone and returned to California.[64]

After her father's death, Alexander sought to find meaning in her life. She wrote: "I felt I had to do something to divert my mind and absorb my interest and the idea of making collections of west coast fauna for study gradually took shape in my mind."[65] In 1905, she met U.S. Biological Survey Director C. Hart Merriam, Florence Merriam Bailey's brother and cousin of her mentor John Merriam. Hart Merriam was seeking specimens from the West Coast and Alaska. In February 1906, Alexander sent him a letter in which she outlined her plans for collecting big game in Alaska that summer, requesting advice about the fieldwork and sharing her idea of establishing a natural history museum on the West Coast. In May, Alexander, her friend Edna Wemple, and a government collector spent 4 months in the Alaskan wilderness, collecting marmots, ermines, squirrels, porcupines, minks, sea otters, moose, and brown bears.[66]

Alexander planned a second expedition to Alaska in 1907, to the Sitkan District. While arranging the trip, she encountered a Throop Polytechnic Institute[67] biology instructor named Joseph Grinnell. That meeting began a long, mutually beneficial association. Grinnell loved fieldwork. He had spent 2 years in Alaska collecting birds and recording natural history observations. He shared experiences and advice with Alexander.

After Alexander returned, Grinnell visited and examined her specimens (476 mammals, 532 birds, and 33 sets of eggs).[68] Alexander shared with him her vision of founding a museum of vertebrate zoology. The two began a regular correspondence. Grinnell thought the museum should be located at Stanford. Alexander insisted the museum be located on the Berkeley campus. In fact, Alexander had already written to the President of Berkeley:

> Dear President Wheeler.— Should the University of California within the next six months erect a galvanized iron building furnished with electric light, heat and janitor's services and turn it over to my entire control as a Museum of Natural History for the next seven years, I will guarantee the expenditure of $7000—yearly during that time for field and research work relating exclusively to mammals, birds, and reptiles of the West Coast, with the understanding that the University of California would be in no way responsible for the management of the funds for carrying out the work, or selection of collectors.[69]

The funds offered by Alexander would pay salaries for museum staff, buy supplies and equipment for storing specimens, and cover the costs of preparing specimens and publishing results. The nucleus of the collection would be the 1,300 specimens then housed in the Department of Zoology and Alexander's personal collection of

---

[63] Bonta (1991), p. 51.
[64] Stein (2001), p. 46.
[65] As quoted in Stein (2001), p. 47.
[66] Stein (2001), p. 62.
[67] Now California Institute of Technology, "Caltech".
[68] Stein (2001), p. 75.
[69] Stein (2001), p. 78.

nearly 3,500 specimens. Her letter stated that specimens obtained during the course of research at the museum would become property of the university and that all "materials would be and remain accessible to any qualified student within or without the university wishing to carry on special lines of investigation."[70] Her statement followed the policy generally accepted in the East of loaning cataloged specimens to researchers, contrary to the restrictive loan policies at the California Academy of Sciences in San Francisco. She indicated that Joseph Grinnell would be appointed director of the museum. If Wheeler would not accept her entire proposal, she would withdraw the offer. John Merriam supported this initiative and wrote to Wheeler, pointing out that the museum would make the university a recognized center for the study of reptiles, birds, and mammals. On 23 March 1908, Alexander telegrammed Grinnell: BEGIN WORK NEW MUSEUM TODAY YOUR APPOINTMENT AS DIRECTOR WILL BEGIN FROM APRIL FIRST.[71] Grinnell served as director of the Museum of Vertebrate Zoology for 32 years.

With the museum established, Alexander was ready to return to Alaska, this time to Prince William Sound, but she needed to find a woman to accompany her. As John Anderson pointed out in *Deep Things Out of Darkness*, "one can only marvel at a world in which it was acceptable for a woman to travel into the wild to hunt bears but unacceptable for her to travel as a sole woman in the company of men."[72] Alexander invited Louise Kellogg, a woman 13 years younger, who professed to be ready for adventure. The 1908 expedition was a success. Kellogg was a crack shot, she enjoyed herself as much as Alexander did, and the party collected two bears. Alexander wrote that the greatest discovery during the trip was Kellogg. That fieldtrip began a 42-year personal and professional relationship that ended only with Alexander's death in 1950.

Louise Kellogg (1879–1967) grew up in Oakland, not far from Annie Alexander's family home. She learned from her father how to hunt and fish in the marshes surrounding San Francisco Bay. Kellogg graduated from the University of California in 1901 with a degree in classics, and soon afterward got a job teaching school. Her 1908 expedition to Alaska with Alexander was her first exposure to fieldwork.

For more than 30 years, Alexander and Kellogg collected vertebrates and fossils as a team. Then, during their last decade of fieldwork together, they redirected much of their energy to collecting plants. Doing so allowed them to explore familiar landscapes with a fresh eye.[73] Over their lifetimes, the two collected 6,744 mammals, birds, reptiles, and amphibians for the Museum of Vertebrate Zoology; 17,851 plant specimens for the University Herbarium; and thousands of fossils for the Museum of Paleontology, which Alexander had founded in the 1920s.[74] Researchers found the women's specimens invaluable because they were meticulously prepared and accompanied with detailed field notes. Historian Robert Kohler wrote:

---

[70] As quoted in Stein (2001), p. 78.
[71] As quoted in Stein (2001), p. 82.
[72] Anderson (2013), p. 239.
[73] Stein (2001), p. 274.
[74] Bonta (1991), p. 55.

For Alexander and Kellogg, field collecting provided the freedom to pursue an unconventional lifestyle, do skilled work that they liked and did well, and contribute to science, all the while enjoying outdoor life. Although Grinnell encouraged Alexander to work up her collections herself—to take a step toward career science—she never did. It was not that she doubted her intellectual capacity ... or felt constrained by gender roles. Rather, it was that everything she did she had to do in a first-rate way, and she knew that to do first-rate science she would have to submit to formal academic training, which was not to her liking. Besides, as a hunter and collector she already had just about everything she wanted.[75]

In 1911, Alexander purchased 525 acres of undeveloped land on remote Grizzly Island, about 40 miles northeast of Oakland.[76] The land would ensure financial security and independence for Kellogg in later life. They grew hay and bred milking shorthorn cattle. Later they sold their herd and switched to growing asparagus—a good choice because asparagus was a spring crop, freeing summers and autumns for field trips and winters for visits to Hawaii and for desert explorations. Their last great offshore adventure was a plant-collecting trip to Baja California in 1947. Alexander (age 79) and Kellogg (age 66) were joined by 40-year-old Annetta Carter, who worked at the University Herbarium. The three women returned with specimens representing more than 700 species, including undescribed taxa and many range extensions.

Alexander suffered a stroke in November 1949. She lapsed into a coma and died 10 months later. Throughout her life, Alexander had been aware that many women felt stymied by the limits set upon them by men. She also knew that financial independence and a determined personality had empowered her to follow her passion. She shunned publicity, did not want to have buildings named after her and didn't particularly want to have species named after her. As reflected in the title of Alexander's biography by Barbara Stein, Annie Alexander lived life "on her own terms."[77] Kellogg lived another 17 years. She made her last expedition, to Baja California, at the age of 80. Kellogg left an endowment for the University Herbarium, with funds to be used to support botanists working with the collection. In 1991, the women's beloved ranch became part of the Grizzly Island Wildlife Area.

Mary Cynthia Dickerson (1866–1923) began life in a small town in central Michigan, where her father was a grocer.[78] She graduated with a bachelor's degree from the University of Chicago in 1897. For 8 years, she was head of zoology and botany at Rhode Island State Normal School (later renamed Rhode Island College), where she taught nature study and honed her skills as a naturalist while making observations for the two books she would write.

In 1901, Dickerson published her first book, *Moths and Butterflies*. She also contributed to *Country Life in America*, a lavish folio publication designed for wealthy gentleman farmers, in which she focused on small woodland animals—toads, mice, squirrels, muskrats, and shrews, charting their lives through the seasons. Her second book, *The Frog Book* (1906), contained over 300 of her photographs; the narrative

---

[75] Kohler (2006), p. 219.
[76] Stein (2001), p. 120.
[77] Stein (2001). The title of the biography is *On Her Own Terms: Annie Montague Alexander and the Rise of Science in the American West*.
[78] Fabian (2013).

was derived from field notes she made over a period of 10 years. Dickerson wrote the book "with the direct aim of making these forms known and appreciated by people interested in nature."[79] The book became a classic, valued by amateurs and professional biologists alike. It is still treasured for its illustrations and accurate descriptions of the appearance, life histories, and habits of North American frogs.

In 1908, Dickerson was hired by the American Museum of Natural History in New York City to work in the Department of Woods and Forestry. The following year, the Department of Ichthyology and Herpetology was established, with three ichthyologists and Dickerson as the sole herpetologist. She now had a three-pronged job: exhibition work, collection-building, and establishing a library and center for herpetological research.[80] She corresponded with collectors and sought new ones, oversaw the construction of dioramas relating to amphibians and reptiles, and edited expedition reports. From 1910 to 1920, she edited the *Museum Journal* (later renamed *Natural History* magazine). She described over 20 new species of reptiles. In February 1920, the departments split, and Dickerson became the first curator of the Department of Herpetology. Her tenure in that position was short-lived, however. She developed mental illness, and by late December of that year, she was committed to an asylum on Wards Island. She died there 3 years later, at age 57, having established the American Museum of Natural History as a major research center for herpetology (Figure 3.6).

Edith Marion Patch (1876–1954) was born in Worcester, Massachusetts, the youngest of six children. She kept caterpillars, toads, and green snakes as pets, and she loved exploring her natural surroundings. A childhood incident foreshadowed her future career. She read a story about a cabbage caterpillar that metamorphosed into a yellow butterfly. She knew better: cabbage butterflies are white. Worse, the accompanying illustration was a night-flying moth. She vowed that someday she would write biologically accurate nature books for children.[81]

In 1884, the family moved to Minnesota, where Patch became fascinated with the plants and animals on her family's 10 acres of prairie. She was particularly intrigued with the monarch butterfly life cycle. During her senior year in high school, Patch wrote a prizewinning essay about monarchs. With the $25 award, she bought the book she had most wanted—John Henry Comstock's *Manual for the Study of Insects*, illustrated by Anna Comstock (profiled in Chapter 4). She entered the University of Minnesota, initially interested in English. During her university years, she won prizes for her sonnets and romantic stories, but she was increasingly captivated by insects. She graduated in 1901 with a Bachelor of Science degree.

Patch began the discouraging process of applying for entomology positions, during which she was told repeatedly that entomology was not a suitable profession for women.[82] She taught high school English for 2 years and continued to search for an entomology job. One day she received a letter from Dr Charles Woods, director of the Maine Agricultural Experiment Station, offering her a 1-year, unpaid position to

[79] Dickerson (1969), p. xiii.
[80] Myers (2000); http://hdl.handle.net/2246/1599; accessed 22 February 2018.
[81] Bonta (1991), p. 174.
[82] Bonta (1991), p. 175.

FIGURE 3.6   Mary Cynthia Dickerson (1866–1923). Image in the public domain.

organize a Department of Entomology at the University of Maine. After that year she might be appointed head of that department. She accepted. Woods arranged for her to teach entomology at the university to provide some income. Ridiculed for hiring a woman, one of Wood's responses was "So far as the people on my staff are concerned, I am not at all concerned whether they are attired in trousers or skirts, just as long as they do the work."[83] According to Patch:

> When one of the leading Bureau of Entomology men remarked to Doctor Woods ... "I hear you have appointed a woman as entomologist. Why on earth did you do that? A *woman* can't catch grasshoppers," he received the drawled reply, "It will take a lively grasshopper to escape Miss Patch."[84]

Patch organized a Department of Entomology at the University of Maine, became head of that department in 1904, and worked there until she retired in 1937. In 1910, she received her master's degree from the University of Maine. She fulfilled her dream of working with John Henry Comstock and received her PhD from Cornell in 1911. She amassed one of the most complete collections of aphid species in the world, for the Maine Agricultural Experiment Station. In 1938, Patch published *Food Plant*

---

[83] https://entomologytoday.org/2015/08/13/remembering-edith-patch-the-first-female-president-of-the-entomological-society-of-america; accessed 1 June 2018.
[84] Bonta (1991), pp. 175–176.

Catalogue of the Aphids of the World, still one of the most comprehensive books on aphids and their host plants.

Patch considered herself a combination scientist-naturalist, both specialist and generalist.[85] She divided her day by spending mornings as a professional entomologist and afternoons as a field naturalist.[86] She understood the power of story to communicate both scientific ideas and the wonder and beauty of nature.[87] She fulfilled her vow to write biologically accurate children's books—17 of them, ranging from *Dame Bug and Her Babies*, a collection of stories about insect mothers and their babies, to *Little Gateways to Science*, a story about the life histories of a dozen birds and the negative effects humans can exert on nature.[88] Patch's enthusiasm for discovery is evident throughout her children's books, in which she encouraged the reader to engage in nature exploration.

In 1924, Patch became director of the Maine Agricultural Experiment Station, the first female to lead a U.S. agricultural station.[89] Six years later, she earned another first when she became the first woman president of the Entomological Society of America. She foreshadowed Rachel Carson's book *Silent Spring* when in 1936 she gave a speech in a Maine Agricultural News Radio Program about the dangers of insecticide overuse.[90] Not long afterward, she spoke of the effects of chemical insecticides on nontargeted insects during an address to the Entomological Society of America. Patch left behind more than 100 professional and popular writings. She encouraged children to love nature, and she provided inspiration for other women aspiring to become entomologists.

Patch was determined to get a job in entomology at a time when it was rare for women to study insects. Not only was she successful, she was accepted by her male peers, as reflected in the following story from *Entomology Today*, the news bulletin of the Entomological Society of America:

> Patch was expected to adhere to certain societal etiquettes, only some of which she followed. But her polite, often wordless deviation from the norms of her time helped pave the way for the success of women in science. When Patch was discouraged from attending an after-dinner address during a meeting of the Entomological Society of America because the men would be smoking (women were not allowed to be in the presence of a man while he smoked during this time), she figured out where the meeting was, walked in and quietly took a seat. The smoke-filled room fell silent as the men looked side-to-side, eyebrows raised. Within seconds, every cigar and pipe in the room had been put out. She was present at all subsequent meetings.[91]

---

[85] Bonta (1991), p. 178.

[86] Bonta (1991), p. 178.

[87] https://entomologytoday.org/2015/08/13/remembering-edith-patch-the-first-female-president-of-the-entomological-society-of-america/; accessed 1 June 2018.

[88] https://entomologytoday.org/2015/08/13/remembering-edith-patch-the-first-female-president-of-the-entomological-society-of-america/; accessed 1 June 2018.

[89] Bonta (1991), p. 177.

[90] https://entomologytoday.org/2015/08/13/remembering-edith-patch-the-first-female-president-of-the-entomological-society-of-america/; accessed 1 June 2018.

[91] https://entomologytoday.org/2015/08/13/remembering-edith-patch-the-first-female-president-of-the-entomological-society-of-america/; accessed 1 June 2018.

Of the 11 women profiled for this time period, some were independently wealthy (e.g., Eastwood and Alexander) while others struggled financially (e.g., Chase). Some were highly educated (e.g., Brandegee and Patch) while others received minimal formal education (e.g., Chase). Six were married; of these, Britton, Brandegee, and Bailey all worked extensively in the field with their husbands. None had children. All faced challenges pursuing their scientific passions during times when their fields were dominated by men. Yet, with the exception of Beatrix Potter in England, all earned the respect of their male colleagues during their lifetimes. Many received honorary academic degrees, awards, and were elected Fellows of scientific societies.

The proof offered by women of the late Age of Discovery (Chapter 2)—that women had both the intellectual capacity and physical stamina to do not only science but also field-based science—carried forward into the Gilded Age. This appears to have been especially true in the United States, where professional opportunities were being created in the East by women's colleges and in the West by the freedom of the frontier. These trends would be accelerated during the Interwar Years (Chapter 4), by what was essentially a combination of formal education and opportunity—the formation of Midwestern Land Grant Universities and their early emphasis on the new science of ecology.

REFERENCES

Anderson, J. G. T. 2013. *Deep things out of darkness: a history of natural history*. Berkeley, CA: University of California Press.

Bonta, M. M. 1991. *Women in the field: America's pioneering women naturalists*. College Station, TX: Texas A&M University Press.

Bonta, M. M. 1995. *American women afield: writings by pioneering women naturalists*. College Station, TX: Texas A&M University Press.

Dickerson, M. C. 1969. *The frog book: North American toads and frogs*. Reprinted. Toronto, Ontario: Dover Publications.

Fabian, A. 2013. Charming toads. *Michigan Quarterly Review* 53(1). http://hdl.handle.net/2027/spo.act2080.0052.102; accessed 22 February 2018.

Kofalk, H. 1989. *No woman tenderfoot: Florence Merriam Bailey, pioneer naturalist*. College Station, TX: Texas A&M University Press.

Kohler, R. E. 2006. *All creatures: naturalists, collectors, and biodiversity, 1850–1950*. Princeton, NJ: Princeton University Press.

Lear, L. 2007. *Beatrix Potter: a life in nature*. New York: St. Martin's Press.

Lipscomb, D. 1995. Women in systematics. *Annual Review of Ecology & Systematics* 26:323–341.

Moring, J. 2002. *Early American naturalists: exploring the American West 1804–1900*. New York: Cooper Square Press.

Myers, C. W. 2000. A history of herpetology at the American Museum of Natural History. *Bulletin of the American Museum of Natural History*, Number 252. American Museum of Natural History, New York.

Norwood, V. 1993. *Made from this Earth: American women and nature*. Raleigh, NC: University of North Carolina Press.

Ogilvie, M. B. 1986. *Women in science: antiquity through the nineteenth century. A biographical dictionary with annotated bibliography*. Cambridge, MA: The MIT Press.

Rudolph, E. D. 1982. Women in nineteenth century botany: a generally unrecognized constituency. *American Journal of Botany* 69:1346–1355. https://www.jstor.org/stable/2442761.

Shearer, B. F., and B. S. Shearer, eds. 1996. *Notable women in the life sciences: a biographical dictionary*. Westport, CT: Greenwood Press.

Shushkewich, V. 2012. *More than birds: adventurous lives of North American naturalists*. Toronto, Canada: Dundurn.

Stein, B. R. 2001. *On her own terms: Annie Montague Alexander and the rise of science in the American West*. Berkeley, CA: University of California Press.

Swallow, P. C. 2014. *The remarkable life and career of Ellen Swallow Richards: pioneer in science and technology*. Hoboken, NJ: John Wiley & Sons.

Swanson, R. L. 2013. Clean up our home: Ellen Swallow Richards' human ecology and emerging environmental ideologies, 1890–1915. Honors Program Theses. 50. https://scholarworks.uni.edu/hpt/50; accessed 30 January 2022.

4 1917 to 1945 (War and Interwar Years)

The discipline of ecology created tremendous scientific opportunities for women. Ernst Haeckel coined the term "oecologie" in 1866, but it took another half century for the idea of a science centered around the interactions of organisms with their environment to mature to the point of having practitioners, professional societies, meetings, and journals. Ecology arose because once the incredible diversity of species became understood, field biologists became curious about how these species interacted. While natural historians passively described nature and collected specimens, ecologists were aggressively quantitative. Aldo Leopold recognized the difference. After interviewing an older candidate to supervise the prairie restoration at the University of Wisconsin's Arboretum, Leopold wrote:

> He is a kindly elderly gentleman of rather wide experience in horticulture and soils, with a good botanical background, but no ecology, since there was no such science in his day. … As a test of his command of ecological science: he had never heard of a quadrat.[1]

The activities of most of the women field biologists profiled here have, up to this point, fallen under the category of natural history—the description of nature. But the opening of land grant institutions beginning in the 1860s and later the interests of their faculty in this new discipline of ecology created increased opportunities in state-run coeducational institutions for women wishing to make science a career. Other disciplines based in ecology, such as wildlife biology, conservation biology, and restoration ecology, lay in the future, but during the Interwar Years, women interested in field-based science were becoming ecologists.

In the wake of America's first ecologist the Illinoisan Stephen Forbes, four great early ecological "schools" developed and grew: the Nebraska/Minnesota school of plant ecology, the Chicago schools of plant and animal ecology, and the Wisconsin school of limnology.[2] It was no accident they flourished in the Midwest. The University of Chicago gave ecology a central position in its curriculum. Cornell University, the only Ivy land grant, also embraced the field of ecology. At these schools, women were accepted and nurtured, as this open mindedness to new disciplines such as ecology signaled an open mindedness to the types of students the land grants and Chicago admitted. Novelist Willa Cather, who attended the University of Nebraska, was deeply influenced by the ecologists there, a connection reflected in passages such

---

[1] Court (2012), p. 103.
[2] Egerton (2014) combines the Chicago plant and animal schools. We split them because the trajectories of their descendants are so distinctly different. For example, Chicago's Shelford collaborated not with the descendants of Chicago's Cowles but rather with Nebraska's Clements to write *Bio-Ecology*.

DOI: 10.1201/9781003311508-5

as "the grass was the country," and ... "its roots ran deep and shaped the heartland of a nation."[3]

Just as women had been drawn to natural history, they became interested in ecology. At the University of Illinois, Forbes' institution, 100% of pre-1920s PhD degrees in ecology and 50% of 1920–1929 PhDs in ecology were awarded to women.[4] At Cornell, women received 50% of the pre-1920s PhD degrees in ecology.[5] (Unfortunately, absolute numbers on which these percentages are based were not provided.) In this chapter, we profile women field biologists who were educated and shaped by these four "ecological schools," as well as selected other field biologists who worked during this time.

THE CHICAGO PLANT ECOLOGISTS

Harriet George Barclay (1901–1990) was born in Minnesota. After receiving her BS and MA degrees in botany from the University of Minnesota, she completed her PhD in plant ecology working with botanist and pioneer ecologist Henry Cowles at the University of Chicago.[6] Soon after graduating, she married fellow former graduate student B. D. Barclay. The couple obtained appointments in the Botany Department at the University of Tulsa, which she would chair and where she taught until her retirement in 1971. She enjoyed taking students to the field, and it was estimated that she spent 18 summers at the Rocky Mountain Biological Laboratory (RMBL), where she taught courses in plant ecology and floral identification.[7] Barclay was a gifted teacher.

Barclay's interest in high altitude plants led her later in life to study alpine plants in South America. She also explored Africa, Australia, and Asia. Barclay was known for her paintings and photographs of her study sites. She led efforts to identify and preserve some of Oklahoma's unique natural areas, such as the 83-acre Redbud Preserve north of Tulsa. In 1959, she was named Woman of the Year by the American Women in Radio and Television, and in 1971, she was named the Conservationist of the Year by the Oklahoma Wildlife Federation. In 1976, she was inducted into the Oklahoma Hall of Fame.

Mildred E. Faust (1899–1988) was born in Emporia, Kansas. She received her BS from Penn College of Iowa and her PhD in 1933 with Henry Cowles at the University of Chicago. She was a professor of botany at Syracuse University for 39 years and had an adjunct position across the street at the State University of New York's College of Environmental Science and Forestry. Faust created an exhaustive floral list for Onondaga County and provided key contributions to the flora of New York State. She is best known for her collaboration with Edward Knobel on *The Field Guide to the Grasses, Sedges, and Rushes of the United States* and for her 1936 paper "Germination of *Populus grandidentata* and *P. tremuloides*, with particular

---

[3] Court (2012), pp. 218–219, Cather (1918), p. 13.
[4] Langenheim (1996).
[5] Langenheim (1996).
[6] From Dr Jean H. Langenheim, circa 1985 at https://esa.org/history/harriet-george-barclay-tributes-to-a-teacher; accessed 18 March 2019.
[7] Langenheim notes "she provided inspiration for generations of students through her infectious enthusiasm for plant ecology."

reference to oxygen consumption." Faust continued the Chicago tradition of relying on field trips to demonstrate ecological concepts to students, often taking students by train (another Chicago tradition) to Clark Reservation State Park. She was known for her witty, energetic outings to explore ferns and spring wildflowers. Faust has a trail named after her at Clark Reservation State Park, and in 1979 the Syracuse Botanical Club established the Mildred E. Faust Herbarium.

Henry Cowles heavily influenced botanist and ecologist Emma Lucy Braun (1889–1971), who received her PhD at the University of Cincinnati in 1914. Braun was born in Cincinnati, Ohio.[8] "She was born a Victorian, and she died a Victorian in 1971 at 82 years of age. Times change, but she did not change."[9] She received all her degrees at the University of Cincinnati and served on the faculty there for the rest of her life.[10]

Braun was 5 years younger than her sister Annette, who would become a distinguished entomologist. Their parents encouraged the girls' interest in nature by taking them by street car to a nearby woods to explore. While in high school, Braun began creating her herbarium—a lifelong pursuit—by collecting and pressing local plants.[11] After their parents died, the Braun sisters lived in a limestone house on two wooded acres. They lived a spartan life, eating simply and limiting social engagements. Braun's faculty salary covered their expenses; Annette kept house. In the field, Annette was Braun's partner, and after the data were collected, she assisted in editing Braun's publications.[12]

Braun was demanding and always in complete control. Although she rarely praised or thanked anyone, and did not handle criticism well, her students respected her vast knowledge, and several became professional botanists.[13] She advised many female graduate students.[14] She wrote 4 books and published 180 articles in 20 journals.[15]

Braun was one of America's first female academic field biologists. She taught her own courses, advised her own graduate students (uncommon for a female faculty member at that time), and developed her own research. Women before her often had their ideas subsumed by men. For example, Harriet Bell Merrill's work went unacknowledged by Edward Birge, and the intellectual contributions of Edith Clements and Rosa Eigenmann were incorporated into the works of their husbands. Braun challenged some of the great botanists of her time—including Henry Gleason, Bohumil Shimek, Ed Deevey, and Paul Sears—on the basis that their conclusions were drawn from incomplete data sets.

---

[8] Most of this account comes from Stuckey (2001).

[9] Durrell (2001).

[10] BA in 1910, MA in geology in 1912, and PhD in botany in 1914. From 1914 to 1917, she was an assistant in botany; from 1917 to 1923 she was an instructor in botany; from 1923 to 1927 she was an assistant professor of botany; from 1927 to 1946 she was an associate professor of botany; from 1946 to 1948 she was a professor of plant ecology; and from 1948 until her death in 1971 she was an emeriti professor of plant ecology.

[11] Durrell (2001).

[12] Durrell (2001).

[13] Durrell (2001).

[14] Stuckey (1973); these included Sylvia Geisler (1926), N. Mildred Irwin (1929), Alice P. Withrow (1932), Dorothy Parker (1936), and Isabel Thompson (1939).

[15] Stuckey (1973).

FIGURE 4.1   Emma Lucy Braun; image in the public domain.

Her specialty was phytogeography (distribution of plants) and the factors determining these distributions. As she described,

> The interpretation of existing phytogeography must be based on evidence from a number of unlike sources: from the distribution pattern itself, the relationship of species and communities, their geographic extent or degree of continuity of disjunction; from the physical features of the environment, which alone or in part may explain existing distribution; from hybrids and population variation; and from fossils.... Much of the information of significance for phytogeographic studies is hidden in a mass of irrelevant material.[16]

She also placed great emphasis on glacial relict populations (Figure 4.1).[17]

From such insights, Braun developed her theory about the formation of vegetation characteristics and composition in the eastern deciduous forests in general and southwestern Ohio in particular:

> Each of the five or six great glacial advances forced a southward migration of vegetation. ... Arctic and coniferous belts intervened between the ice and the more southern deciduous forest. These [vegetation belts] were not wide, for the effects of glacial refrigeration do not seem to have extended far beyond the limits of the ice cap.[18]

---

[16] Braun (1955).
[17] Braun (1928).
[18] Braun (1928).

This statement—that the once-northern vegetation belts were narrow south of the glacial boundary—was her most controversial.

Braun was a charter member of the Ecological Society of America, established in 1915. She became the first woman elected president of the society in 1950. To honor her, Ecological Society of America established the E. Lucy Braun Award for Excellence in Ecology, given at the society's annual meeting to a student for the outstanding poster presentation.

THE CHICAGO ANIMAL ECOLOGISTS

Victor Shelford trained at the University of Chicago under Cowles, and following his graduation, Stephen Forbes invited him downstate to anchor the University of Illinois' ecology program. Shelford's earliest female PhDs have been nearly lost to history, despite publishing numerous papers in high-profile journals. For example, Minna E. Jewell (1892–1985) received her PhD in 1918 from the University of Illinois. Following her graduation, she taught at Kansas Agricultural College (now Kansas State University) and Thornton Junior College (now South Suburban College) in Harvey, Illinois. Her research interests included the biology of northern bog lakes and prairie streams, groundwater, and the ecology and systematics of freshwater sponges. Her 1935 paper on sponges was based on collections made from 103 lakes and 15 streams in northeastern Wisconsin. Although she published many of her results in *Ecology* and *Ecological Monographs*, she received little recognition for her groundbreaking work from either ecologists or limnologists (freshwater biologists). She foreshadowed Dr Ruth Patrick (profiled on pages 98–100) in 1922, when she presented a paper entitled "The Sangamon River—A study in stream pollution."

Shelford's second female PhD graduate was Martha Wheatley Shackleford (1898–1982), who received her PhD working on animal assemblages of the prairie peninsula. She published her dissertation research in the journal *Ecology*.[19] After graduation, Shackleford became head of the Department of Zoology at the Oklahoma College for Women[20] in Chickasha. She continued her research through a series of papers published in the *Oklahoma Academy of Science* on prairie invertebrate community composition. Her research was fieldwork intensive. For example, for her 1931 paper,[21] she conducted 1,250 insect net sweeps.

Jane Claire Dirks-Edmunds (1912–2003) was born in the Ozark Mountains of Arkansas, the youngest of ten children. When she was 12, her parents moved to Washington State, finally settling in Umpqua Valley, Oregon. Dirks-Edmunds attended Linfield College (now University) from 1932 to 1937, where she received her BS in biology. She earned her PhD at the University of Illinois under Shelford; her dissertation was published in *Ecological Monographs*.[22] Following graduation, Dirks-Edmunds returned to Linfield—the first female PhD hired by the institution. In 1944, she spent three semesters at Whitworth College (now University) as the

---

[19] Shackleford (1929).
[20] Today, the University of Science and Arts of Oklahoma [USAO].
[21] Shackleford (1931).
[22] Dirks-Edmunds (1947).

FIGURE 4.2   Jane Dirks-Edmunds; image in the public domain.

head of the biology department. She returned to Linfield in 1946 and taught until her retirement in 1974 (Figure 4.2).

Dirks-Edmunds is best known for her research on a single hectare plot on Saddleback Mountain in Oregon's Coastal Mountain range. The site had been logged in the early 1900s and then logged a second time during the Second World War. The area became subjected to wildfires and remained in early successional stages, never recovered as a mature forest, and was subject to fires afterward. She, along with her mentor James Macnab, studied the soil microbiota of this former Douglas fir and hemlock community. Dirks-Edmunds continued this research after Macnab retired. She studied the Saddleback Mountain plot from 1935 to 1969. In 1983, Dirks-Edmunds summarized her research in her book, *Not Just Trees*.[23]

THE NEBRASKA/MINNESOTA PLANT ECOLOGISTS

The Nebraska/Minnesota School of Plant Ecology was built around Frederic Clements and his appointments at the Universities of Nebraska and Minnesota, and the Carnegie Institute. Two women, Edith Schwartz (who would marry Clements) and Francis Louise Long, received PhDs working with him.

---

[23] Dirks-Edmunds (1947).

Edith Schwartz (1874–1971) was born in Albany, New York. Shortly afterward her family moved to Omaha, Nebraska.[24] She attended the University of Nebraska and during her senior year met Clements. They married in 1899 and honeymooned in a cottage at Pikes Peak, Colorado. While there, they hatched the idea for a summer research laboratory in the Rocky Mountains—what would become the Alpine Laboratory at Pikes Peak.

After marrying Clements, she entered the PhD program in botany. She studied the microscopic anatomy of leaves and received her doctorate in 1907, the first woman to earn a PhD from the University of Nebraska. The couple began working together, but not equally, as she would describe in her memoir *Adventures in Ecology: Half a Million Miles: From Mud to Macadam* (1960): "He furnished the brains and I the manual dexterity [which] led to a division of labor. ... [Me] sitting on the running-board of the current automobile and taking his dictation of field notes."[25] The couple worked together throughout their lives, with Schwartz devoting much of her effort to furthering her husband's career. For many years after he died, she continued working on their joint manuscripts.

Frances Louise Long (1885–1946) was born in Madison, Nebraska.[26] She attended the University of Nebraska and received her BA and BS degrees in 1906. She followed Clements to the University of Minnesota, where she received her MA (1914) and PhD (1917). After graduation, she joined Clements at the Carnegie Institution for Science in Washington, D.C., and worked with him at his Alpine Laboratory at Pikes Peak. Her research involved a wide range of topics including the effects of light, desiccation, age, and low nutrient availability on plant growth and function. She was also interested in pollination, and, like Clements, in improving research techniques. Little biographical information exists on many of Clements' students,[27] and this is true of Long. We know that she was a talented experimental biologist and facile with laboratory instruments.[28]

These progressive ecology-oriented schools began hiring the women field biologists they trained. One was Josephine Elizabeth Tilden (1869–1957).[29] Born in Davenport, Iowa, Tilden's family moved to the Twin Cities, where she excelled in high school and was admitted to the University of Minnesota. Before graduating with her bachelor's degree in 1895, she spent a summer in Yellowstone, where she was interested in blue-green algae and their associations with hot springs. Tilden was among the first scientists—male or female—to collect algae from these unique habitats. Following graduation, she was invited to join the University of Minnesota faculty to continue her research, with one stipulation—that she stay at least 5 years and provide a research plan. She proposed studying algae of the Pacific Ocean, the administration agreed, and in 1895 Tilden became the first female faculty member appointed at the University of Minnesota.

---

[24] Much of this account has been extracted from Oberg (2019); accessed on 14 July 2020.
[25] Clements (1960). While Hagen (1993) calls this book "light-hearted," we have a different opinion. Much of the book is about the unreliable cars and bad roads of the 1920s (it was published in 1960), and there are several extraordinarily insensitive comments about race, which might be forgiven except for her husband's outspoken advocacy of eugenics.
[26] www.plantphysiol.org; accessed 14 July 2020.
[27] Hagen (1993); https://academic.oup.com/plphys/article/21/3/371/6073249; accessed 30 January 2022.
[28] Hagen (1993).
[29] Hansen (1996).

The following year she began collecting algae in Washington and British Columbia. Tilden's strategy involved asking local fishermen the location of rich algal beds. On the outer coast of Vancouver Island, she discovered "a great sandstone shelf in which boulders have ground innumerable cistern-like pot-holes varying in size from mere tea cups to great wells 30 feet across and 20 or more in depth."[30] She named the area Botanical Beach and wrote:

> The algae covering the exposed shore were beyond my wildest dreams. We remained four days with only a two-quart jar of cooked beans to eat and tea to drink, collecting algae every daylight moment. Wet to the skin, with no shelter from the rain, those four days were the happiest I have ever spent.[31]

The property was owned by homesteader Tom Baird. Tilden asked him about establishing a marine field station there, and Baird responded, "If you like the area so much, I'll give you a deed to my best 4 acres—take your pick."[32] Tilden floated the idea back in Minnesota, and in 1901 it came together. Four universities—Minnesota, Colorado, North Dakota, and Iowa—shared the Minnesota Seaside Station on Vancouver Island, which ran for 5 years with student and faculty funding, then with Tilden's funding. In 1907, the station was closed and abandoned.

Tilden had her professional disagreements. She blackballed botanist William Setchell because of his "unpleasant spirit" and "ungentlemanly tone" in criticizing her dried specimens.[33] DeAlton Sanders, the botanist on the 1899 Harriman Expedition to Alaska, wrote "I believe [Tilden's] work for pure superficiality cannot be equaled."[34] Botanist William Farlow at Harvard wrote "Miss Tilden's specimens are worse than any I ever saw ... A good part of the ... numbers I examined are incorrectly named."[35]

Despite this criticism of her taxonomic skills, Tilden flourished at the University of Minnesota. In 1903, she was granted her MS degree and promoted to assistant professor; in 1910, she became full professor. Her accomplishments included writing *The Algae and Their Life Relations* (1935), the first textbook on phycology; two books on collections of dried algal specimens—*American Algae* and *South Pacific Algae*; a bibliographic card index to the world literature—*Index Alagrum Universalis*; and the first American guide to the blue-green algae—*The Myxophyceae of North America and Adjacent Regions* (1910).

In 1934–1935, Tilden took ten students on a field expedition to the South Pacific. By the time they reached New Zealand, their funds had run out, so she borrowed several thousand dollars from the university. She was certain that sales from the exsiccate (they brought back over 51,000 algal and 52,000 vascular plant specimens) would cover the loan. When they did not, she was forced to retire from the University of Minnesota. Tilden then moved to Florida and established a commercial citrus grove. Financial problems dogged her throughout her life, and she died a pauper.

---

[30] Hansen (1996), p. 186.
[31] Hansen (1996), p. 187.
[32] Hansen (1996), p. 187.
[33] Silva (2009).
[34] Silva (2009).
[35] Silva (2009).

THE WISCONSIN LIMNOLOGISTS

Harriet Bell Merrill (1863–1915) was born in Stevens Point, Wisconsin.[36] Her father was a lumber businessman. Her mother, Anna Comstock Emmons (apparently not related to Anna Botsford Comstock, discussed later in this chapter), was a school-teacher from upstate New York. Merrill graduated from the University of Wisconsin summa cum laude in 1890 with her BS; 3 years later she received her MS, also from Wisconsin, working with Edward A. Birge on zooplankton in the order Cladocera (e.g., *Daphnia* spp.). Following graduation, she did research at the Marine Biological Laboratory at Woods Hole. In 1890, she became director of the Physiology and Biology Departments at Milwaukee's East and South high schools. Seven years later, she became head of the science department at Milwaukee Downer College, where she taught chemistry, biology, zoology, anatomy, physiology, and psychology.

From 1902 to 1903, and again from 1907 to 1909, Merrill conducted expeditions throughout South America and the Caribbean. She was small, always traveled alone, and made a fashion statement by wearing men's boots with her skirts hiked up, loaded down with heavy cameras and field collecting equipment. She became legendary as "the courageous American woman."[37] She kept a detailed journal and wrote volumes of letters to her friends, vividly describing the countryside, Indigenous cultures, and her experiences. Her second trip to South America focused on zooplankton, and from her collections of ~700 samples, she described 82 taxa—several of them new to science. She also collected Indigenous artifacts, which she deposited in the Milwaukee Public Museum. The Milwaukee Sentinel published her letters from the field. She returned to the University of Wisconsin as an assistant professor in zoology and gave public lectures at Cornell, Chicago, and the University of Wisconsin. She had health problems for most of her adult life, but they never slowed her down. At the time of her death from myocarditis at age 52, she was enrolled in a PhD program at the University of Illinois.

Merrill has only now begun to achieve recognition.[38] Birge considered her letters, notebooks, and microscope preparations important enough to be saved, but he did not identify her contributions in his publications. (During this time, Birge was considered the world's expert on zooplankton). After Birge's death in 1950, limnologist David Frey at Indiana University acquired Birge's cladoceran material. He found notebooks and slide preparations of South American species, but it was only after one of Frey's students discovered Merrill's letters to Birge that he realized these had been collected by Merrill. As Frey described,

> Merrill never published anything about these collections, but her notes indicate a very high level of understanding of the Cladocera. The notebooks contain many records of first collections of particular species in South America and the presence of species not yet described ... She was potentially connected with Birge for 27 years, yet there is no mention of her anywhere.[39]

---

[36] https://writingroughshod.com/2016/08/16/just-the-facts-maam-resources-on-harriet-bell-merrill; accessed 20 June 2019.

[37] Beckel (1990); accessed 9 July 2020.

[38] Beckel (1990); accessed 9 July 2020.

[39] Beckel (1990), pp. 3–4.

THE CORNELL LEGACY

During the late 1800s and early 1900s, women in the United States were finally being offered faculty positions in coeducational academic institutions. When women were hired, however, they usually were not promoted.[40] One early exception was Anna Botsford Comstock, who rose to full professor at Cornell University.

Nature educator extraordinaire, Comstock (1854–1930) was born to a prosperous Quaker farming couple in western New York. Her parents gave her superb education opportunities and encouraged her to be a freethinker; her mother instilled in her a love of nature. At 17, she was teaching school full time. After someone suggested she enroll at the new school that had recently opened its doors to women, she enrolled at Cornell in 1874. Comstock's favorite class was entomology, taught by John Henry Comstock. He encouraged her to draw, and she discovered that drawing helped her see and appreciate the details of insect morphology.

They married in 1878. Comstock helped keep her husband's lab in order, wrote his business letters, and illustrated his lectures.[41] In 1879, he was offered the position of chief entomologist at the USDA, and they moved to Washington, D.C. The Comstocks returned to Cornell in 1881, where she resumed her education and received her BS degree in 1885. She learned wood engraving to illustrate her husband's next book *An Introduction to Entomology*. After it was published (1888), the quality of its illustrations earned her an international reputation. Her husband founded the Comstock Publishing Company, which published their next book *Manual for the Study of Insects* (1895).

About this same time, Anna Comstock was swept into the Nature Study Movement. Its mantra, borrowed from Louis Agassiz, was "study nature, not books." A program of nature study in New York had begun at Cornell, under the direction of horticulturist Liberty Hyde Bailey. For a year, Comstock surveyed New York public schools to see if they taught nature study. Most did not, and many teachers felt they did not know enough to teach nature study to their students. Comstock and several others created instructional leaflets for teachers. In 1897, Comstock was appointed assistant professor of nature study. Cornell quickly withdrew her appointment, though not her salary and responsibilities, because the Board of Trustees objected to a woman having professorial status—it might damage the university's reputation. She was demoted to lecturer.

In 1909, she began writing a *Handbook of Nature Study*, designed as a guide for teachers.[42] Every publisher she approached rejected the manuscript because of its massive size. Her husband felt that the book would be expensive to produce and that it would lose at least $5,000. Nonetheless, Comstock Publishing Company published the book in 1911. It was 938 pages, with 232 planned lessons, suggested field trips, experiments, and questions to engage students. The book sold well. Comstock was given the title of assistant professor in 1913, this time to keep. Six years later she was promoted to full professor. She retired in 1921, the first female professor at Cornell (Figure 4.3).

---

[40] Lipscomb (1995), p. 333.
[41] Bonta (1991), p. 156.
[42] Norwood (1993), p. 80.

FIGURE 4.3   Anna Botsford Comstock; image in the public domain.

In 1923, the League of Women Voters voted Comstock one of the twelve great-est women in America within their chosen fields.[43] When she told her husband that she felt unworthy of the honor, he assured her that she deserved the honor based on *Handbook of Nature Study* alone. The book had already sold over 40,000 copies, bringing the love of nature to countless numbers of children.

Anna Comstock was one of the first educators to encourage teachers to take their students *outside* to study nature. The *Handbook* has now gone through 25 editions and has been translated into eight languages; it is still in print. Her colleague Liberty Hyde Bailey wrote: "Anna Botsford Comstock blessed us all. She leaves a fragrant memory of high achievement, noble service, unselfish cooperation, constructive counsel, inspired teaching, loving kindness and unforgettable companionship. Her life was a poem."[44]

Emmeline Moore (1872–1963) was born in Batavia, New York, where she grew up on a farm. She graduated from Geneseo Normal School,[45] and then taught classes to earn money to cover her tuition and living expenses at Cornell University. A year after she earned her BA in 1905, she earned her MA at Wellesley College. Following

---

[43] Bonta (1991), p. 165.
[44] As quoted in Bonta (1991), p. 166.
[45] Now SUNY Geneseo.

teaching positions in the United States and South Africa, she returned to Cornell and earned her PhD in biology (1916). Moore taught biology at Vassar College and was promoted to assistant professor. Her passion was research, however. She began working at the U.S. Bureau of Fisheries during summers and eventually parlayed this part-time work into a full-time job as chief aquatic biologist and director of the New York State Biological Survey. She studied rivers, lake pollution, and fish diseases while working with the New York Conservation Department. Moore produced some of the best early state surveys of aquatic resources in the state. The respect she earned from the quality of her research resulted in her being elected the first woman president of the American Fisheries Society (AFS). She also published technical papers on fish culture and fish diseases. After her retirement, she continued research at the Yale Laboratory of Oceanography.

Moore was twice honored with the Walker Prize from the Boston Society of Natural History and was awarded an honorary degree from Hobart College. A New York oceanographic research ship was named in her honor. The AFS established the Emmeline Moore Prize. The prize is a prestigious career achievement award to recognize AFS members who display a strong commitment to diversity issues and to those who encourage greater involvement of underrepresented groups in fisheries science, education, research, or management.

In 1937, only a handful of biologists were banding birds. One of them was Amelia Marguerite Baumgartner (1909–2004).[46] Baumgartner was born in Rochester, New York, and received her BS at the University of Rochester. She spent three summer sessions at the Allegeny School of Natural History, where her interests in birds and in nature in general were reinforced. She enrolled in graduate school at Cornell, where she did her bird-banding work and may have been the first woman to receive her PhD in ornithology.[47] At Cornell, she met and married fellow graduate student, Frederick Baumgartner. After she graduated, Baumgartner taught at Oneonta State Normal School. Later, Frederick was offered a job at Oklahoma A&M College (later, Oklahoma State University), and the couple moved to Stillwater. They raised four children, and she wrote nature articles for the *Sunday Oklahoman* and edited *The Scissortail*, the quarterly publication of the Oklahoma Ornithological Society. In 1965, her husband took a faculty position at the University of Wisconsin, Stevens Point. She guest-lectured at Stevens Point and continued to write for the local newspaper, the *Stevens Point Daily Journal*, titling her often lengthy nature columns "Across the Footbridge."[48]

In addition to highlighting natural history phenomena, Baumgartner's columns covered controversial topics such as the increasing use of herbicides. Both Baumgartners were elected presidents of the Citizen's Natural Resources Association, and through this organization fought for a ban on dichloro-diphenyl-trichloroethane

---

[46] In an online reminiscence (https://www.youtube.com/watch?v=ZWgoW63lhTw; accessed 18 February 2018), Fran Hamerstrom recalls that during the time of her discussions to join Aldo Leopold's laboratory only a few women were banding birds, and one was Baumgartner.

[47] http://www.friendsofthelittleploverriver.org/assets/Uploads/Miscellaneous/1975-7-11-SPJ.pdf; accessed 18 February 2018.

[48] Berry (2014), p. 55.

(DDT) application in Wisconsin, making it the first state to ban DDT and leading to the eventual federal ban on this pesticide in the United States.[49]

After their retirement, the Baumgartners returned to Oklahoma, where they published *Oklahoma Bird Life*.[50] They started a nature school, which they named "Little Lewis Whirlwind," after the Cherokee man who owned the land. In announcing their intention to open the school, they said;

> In naming our new home for its first owner, we have dedicated it to the dream of a sanctuary where all the Lewis Whirlwinds, of whatever age, ethnic background, occupation or inclination, may find inspiration and strength in the eternal values of the natural world.[51]

ELSEWHERE IN THE UNITED STATES

The legacy of botanists Kate Brandegee and Alice Eastwood at the California Academy of Sciences continued in the early 1920s with Eastwood's protégé, Ynés Enriquetta Julietta Mexía (1870–1938). Mexía described herself as "a nature lover and a bit of an adventuress."[52] Eastwood referred to Mexía as "one of the most remarkable women of the world."[53] In contrast, others described her as "spoiled, egotistical, quick-tempered, and overbearing."[54] At the time of her birth in Washington, D.C., Mexía's father was a Mexican diplomat. Her mother left him when Mexía was three and took her seven children to live in Texas.

In 1888, Mexía moved to Mexico City to live with her father. In 1897, she married a German-Spanish merchant. He died 7 years later.[55] After her father died, Mexía inherited his hacienda and started a poultry and pet stock-raising business. Four years later she remarried, this time to one of her employees. Within a year he had ruined her financially. Mexía suffered a nervous breakdown and sought medical help in San Francisco. She sold her business, divorced her husband, and remade herself in San Francisco.

In 1920, she began hiking with the local Sierra Club and discovered a passion for nature. At age 51, she enrolled at Berkeley and gravitated toward botany. Eastwood taught her botanical collecting, and in 1925, Mexía made her first collecting trip to western Mexico. She returned with 500 species of plants, some new to science. She loved collecting, so she decided to make a living collecting and selling plant specimens to finance future collecting trips. In 1926, she returned to Mexico. Seven months later, she had collected 33,000 specimens, including 1 new genus and 50 new species. She collected on the slopes of Alaska's Mount McKinley, and then she went to Brazil.

---

[49] Berry (2014).
[50] Baumgartner and Baumgartner (1992).
[51] http://www.friendsofthelittleploverriver.org/assets/Uploads/Miscellaneous/1975-7-11-SPJ.pdf; accessed 18 February 2018; Berry (2014), p. 61.
[52] Bonta (1995), p. 136.
[53] Bonta (1995), p. 137.
[54] Bonta (1995), p. 137.
[55] Bonta (1995), p. 136.

In 1931, Mexía set off on the greatest adventure of her life—a collecting trip that took her 3,000 miles up the Amazon River to the Pongo de Manseriche, a 7-mile gorge in northwestern Peru known as the "Iron Gateway of the Amazon."[56] The rains came, and for 3 months she was marooned above the Pongo, collecting botanical specimens, bartering trade goods with the local people, and eating toucans, parrots, and monkeys.[57] Finally, she convinced the local Aguaruna Indians to build a balsa raft, and for 2 weeks she, her three guides, and her collection of plants and animals floated back to Iquitos. During the journey, she prepared specimens and wrote notes. From Iquitos, she sent her specimens back to California and she continued across the continent by hydroplane, airplane, mule, automobile, and railroad.[58] Nearly 2.5 years later, she returned to California with 65,000 plant specimens. Her last collecting trip, at age 67, was to southwestern Mexico. She cut the trip short and returned to California with her 13,000 specimens, suffering from chest pain. She died of lung cancer less than 2 months later.

In 10 years of collecting plants, Mexía had covered more geographical area and collected more plant specimens than any other woman collector of her era—over 137,000 specimens. She wrote extensively about her travels and lectured in the San Francisco area, entertaining audiences with stories of her adventures. Many of her specimens are housed at the California Academy of Sciences. She earned the reputation from men and women alike of being a dedicated and hardy field worker.

Farida Wiley (1887–1986) shared her love of nature with many thousands of people. She grew up on a horse farm in Sidney, Ohio. By age 12, she was sending reports on nesting bird data to the U.S. Biological Survey.[59] Her formal education ended with high school, after her parents died. In 1919, she moved to Long Island, where she landed a job teaching botany to blind students at the American Museum of Natural History. That led to full-time work in the education department, where she taught natural history for 60 years. In all those years, she never called in sick.[60] She never married. In 1953, she received the Silver Medal from the museum for "over 50 years as a distinguished naturalist and teacher."[61]

Wiley also taught at Pennsylvania State College, the Maine Audubon Camp, and a New York University branch campus on Long Island.[62] Her book *The Ferns of the Northeastern United States* was published in 1936, and she wrote and edited other books.[63] She met naturalist and nature essayist John Burroughs at his home in the Hudson Valley the year before he died. Later she helped to establish the John

---

[56] Bonta (1991), p. 109.

[57] Shearer and Shearer (1996), p. 286.

[58] Shearer and Shearer (1996), p. 287.

[59] https://www.washingtonpost.com/archive/lifestyle/1986/12/26/miss-wiley-grand-birder-of-central-park/716901e9; accessed 28 February 2018.

[60] https://www.washingtonpost.com/archive/lifestyle/1986/12/26/miss-wiley-grand-birder-of-central-park/716901e9; accessed 28 February 2018.

[61] www.nytimes.com/1986/11/18/obituaries/farida-a-wiley-is-dead-at-99-naturalist-and-bird-watcher.html; accessed 28 February 2018.

[62] www.nytimes.com/1986/11/18/obituaries/farida-a-wiley-is-dead-at-99-naturalist-and-bird-watcher.html; accessed 28 February 2018.

[63] Including *The Story of Landscape* (1952) and *Theodore Roosevelt's America* (1962).

Burroughs Memorial Association, and she edited a selection of his writing (*John Burroughs' America*, 1951).

Wiley was best known as the "Grand Birder of Central Park," the lady who, good weather or bad, led people twice-weekly on early-morning nature walks through The Ramble in Central Park every spring and fall. She began leading walks in her 50s and led her last at the age of 94. She sometimes had as many as 100 followers on these walks. At one point, Wiley counted 150 species of birds in The Ramble. Over time, she watched the diversity decline to 50 species. She feared the cause was habitat loss and the effect of pesticides.

Some eastern women with access to nature, but not to the big-time academic programs for graduate work, made their own way in their own style. Perhaps the best example is Margaret Morse Nice (1883–1974), who, more than any other woman in North America, developed the field of animal behavior and became famous for it. Nice was born in Amherst, Massachusetts.[64] Her father was a professor of history at Amherst College. Her mother had studied botany at Mount Holyoke Seminary. She began taking notes on birds when she was eight. Two years later, she received Mabel Osgood Wright's *Bird-Craft*, and for the first time she could study colored bird images. The following Christmas, she requested "any book of [naturalist] Olive Thorne Miller."

Nice entered Mount Holyoke College in 1901. During her junior year she bought a .22 caliber rifle and a pistol, so she could walk in the woods unescorted. She bought Frank Chapman's *Handbook of the Birds of North America* but found that bird identification did not interest her. She could make few connections between her college courses and her experiences with nature. Reflecting on this, she wrote:

> I believe that I was the only one of my class of 150 who graduated without definite plan for earning a living, either at once or in the future. ... I saw no future in laboratory zoology. ... I was to return to Amherst to be a daughter-at-home as my parents wished.[65]

A year later "fate took a hand" when she attended a dynamic lecture by Clark University professor Clifton Hodge, in which he recounted his studies on animal behavior.[66] She met Hodge, he offered her a project studying the behavior of Bobwhite Quail, and she found purpose in life.

In 1907, Nice joined the American Ornithologists' Union and attended their 1908 meeting in Cambridge, Massachusetts. Two years later, she married Clark University faculty member Leonard Blaine Nice and quit her PhD program. Four years later, her husband was offered a faculty position at the University of Oklahoma; the couple and their two daughters moved to Norman. Nice remained committed to documenting behavior, both of nonhuman animals and children. In 1915, she published "Development of a child's vocabulary in relation to environment,"[67] based on observations of her daughter Constance. She remained interested in the development of human language for the next decade, and although she never mentioned it, her

---

[64] Much of this account in taken from Nice (1979).
[65] Nice (1979), p. 21.
[66] Author of the 1902 book, *Nature Study and Life*.
[67] Pedagogical Seminary 22:35–64.

FIGURE 4.4   Margaret Morse Nice; image in the public domain.

aptitude with both the acquisition of language in humans and the subtleties of bird calls suggests she had an unusually keen ear.[68]

By the fall of 1918, Nice had four children and was deeply frustrated intellectually. She wrote "[I] resented the implication that my husband and children had brains, and I had none. He taught; they studied; I did the housework."[69] A few months later she wrote in her notebook: "Research is a passion with me; it drives me; it is my relentless master."[70] In 1919, Nice bought Florence Merriam Bailey's *Birds of the Western United States* and realized that many of the regional species were misidentified. She began going out with the goal of correcting these errors, and in the process discovered Song Sparrows close to her home.

In Norman, she developed her now famous nest-watching skills working with Mourning Doves in her backyard. After the family moved to Columbus, Ohio, nest-watching became her primary interest (Figure 4.4). In 1928, she banded her first Song Sparrow. The following year, she began mapping Song Sparrow territories—the work that would bring her international acclaim.

The key to her Song Sparrow project was banding birds, which allowed her to identify and follow individuals. Among her early discoveries was that in any given

---

[68] Similar to the legendary field biologist Ted Parker (Lannoo, 2018).

[69] Nice (1979), p. 41.

[70] Nice (1979), p. 41.

year about 50% of the nesting males and 10% the nesting females did not migrate. She found that males have a repertoire of 6–24 songs. Song repertoires appear to be learned by young males in the late summer or early fall. In late winter, they begin singing a wide variety of these learned songs, then hone them down to develop their unique repertoire, which the next spring they use to advertise and defend territories. She published her Song Sparrow monograph in two parts in the German journal *für Ornithologie,* as Zur Naturgeschichte des Singammers. Evolutionary biologist Ernst Mayr called it "the finest piece of life history work, ever done."[71]

In 1938, she spent a month in Austria with Konrad Lorenz, where she worked on European Redstarts. The following year she published her second book *The Watcher at the Nest,* with ten chapters devoted to Song Sparrows. In 1943, she published her third book *Behavior of the Song Sparrow and Other Passerines.* That same year the American Ornithologists' Union awarded her the Brewster Medal for her Song Sparrow monograph. In 1945, Nice assisted Frances Hamerstrom in sending food and clothing to European scientists suffering from the effects of the Second World War. Nice died before her inspirational autobiography *Research is a Passion with Me* was published. In 1997, the Wilson Ornithological Society honored Nice by establishing an award to recognize a recipient's lifetime of contributions to ornithology: the Margaret Morse Nice Medal. Two field ornithologists profiled in Chapter 7 have received this prestigious award—Frances James (1999) and Rosemary Grant (along with Peter Grant in 2012).

During the first half of the twentieth century, the Neotropics attracted naturalists and biologists game to explore new landscapes. But it was a man's world. After construction of the Panama Canal, travel grants were awarded to men to undertake a biological survey of the zone. The 1911–1912 biodiversity survey of the Panama Canal Zone was carried out entirely by male scientists.

In 1923, the governor of the Canal Zone set aside Barro Colorado Island as a nature preserve. Within a year, the Barro Colorado Island Biological Laboratory had been constructed. For the first 20 years of its existence, the lab was private, funded by dues from museums, universities, and donations from philanthropists. Women hoping to conduct research at the station, as for example they had been able to do at the Marine Biological Laboratory in Woods Hole, Massachusetts, found themselves unwelcome.

Early on, it was proposed that a women's dormitory be built at the Barro Colorado Laboratory. David Fairchild, a prominent North American naturalist, reacted strongly against this proposal:

> I am myself a married man and as fond of my wife as any man could be but the question is not one of convenience; it is as to whether it is possible, if we let women in, to keep there at Barro Colorado the one thing which to me is absolutely sacred. I do not believe the same curiously stimulating atmosphere can be maintained in a body of men and women that can be in a body of men alone. ... Let us keep a place where real research men can find quiet, keen intellectual stimulation, freedom from any outside distractions, and if possible as simple as any well equipped institution can be made.[72]

---

[71] Nice (1979), p. 127.
[72] Henson (2002).

In addition to the concern that women would lower the intellectual level of discourse at the station, some men worried that moral promiscuity was likely if women were allowed to live on the island.

For the first two decades after the station opened in 1924, women were allowed to conduct research on the island, but they had to live on the mainland and travel to their study sites by boat. Early morning bird observations and nighttime collecting were impossible, and women missed out on the camaraderie and intellectual stimulation of evening scientific discussions. The lab was an enclave of North American male scientists from American institutions such as the Museum of Comparative Zoology at Harvard, the Smithsonian Institution, the American Museum of Natural History, The Johns Hopkins University, and the University of Chicago. Not until after the Second World War when the lab became part of the Smithsonian Institution were facilities built for women on the island.

How did women face these challenges? Some persisted and worked on Barro Colorado Island; others chose to work elsewhere in the tropics. One woman in the latter category who was shut out of botanical exploration soon after construction of the Panama Canal was Agnes Chase, profiled in Chapter 3.[73]

Albert Hitchcock, Chase's mentor and supervisor at the USDA, had recommended that Chase join the expedition to Panama for botanical work, as she had proven herself one of the best researchers in systematic botany and was a highly competent field worker. He would support her to the tune of $54; she would use her own funds for the remainder needed. Frederick W. True, Assistant Secretary of the Smithsonian, responded:

> I regret to say that I am unable to recommend that Mrs. Chase be sent to the Canal Zone in connection with the Biological Survey, both because I doubt whether the sum mentioned by Professor Hitchcock would be sufficient for her expenses and because I doubt the advisability of engaging the services of a woman for the purpose.[74]

Chase would not be deterred from collecting grasses in the tropics. In 1913, after she was prevented from joining the 1911–1912 expedition, she collected grasses in Puerto Rico. In 1924–1925 she spent 8 months collecting in Brazil, a trip that netted over 500 species of grasses. After that, she made several more collecting trips to South America. On her second trip to Brazil, she was accompanied for part of the time by Ynés Mexia (profiled earlier in this chapter). Chase never visited the Barro Colorado Island Laboratory.

Chase used primarily her own funds for fieldwork, usually traveled alone, and developed a correspondence network of Latin American scientists who welcomed her with hospitality, logistical support, and advice. She forged strong bonds with her Latin American colleagues, informal ties that were more egalitarian than those attempted by many of her North American male colleagues.[75] She facilitated the careers of many young naturalists from Latin America by inviting them to work with her in Washington, D.C., and boarding them in her home. She sought to open

---

[73] The discussion of Chase and other women scientists working in Latin America from 1900 to 1950 is taken from Henson (2002).

[74] Henson (2002).

[75] Henson (2002).

doors for young women, doors that had been closed to her. We use this example of Chase not only to highlight the discrimination against women in field biology during this era but also to celebrate the dedication and determination of women who broke through male-constructed barriers.

Fieldwork is not restricted to natural history and the biological sciences, of course. Geologists, paleontologists, archeologists, and anthropologists also work outdoors and are exposed to the same sorts of challenges as field biologists. We would be remiss if we ignored one of the first, and most famous (and controversial), of the female field scientists during this time period—cultural anthropologist Margaret Mead.

Margaret Mead (1901–1978) grew up restless and impatient. In the fall of 1919, she attended DePauw University in Indiana. She found herself far ahead of her classmates, so after her freshman year she transferred to Barnard in New York City.[76] Mead graduated in 1923 with honors and later earned her master's degree (1924) and PhD (1929) from Columbia University.

Mead is best known for the conclusions drawn from her groundbreaking field research in anthropology. *Coming of Age in Samoa*[77] was her first and best-known book. In it, she emphasized there were various ways for humans to do almost anything and that these different means were as much culturally based as they were biologically determined.[78]

Mead had a complicated personality. She made her own rules and then broke them. She had three marriages, a daughter, and many relationships. She once claimed she made a new, important friend every 2 or 3 months.[79] She was focused on people and gave to them her time and attention. "I never had enough companionship," Mead once said of her childhood, [and perhaps] she never had enough companionship throughout her entire life. She never stopped searching for new people and fresh ideas.[80]

Mead's interests centered on issues related to gender and development. Children interested her more than adults did, and she emphasized women's issues over men's. She was attracted to the ordinary, unsung, and mundane. She had no patience for close-mindedness and did not suffer fools.[81] She was called "a monumental woman," "one of those" who "left their footprints."[82]

Mead's sin, she once said, was greed for new experiences. All her appetites were hearty. By conventional standards she moved around too much, ate and drank too much, and wrote too much. Her bibliography lists 39 books, 1,397 other publications, and 43 records, tapes, films, and videotapes. She had 28 honorary degrees and won 40 awards.[83] As Robin Fox wrote in *The New York Times*, she was "part of that long, informed, liberal, questing, progressive, courageous, challenging tradition of critical human inquiry that is so essential a part of eternal vigilance."[84] Mead's legacy

---

[76] Howard (1984), pp. 38–39.
[77] Mead, M. 1928. *Coming of Age in Samoa*. William Morrow and Company.
[78] Howard (1984), p. 13.
[79] Howard (1984), p. 13.
[80] Howard (1984), pp. 13–14.
[81] Howard (1984), pp. 14–15.
[82] Howard (1984), p. 15.
[83] Howard (1984), p. 439.
[84] Howard (1984), p. 441.

FIGURE 4.5   Margaret Mead; image in the public domain.

includes making cultural anthropology accessible to the general public through her books and lectures and mentoring many young anthropologists (Figure 4.5).

The women profiled here represent an unprecedented range of disciplines and depth of knowledge, attributable in large part to the work women did during the First World War and the postwar release that resulted in the freedom characterizing the roaring 20s. As we are about to see, this movement started by the First World War would be accelerated by the Second World War.

REFERENCES

Baumgartner, F. M., and A. M. Baumgartner. 1992. *Oklahoma bird life*. Norman, OK: University of Oklahoma Press.

Beckel, A. 1990. Harriet Bell Merrill: early limnologist recognized after 75 years. Limnology News No. 4, Spring. Available at: Braun, E. L. 1955. The phytogeography of unglaciated eastern United States and its interpretation. *The Botanical Review* 21:297–375.

Berry, B. 2014. *Banning DDT: how citizen activists in Wisconsin led the way*. Madison, WI: Wisconsin Historical Society Press.

Bonta, M. M. 1991. *Women in the field: America's pioneering women naturalists*. College Station, TX: Texas A&M University Press.

Bonta, M. M. 1995. *American women afield: writings by pioneering women naturalists*. College Station, TX: Texas A & M University Press.

Braun, E. L. 1928. Glacial and post-glacial plant migrations indicated by relic colonies of southern Ohio. *Ecology* 9:284–302.

Braun, E. L. 1955. The phytogeography of unglaciated eastern United States and its interpretation. *The Botanical Review* 21:297–375.

Cather, W. S. 1918. *My Antonia*. Boston: Houghton Mifflin; repr., Read How You Want Classics Library.

Clements, E. S. 1960. *Adventures in ecology: half a million miles … from mud to macadam*. New York: Pageant Press.

Court, F. C. 2012. *Pioneers of ecological restoration: the people and legacy of the University of Wisconsin Arboretum*. Madison, WI: The University of Wisconsin Press.

Dirks-Edmunds, J. C. 1947. A comparison of biotic communities of the Cedar-Hemlock and Oak-Hickory Associations. *Ecological Monographs* 17:2352–2360.

Durrell, L. 2001. Biographical accounts of Lucy Braun (1889–1971). Introduction: excerpts from "Memories of E. Lucy Braun". In *E. Lucy Braun (1889–1971: Ohio's foremost woman botanist. Her studies of prairies and their phytogeographical relationships. An anthology of papers*, ed. R. L. Stuckey, pp. 1–4. Columbus, OH: The Ohio State University.

Egerton, F. N. 2014. History of ecological sciences, part 50: formalizing limnology, 1870s to 1920s. *Bulletin of the Ecological Society of America* 95:131–153.

Hagen, J. B. 1993. Clementsian ecologists: the internal dynamics of a research school. *Osiris* 8:178–195.

Hansen, G. I. 1996. Josephine Elizabeth Tilden (1869–1957). In *Prominent phycologists of the 20th century*, eds. D. J. Garbary and M. J. Wynne, pp. 184–193. Hantsport, Nova Scotia: Lancelot Press Limited.

Henson, P. M. 2002. Invading Arcadia: women scientists in the field in Latin America, 1900–1950. *The Americas* 58:577–600. https://www.jstor.org/stable/1007799.

Howard, J. 1984. *Margaret mead: a life*. New York: Simon and Schuster.

Langenheim, J. H. 1996. Early history and progress of women ecologists: emphasis upon research contributions. *Annual Review of Ecology and Systematics* 27:1–53.

Lannoo, M. J. 2018. *This land is your land: the story of field biology in America*. Chicago: University of Chicago Press.

Lipscomb, D. 1995. Women in systematics. *Annual Review of Ecology & Systematics* 26:323–341.

Nice, M. M. 1979. *Research is a passion with me*. Toronto: Consolidated Amethyst Communications, Inc.

Norwood, V. 1993. *Made from this Earth: American women and nature*. Raleigh, NC: University of North Carolina Press.

Oberg, J. H. 2019. Founders of plant ecology: Frederic and Edith Clements. University of Nebraska State Museum, Publications of UNSM Staff and Affiliates. Available at digitalcommons.uni.edu/unsmaffil/1; accessed 17 November 2021.

Shackleford, M. W. 1929. Animal communities of an Illinois prairie. *Ecology* 10:126–154.

Shackleford, M. W. 1931. Autumnal herb societies of an Oklahoma prairie in 1927, 1928, and 1930. *Proceedings of the Oklahoma Academy of Science* 11:13–14.

Shearer, B. F., and B. S. Shearer, eds. 1996. *Notable women in the life sciences: a biographical dictionary*. Westport, CT: Greenwood Press.

Silva, P. C. 2009. Historical, nomenclatural, and distributional notes on two Pacific Coast kelps: *Lessoniopsis littoralis* and *Pleurophycus gardneri* (Phaeophyceae, Laminariales, Alariaceae). *Madroño* 56:112–117.

Stuckey, R. L. 1973. E. Lucy Braun (1889–1971), outstanding botanist and conservationist: a biographical sketch, with bibliography. *The Michigan Botanist* 12:83–106, reprinted in 2001, In *E. Lucy Braun (1889–1971: Ohio's foremost woman botanist. Her studies of prairies and their phytogeographical relationships. An anthology of papers*, ed. R. L. Stuckey, pp. 5–28. Columbus, OH: The Ohio State University.

Stuckey, R. L. 2001. *E. Lucy Braun (1889–1971: Ohio's foremost woman botanist. Her studies of prairies and their phytogeographical relationships. An anthology of papers*. Columbus, OH: The Ohio State University.

5 1945 to 1972 (Postwar Years)

Women's role in filling the labor shortage during the Second World War (i.e., the iconic image of Rosie the Riveter[1]) forever changed men's ideas of what women could do. While returning servicemen displaced women factory workers after the war, in the United States and Britain society didn't forget, and women made substantial strides toward becoming not only accepted but also admired professionals. In field biology it was no different. The years between 1945 and 1972 saw a large number of women rise to prominence and become role models for women in the Civil Rights Era.

Miriam Rothschild (1908–2005) was one of Britain's leading twentieth-century naturalists. She was born into extreme wealth, the daughter of the Honorable (Nathaniel) Charles Rothschild of the Rothschild banking dynasty. Her father shared his love of nature with her. As a 4-year-old, Rothschild collected insects and kept a tame quail. She was educated at home by a governess. Later, she took zoology classes at Chelsea Polytechnic,[2] but she never earned a degree. In 1943, she married water polo champion George Lane. They had four children.[3]

Rothschild published over 350 scientific papers and several books. Her guiding force was observation. She felt that a good observer not only noticed things but considered them. Among her discoveries, she demonstrated that oleander aphids and monarch butterflies defend themselves by sequestering chemicals from the toxic plants they eat; the reproductive cycle of rabbit fleas is under the direct control of their hosts' hormones; wood pigeons are a major source of avian tuberculosis in cattle; insects produce odors to protect themselves; odors emitted by toxic butterflies are mimicked by other species; and warning coloration in insects is directed against predators that hunt by sight.[4] Rothschild studied how fleas jump (and famously said "fleas jump as high for fleas as the Empire State Building would be for humans").[5] Believing wildflowers should return to gardens and pastures to attract butterflies, she collected, propagated, and sold their seeds. She advised Prince Charles on his wildflower meadow, and Lady Bird Johnson on her program to beautify American roadsides. Rothschild was enthusiastic, determined, and well-informed. She was also unapologetic, outspoken, and cantankerous. *The Times of London* said about her achievements and personality: "Imagine Beatrix Potter on amphetamines."[6]

---

[1] Colman (1995).
[2] Now part of King's College London.
[3] https://jwa.org/encyclopedia/article/rothschild-dame-miriam; accessed 27 February 2018.
[4] https://jwa.org/encyclopedia/article/rothschild-dame-miriam; accessed 27 February 2018.
[5] Martin (2005).
[6] Martin (2005).

DOI: 10.1201/9781003311508-6

Rothschild was highly respected by scientists. Her many awards included Fellow of the Royal Society, first woman trustee of the British Museum of Natural History, and the first female President of the Royal Entomological Society. Oxford, Cambridge, Northwestern, and five other universities awarded her honorary doctorate degrees. In 2000, she was conferred the title of Dame Commander of the British Empire for her services to Nature Conservation and Biochemical Research. As a pioneer female entomologist, Rothschild inspired many young women entomologists in Europe, North America, and beyond.

Botanical artist Margaret Mee (1909–1988) followed in the footsteps of Maria Sibylla Merian and Marianne North. Mee painted some of the most celebrated botanical works of the twentieth century, and her paintings are considered some of the finest records ever made of Amazonian plants and their habitats.[7] Mee was born in Chesham, Buckinghamshire, not far from London. Her Aunt Nell, an illustrator of children's books, encouraged Mee to develop her artistic talent. After the Second World War, she enrolled at St. Martin's School of Art in London. There she met and married Greville Mee, a graphic artist.

In 1952, the couple moved to São Paulo, Brazil. Shortly thereafter, at age 47, Mee began her work as a botanical artist. She made 15 expeditions by houseboat and canoe, often for months at a stretch, usually alone except for local guides, to paint orchids, bromeliads, and other plants of the Amazon rain forest.[8] Her diaries reveal acceptance and even humorous descriptions of intense heat and humidity, near starvation, blood-sucking flies and biting ants, malaria and hepatitis, capsizing boats, and threats by drunken gold prospectors, intermingled with expressions of passion for the forest with its magical sounds, sights, and smells.[9] She published *Flowers of the Brazilian Forests* (1968), *Flowers of the Amazon* (1980), and *Margaret Mee In Search of Flowers of the Amazon Forest* (1988).[10]

On her last expedition, in 1988, she painted a flower she had first seen in 1965 along the Rio Negro—the night-flowering *flor da lua*, the Amazon moonflower (*Selenicereus wittii*), an epiphytic cactus. This elusive scarlet and green cactus grows on tree trunks in flooded forest. Flowers of the cactus had never before been painted or photographed in nature.[11] After searching for over 20 years, she finally found a plant in bud near Manaus. She painted the bud, waited for it to open, then painted the pure white blossom by flashlight.

Mee discovered at least eight undescribed species of plants; four were named for her. Some of her paintings represent species that may now be extinct. Her paintings gave her a platform on which to campaign publicly for a halt to rainforest destruction and expose the corporate activities responsible. She also called international attention to the cause of Indigenous peoples. She received Order of the British Empire for services to Brazilian botany and Brazil's Order of the Southern Cross. Her spirit lives on in The Margaret Mee Fellowship Programme, which enables Brazilian artists to study at Kew Gardens, in London. On the eve of an exhibition at Kew, Mee

---

[7] Lewis-Jones and Herbert (2016).
[8] Lacey (2005).
[9] Lacey (2005).
[10] Lacey (2005).
[11] Ryle (1988).

was killed in a car accident. Remembered for her tireless efforts to preserve the Amazon rain forest, her ashes were scattered in Brazil, beside the moonflower she illustrated.[12] Mee's paintings have inspired other artists to portray the beauty of tropical forests. Her conservation work has spurred others to fight to protect the Amazon rain forest, and her adventurous spirit has inspired other women to study the natural history of exotic landscapes.

In North America, women interested in field biology often took an academic route, and many continued to be interested in botany. Recall (from Chapter 4) that Henry Cowles was a botanist and pioneer ecologist at the University of Chicago. Cowles trained many students who went on to become prominent ecologists themselves. One of these was W. S. Cooper, who trained Henry J. Oosting. Elsie Quarterman and Catherine Keever received their PhDs the same year (1949) at Duke University, working with ecologist Oosting.[13] Quarterman and Keever collaborated extensively. One reviewer of Quarterman and Keever's National Science Foundation proposal to study southeastern hardwood forests wrote "If these two fool women want to work among the chiggers, ticks, and cottonmouths of the southeastern forests, I say let 'em do it!"[14]

Elsie Quarterman (1910–2014) was born in Valdosta, Georgia. She was hired by Vanderbilt University, and in 1964 became the first woman Department Chair at Vanderbilt.[15] Quarterman studied the endemic plants of cedar glades, including aspects of the plants' life histories and community-level relationships (Figure 5.1). She rediscovered the native Tennessee coneflower,[16] a cedar glade endemic. In her honor and that of her collaborator, Catherine Keever, the southeastern chapter of the Ecological Society of America established the Quarterman-Keever Award for the best student poster at its annual meeting.

Catherine Keever (1908–2003) was born in rural North Carolina. Her research focused on plant growth and soil building processes in highland areas. She studied Rocky Face Mountain in Hiddenite, North Carolina, and discovered that moss, not lichens as had been assumed at the time, were the first plants to colonize bare rock. Keever taught biology at Winthrop College, Limestone College, the University of Georgia, and finally at Millersville University, in Pennsylvania, until her retirement in 1974. Keever predicted that oak-hickory forests would replace the dying chestnut-oak forests in Southern Appalachia, an observation that was later confirmed.[17]

Estella Leopold (1927–) was born in Madison, Wisconsin, as the youngest daughter of Aldo and Estella Leopold.[18] She graduated with her BS degree from the University of Wisconsin, MS from the University of California, Berkeley, and PhD from Yale; all her degrees were in botany. At Yale, she began extracting pollen and spores from ancient rocks and sediments and comparing this evidence with those of

---

[12] Lewis-Jones and Herbert (2016).
[13] Langenheim (1996).
[14] Ware (2015).
[15] Furlong (2014).
[16] *Echinacea tennesseensis*.
[17] McCormick and Platt (1980).
[18] Flader (2010).

FIGURE 5.1   Elsie Quarterman. (Photo credit: Vanderbilt University Special Collections and University Archives Jean and Alexander Heard Libraries.)

modern species compositions to reconstruct past landscapes. She made palynology (study of plant pollen) her profession (Figure 5.2).

After completing her PhD in 1955, and until 1976, Leopold worked for the U.S. Geological Survey, where she teased apart climate effects on prehistoric plant communities. She worked primarily in the Rocky Mountains. Her science and her family history[19] naturally led her to conservation, and she led the effort to preserve one of her more spectacular field sites, now known as the Florissant Fossil Beds National Monument. Leopold served on the faculty in the Biology Department at the University of Washington for 30 years and is currently Professor Emeritus. She was elected to the National Academy of Sciences in 1974 and awarded the International Cosmos Prize for her conservation work in 2010. She is an elected Fellow of the American Association for the Advancement of Science and of the American Academy of Arts and Sciences.[20]

During an interview with the National Park Service,[21] Leopold stated her concerns about climate change, which are worth repeating.

---

[19] Meine (1988), Lannoo (2018).

[20] https://www.nps.gov/articles/restoring-the-balance.htm; accessed 11 March 2021.

[21] https://www.nps.gov/articles/restoring-the-balance.htm; accessed 11 March 2021.

FIGURE 5.2   Estella Leopold. (Courtesy of the Aldo Leopold Foundation and University of Wisconsin-Madison Archives.)

I'm scared to death. It just seems to me it's so daunting. If we have trouble managing Yellowstone and other Western parks under the present climate—which is tricky enough [here, she is referring to managing fire regimes]—just what do we do when the climate is changing? Some habitats are moving upslope, and the fauna shifts with it. Some habitats lower down may become desiccated, like winter range. What is most alarming to me was when I visited the National Academy computer models at the Aldo Leopold Nature Center in Monona, Wisconsin. … It's absolutely frightening. It appears that between now and 2030 we will probably lose our capacity to raise crops in the Great Plains because of drought. We would lose the corn belt. … I think [young people] need to be advocates. I think they need to be expressing themselves. … that's what is needed. And a lot of folks are not doing it.

Jean Langenheim (1925–2021) was born in Homer, Louisiana.[22] Her family moved to Tulsa, Oklahoma, where she grew up fascinated with natural history. She attended the University of Tulsa and was influenced by botanist Harriet Barclay, who had received her MS under W. S. Cooper and her PhD under Henry Cowles.[23] Because of this influence, Langenheim learned ecology as a field science. In 1945, Langenheim took a summer course at the University of Wisconsin to learn plant physiology.[24] The following year, she took Barclay's Field Botany and Field Ecology courses at the Rocky Mountain Biological Laboratory (RMBL).[25] Barclay was the only female role model Langenheim had during her education.[26] Langenheim used her newfound

---

[22] Unless otherwise indicated, this account is taken from Langenheim (2010).

[23] Langenheim (2010), p. 31.

[24] Langenheim (2010), pp. 33–34.

[25] "Rumble."

[26] Langenheim (2010), p. 36.

familiarity with the Rocky Mountain flora to address questions of plant succession for her dissertation.[27] In 1946, she married Ralph Langenheim, a geology student. Together they did their graduate work at the University of Minnesota.

Langenheim was W. S. Cooper's only female PhD student.[28] From him, she learned to treat ecological issues as the result of "unceasing change."[29] As the Langenheims finished their dissertations, he got a job teaching at Coe College in Cedar Rapids, Iowa. She wrote:

> In general, the Coe College years were not a good time for me. The faculty in the natural sciences were friendly, but some of their wives were less welcoming. They openly questioned why I should bother to complete a Ph.D.; why I did not settle down and have a family, as they had done. ...What I found difficult to reconcile was that these *women* did not allow for any possibilities other than the norm for our gender at that time—the traditional housewife.[30]

In 1952, Langenheim's husband accepted a position in the Department of Paleontology at the University of California, Berkeley. She was invited to work in the herbarium, take field trips with classes, and participate in other activities.[31] In 1955, she taught Field Ecology at RMBL, replacing Harriet Barclay. Langenheim's first position was as a Research Associate at Berkeley. Because of university nepotism policies, she could not receive a salary. She pursued various research and teaching projects as she could, including geological studies in Nevada and at the Grand Canyon in Arizona, and she taught a Field Botany course at Sagehen Creek Field Station in the Sierras.[32] In 1959, Langenheim returned to RMBL to teach Field Ecology.

After her husband was denied tenure at Berkeley, he took a position at the University of Illinois in Champaign. While at Illinois, she wrote "The historic importance of the Universities of Chicago, Minnesota, Illinois, and, later, Wisconsin in the development of American ecology in the mid-twentieth century became clear to me."[33] At Illinois, she participated in the tradition initiated by Victor Shelford (who learned it from Cowles at Chicago) of long class field trips. Langenheim joined several of these excursions and visited the Great Smoky Mountains and the Eastern Deciduous Forest ecosystems so classically described by Lucy Braun.[34]

In 1961, the Langenheims divorced. Soon, an opportunity appeared through Radcliffe. The Radcliffe Institute for Independent Study[35] was established to reduce the discrepancy between the education of American women and their academic accomplishments, and to open new opportunities.[36] Langenheim was accepted and carried out her fellowship at Harvard. She was immediately welcomed, and in turn

---

[27] Langenheim (2010), p. 51.
[28] Langenheim (2010), p. 49.
[29] Langenheim (2010), p. 47.
[30] Langenheim (2010), p. 63.
[31] Langenheim (2010), p. 65.
[32] Langenheim (2010), pp. 124–130.
[33] Langenheim (2010), p. 142.
[34] Langenheim (2010), pp. 146–149.
[35] Today called the Radcliffe Institute for Advanced Studies.
[36] Langenheim (2010), pp. 159–160.

she welcomed the comradery.[37] She began to analyze the chemical composition of amber and was awarded a National Science Foundation (NSF) grant to pursue questions related to angiosperms producing amber.[38] She discovered that amber samples from Mexico came from a tropical flowering tree—upending the assumption that all amber was produced from pine trees.[39]

Langenheim spent the summer of 1964 sampling amber from European museums.[40] The following summer, she continued working on her resin studies in Mexico and Costa Rica.[41] The next year she worked in the Amazon Basin. In 1966, Langenheim accepted a tenure-track position at the University of California, Santa Cruz.[42] She was promoted to associate professor, then to full professor, and became chair of the Biology Department. She retired in 1994 and assumed Emeritus status.[43] Langenheim advised 41 graduate students; 14 were women.[44] She conducted field research on five continents and served as president of four scientific societies. Langenheim died in 2021 at the age of 95, a leading authority on amber and plant resins.[45]

The field of herpetology (study of amphibians and reptiles) was long dominated by men. However, three women born in the late 1800s, Mary Cynthia Dickerson (profiled in Chapter 3), Helen Thompson Gaige, and Doris Mable Cochran, rose to prominence. Helen Thompson Gaige (1890–1976) received her MA degree from the University of Michigan (1910) and became curator at the Museum of Zoology, University of Michigan. Gaige worked mainly on Neotropical frogs. In 1946, her 20 years of service to the American Society of Ichthyologists and Herpetologists (ASIH) earned her the title of Honorary President of Herpetology of ASIH. The ASIH established the Gaige Fund, in honor of her and her entomologist husband Frederick, to assist graduate students in their herpetological research. Doris Mable Cochran (1898–1968), who received her PhD from the University of Maryland (1933), was the first female curator at the Smithsonian Institution. Her research was focused mainly in the West Indies and South America. She wrote a popular book entitled *Living Amphibians of the World* (1961)—a book that sits on the library shelf of most present-day herpetologists. These women paved the way for women in herpetology, including Margaret Stewart.

Margaret M. Stewart (1927–2006) was born in rural North Carolina. She realized her love of nature on her family's 100-acre farm, where she found salamanders in the springhouse, rat snakes in the henhouse, and copperheads in the woodpiles.[46] She received her BS from the Women's College of the University of North Carolina[47] in 1948 and her MA degree from the University of North Carolina at Chapel Hill in

---

[37] Although in 1962, women professors still had to enter the Faculty Club by the side door.
[38] Langenheim (2010), p. 173.
[39] Stevens (2021).
[40] Langenheim (2010), p. 183.
[41] Langenheim (2010), p. 199.
[42] Langenheim (2010), pp. 211–215.
[43] Langenheim (2010), p. 457.
[44] Langenheim (2010), p. 480.
[45] Stevens (2021).
[46] Brown and Breisch (2005).
[47] Now UNC-Greensboro.

1951. There she was introduced to Highlands Biological Station in western North Carolina, where she did her thesis work on food habits and feeding behavior of Marbled Salamander larvae. After teaching for 2 years at tiny Catawba College in Salisbury, North Carolina, she enrolled in a PhD program at Cornell University, where she worked on the life history and natural history of Two-lined Salamanders. In 1956, she took a job teaching field biology and a variety of other courses at the New York State College for Teachers at Albany (now SUNY at Albany), where she spent her entire research career.

Stewart's research interests were varied, but they often centered on the tropics. She loved being outdoors, doing fieldwork. Early in her career she received a Fulbright Fellowship, which allowed her to study in Africa. She published her results in *Amphibians of Malawi* (1967). While in Malawi, she collected an undescribed species subsequently named for her, Stewart's Puddle Frog. [48] In 1966, she began work in Jamaica, studying drought and heat resistance in populations of native and invasive species of frogs. Three years later, she started working in Puerto Rico with coquí frogs, a focus of her behavioral ecology work for 20 years. Stewart famously described the "flying frogs of Puerto Rico." She found that coquís climb tree trunks at dusk to forage in the canopy. Then, at dawn, the frogs "parachute" to the ground and return to diurnal retreats, as it is too windy for them in the canopy during the day. Stewart loved teaching and trained six PhD students and ten MS students.

Stewart was also a conservationist, focusing especially on the Pine Bush, a glacial remnant area between Albany and Schenectady, New York. For 40 years, she took students to Pine Bush to understand the role of fire in natural systems and to see rare plants and animals.[49] The Pine Bush had lost 80% of its habitable land to urban sprawl, endangering protected species, and Stewart was concerned. She served as an "expert witness" in court and served on the Albany Pine Bush Preserve Commission for 12 years, fighting to preserve the land and its remaining rare and endangered flora and fauna.[50]

Stewart was elected president of the Society for the Study of Amphibians and Reptiles in 1979, the first woman to serve as president of a major herpetological society. In 1996, she was elected president of the ASIH, the third woman to lead that society. In 2019, ASIH established an award for mid-career ASIH members, the Margaret M. Stewart Achievement Award for Excellence in Ichthyology or Herpetology. The award honors the memory of Stewart—for her research, service to ASIH, and her contributions as a mentor and role model to ASIH members. The 2020 award went to herpetologist Emily Taylor, an interviewee in Chapter 7. Stewart had a profound impact on the next generation of women herpetologists, including one of the authors of this book (MLC), through her unconditional support and encouragement.

The first influential woman mathematical ecologist Evelyn Chrystalla Pielou (1924–2016) was born in Bognor Regis, England.[51] She served in the British navy during the Second World War, where she met her husband. In 1951, she received her BS degree in botany at the University of London. Throughout the 1950s, working

---

[48] *Phrynobatrachus stewartae.*
[49] Brown and Breisch (2005).
[50] Debakcsy (2019).
[51] Simberloff et al. (2017).

FIGURE 5.3   Evelyn Chrystalla Pielou; image in the public domain.

largely at home with no formal supervision while raising three children, she published several important papers on the statistics of biological patterns. From these, she compiled a dissertation for the University of London and in 1962 was awarded her PhD. She moved to Canada and worked as a research scientist for the Canadian Departments of Forestry and Agriculture, where she published influential papers addressing issues such as randomness in ecological systems.

Pielou was one of the founders of mathematical ecology. She was a pioneer in incorporating statistical rigor into questions of biogeography and developed critical statistical methods for testing hypotheses of spatial arrangements and patterns. Her trademark was intense, data-rich fieldwork, combined with theoretical modeling. Her fieldwork ranged from intertidal marine algae to boreal forests. Pielou was on the faculty at Queens' University in Kingston, Ontario; at Dalhousie University, Halifax; and the University of Lethbridge, Alberta. In 1988, she retired to British Columbia, became a naturalist, led wildlife tours, and wrote five more books. She published ten books in total: five technical scientific books[52] and five popular books.[53] Pielou received the Lawson Medal of the Canadian Biological Association (1984), was the second woman to receive the Eminent Ecologist Award of the Ecological Society of America (1986), and was a Fellow of the Royal Society of Arts. Her obituary in the *Campbell River Mirror* read: "Chris was feisty, free spirited, highly educated and independent. Never afraid to express an opinion, she did not suffer fools gladly. She loved her family dearly and she will be missed" (Figure 5.3).[54]

The first woman wildlife biologist Frances Hamerstrom (1907–1998) was born Frances Flint in Needham, Massachusetts. Her parents were wealthy, and Hamerstrom was predestined to begin life as a debutante and end it as an international

---

[52] Including *Introduction to Mathematical Ecology* (1969) and *Biogeography* (1979).
[53] Including *After the Ice Age: The Return of Life to Glaciated North America* (1991).
[54] *Campbell River Mirror* (2016).

hostess.[55] But from her youngest days she had been drawn to animals and kept wild-life as pets. On 3 May 1916, a day when "the wind roared strong and dry," while Hamerstrom was playing tennis on Martha's Vineyard,[56] a fire started and, driven by the wind, burned over 20 square miles of the island. Before the fire, Martha's Vineyard hosted roughly 2,000 heath hens,[57] afterward, the population was down to 175 and never recovered. In 1932, "Booming Ben," the last surviving heath hen disappeared, and this subspecies of the greater prairie chicken was declared extinct. Hamerstrom admitted to being ashamed that, despite being only 8 years, she never quit the tournament to fight the fire; she suggested the experience left a mark that influenced the rest of her life.[58]

Hamerstrom did not graduate from high school,[59] and she flunked out of Smith College because of her interests in "birds and boys." She met Frederick Hamerstrom (Hammy) at Dartmouth. The couple had opposite personalities but shared interests in nature. They were married secretly. At the Game Conservation Institute in Clinton, New Jersey, they learned to raise game birds. Then they heard Aldo Leopold speak. Inspired, they set out to become wildlife biologists.[60] In 1932, they enrolled at Iowa State University, he worked toward his MS degree and she her BS. Following their graduations, they enrolled at the University of Wisconsin to begin graduate work with Aldo Leopold. Hamerstrom researched winter flocking behavior in chickadees, while Hammy pursued his PhD on the breeding behavior, movement patterns, and parasites of greater prairie chickens. Leopold was interested in Hamerstrom as a researcher because she was expert in attaching and reading bird leg bands. That she was a woman didn't matter to Leopold.[61] She received her degree in 1940; her husband received his in 1942. After the Second World War, they spent 4 years in Michigan, Hammy working for the University of Michigan's Edwin S. George Reserve. In 1949, they returned to Wisconsin to work for the state Department of Natural Resources on their prairie chicken project. The Hamerstroms devoted the rest of their careers to prairie chicken conservation. In the process, they restored four abandoned farmhouses and raised two children.

Hamerstrom had a large, quirky personality. While pregnant with her son Alan, she was told by her obstetrician that she needed to get plenty of sun. Living in rural Wisconsin, she simply took off her clothes. She delighted in telling the story of his unease when she returned for a subsequent check up with no tan lines.[62] When her children were born, she licked them. To age prairie chickens, she bit their heads: "birds of the year 'bit easily' and older birds had tough, entirely resistant skulls."[63] Perhaps the most startling Hamerstrom story is when she permitted their 11-year-old daughter to stay home, by herself, for "a couple of months," while she and her husband traveled to Finland.[64]

---

[55] Hamerstrom (1980), p. 111.
[56] Hamerstrom (1980), p. 3.
[57] *Tympanuchus cupido cupido.*
[58] Hamerstrom (1980), p. 4.
[59] Hamerstrom (1980), p. 29.
[60] Flint (2013), pp. 21–22, McCabe (1987), p. 30.
[61] Hamerstrom (1988), p. 30.
[62] Hamerstrom (1980), pp. 60–63.
[63] Hamerstrom (1980), p. 67.
[64] Hamerstrom (1980), pp. 167–168.

FIGURE 5.4 Frances Hamerstrom; image in the public domain.

The Hamerstroms' work on prairie chickens made them legendary. They not only discovered a landscape pattern that permitted chicken persistence[65] but also received support to secure the necessary land. She published over 100 scientific works, including 10 popular books. Fellow Leopold graduate student Bob McCabe offered "I doubt that many of today's women field biologists or men for that matter could operate comfortably under the difficult conditions that Fran encountered—and do it successfully."[66] She made several appearances on late night television, including the Johnny Carson and David Letterman Shows, advocating for conservation and the outdoors by telling stories and being herself (Figure 5.4).[67]

The challenges faced by women working outdoors, often remotely and alone, are similar whether these women are studying plants, animals, fossils, or Indigenous peoples. Therefore, just as we included the paleontologist Mary Anning in Chapter 2 and the anthropologist Margaret Mead in Chapter 4, we include in this chapter one of the giants in field paleoanthropology—Mary Leakey—for her strength of character, international reputation, and the influence she had on subsequent generations of women. Mary Leakey (1913–1996) was born in London. By the time she was 16, she began attending archeology lectures at University College and the London Museum. R. E. M. Wheeler offered her a spot on the dig he was supervising on the Roman site of Verulamium at St. Albans.[68] More important was her second offer, by Dorothy Hill, to work on the Hembury dig in Devon, one of the earliest key Neolithic sites in Briton.[69] It was there, in 1934, that she met Louis Leakey.

The Leakeys were married on Christmas Eve of 1936. The following year, Louis received an offer to return to Africa to study the Kikuyu people, which he was

---

[65] Forty acres of every square mile (640 acre) managed for grasslands.

[66] McCabe (1987), p. 30.

[67] McCabe (1987), p. 30.

[68] Leakey (1984), pp. 37–38.

[69] Leakey (1984), p. 38.

uniquely positioned to do, since he had learned their language growing up.[70] While Louis studied the Kikuyu, she busied herself with local excavations. He finished his study in 1939 and in 1940 became curator of the Coryndon Museum in Nairobi.[71] In late 1940, their first son Jonathan was born. The family spent the Second World War in Africa, excavating sites in the Rift Valley and on Rusinga Island in Lake Victoria.[72] Their second son, Richard, was born in 1944. After the war, they continued to work in Africa, and in 1948, Mary found the fossilized skull of *Proconsul*, an extinct ape. Their third son Philip was born in 1949.[73]

In 1959, the Leakeys began investigating what they called Bed I sites at Olduvai Gorge. On July 17, she saw a bone projecting from the surface that seemed to be part of a skull. A quick brushing revealed large hominid teeth in an upper jaw. The skull was a new species, which the Leakeys put in the genus *Zinjanthropus*.[74] This find changed everything, not only exciting paleontologists but the public; it made the Leakeys famous.[75] In response, the National Geographic Society supported their research for the next decade. As Mary Leakey noted,

> Once we had resources of that kind at our disposal, we were able to work on a scale appropriate to the archeological sites and the geological problems, and we could import the best specialist advice available as and when we needed it.[76]

In the late 1960s, their son Jonathan discovered the remains of a juvenile and an adult hominid that had large craniums, foot bones suggesting they walked upright, and hand bones suggesting high dexterity. They called this new species *Homo habilis*.[77] About this same time, the Leakeys separated. In 1975, she began excavations at Laetoli,[78] and in 1978, her team uncovered a trio of hominid footprints walking across an ancient ash bed.[79]

In 1982, at age 69, Mary Leakey had a stroke that blinded her left eye. She wrote 2 years later "As any archeologist will know, there comes a time when the main fieldwork and laboratory studies have been completed and the results written up. So it is with me now."[80] She received an honorary doctorate from Oxford, wrote her autobiography, and in December 1996 passed away.

Environmental health scientist Ruth Patrick (1907–2013) was born in Topeka, Kansas. Her father, a successful banker and a naturalist, took young Ruth and her sister on walks and afterward allowed her to look through his microscope at the

---

[70] Leakey (1984), p. 68.
[71] Leakey (1984), pp. 70–77.
[72] Leakey (1984), p. 81–83.
[73] Leakey (1984), pp. 98–99.
[74] Leakey (1984), pp. 120–121; later assigned to the genus *Australopithicus* (with the specific name *boisei*).
[75] Leakey (1984), pp. 121–122.
[76] Leakey (1984), p. 122.
[77] Leakey (1984), pp. 126–128.
[78] Leakey (1984), p. 168.
[79] Leakey (1984), pp. 177–178.
[80] Leakey (1984), p. 210.

woodland soil and stream water they had collected.[81] When she was seven, her father bought her a microscope. Her interest in science survived her teenage years, and in 1926 she entered Coker College, Hartsville, South Carolina. She spent her undergraduate summers immersed in the heavy research atmospheres of Cold Spring Harbor Laboratory and Woods Hole. She received her BS in botany and then attended the University of Virginia, where she received her MS in 1931 and her PhD in 1934, working with diatoms (a type of unicellular algae).

After Patrick married Charles Hodge IV, the couple moved to Philadelphia where Hodge taught zoology at Temple University. Patrick[82] was naturally attracted to the diatom collection at the Academy of Natural Sciences of Philadelphia.[83] She volunteered at the academy and in 1937 was appointed curator of the Leidy Microscopical Collection, still as a volunteer.[84] Two years later, she founded the Department of Limnology, which she chaired. As curator, she grew the Leidy Collection into the Diatom Herbarium. As with Josephine Tilden (profiled in Chapter 4) before her, Patrick gathered and cataloged the scientific literature concerning diatoms and references to newly identified taxa.[85]

Patrick observed that the diatom species composition reflected the health of a stream, and in 1948, she tested this idea with a study of streams in the Conestoga River watershed in Lancaster County, Pennsylvania. Many streams within this drainage received runoff from cropland and pastures; other streams were affected by septic systems, urban sewage effluent, or toxic discharges from industry; still others were relatively pristine. Patrick's team identified the sources of pollution, assessed water chemistry, and quantified the biodiversity of a wide array of organisms, including bacteria, algae, protozoa, rotifers, macroinvertebrates, and fish. To facilitate comparisons across streams, Patrick invented the diatometer, a floating contraption that holds glass microscope slides. Diatoms grow on the slides, which are then collected and examined. Using this device, Patrick established that the diversity of the diatom community varied with different types of pollution. A natural diatom community is characterized by a large number of species with small to moderate population sizes. In contrast, organic pollution typically produces a smaller number of taxa with some having unnaturally large population sizes. Further, toxic pollution produces both a small number of taxa and a small number of individuals per taxon. These same types of patterns held for other taxa. Collectively, the results of Patrick's studies established that as pollution increased, biodiversity decreased, a finding that conservation biologist Tom Lovejoy labeled the "Patrick Principle."

Patrick's breakthrough created a huge demand for her services from various industries and government agencies, including the Atomic Energy Commission and the Army Corps of Engineers. As an example, she assessed the radionuclide contamination of the Susquehanna River following the Three Mile Island accident in 1979. She once estimated that she studied between 800 and 900 streams and rivers around the world.

---

[81] Bott and Sweeney (2014).

[82] She kept her maiden name to honor her father.

[83] Now associated with Drexel University.

[84] Patrick did not become a paid member of the academy staff until 1945.

[85] Termed the Literature Citation File and the New Taxon File, respectively.

She became convinced that knowing the ecology of natural stream ecosystems was critical for restoring them following pollution and contamination. With this in mind, she cofounded the Stroud Water Research Center in 1967, on the banks of White Clay Creek, in Avondale, Pennsylvania. Research there focused on biodiversity, energy flow through stream ecosystems, and watershed-stream linkages. The project quickly produced a new paradigm for flowing water systems called "The River Continuum." Patrick and the Philadelphia Academy established the Benedict Estuarine Laboratory on the Patuxent River in Benedict, Maryland, to assess the impacts of power plants on estuarine environments.[86]

Patrick's consistently applied approach of linking pollution problems to knowledge of stream ecology made her an effective spokesperson for the environment to corporations, government regulators, and the larger public.[87] She advised Presidents Johnson and Reagan on pollution, and she worked with Congress on anti-pollution legislation and participated in developing the 1972 Clean Water Act. She became the first woman to chair the board of trustees at the Academy of Natural Sciences of Philadelphia. She was the first woman and first environmentalist elected to the board of directors of the DuPont Company, in 1975. She thrived on being at the nexus of academia (with its basic science), industry, and government (where science could be applied to solve problems and improve the environment). Her intelligence and experience allowed her to get to the meat of an issue, and her charming manner often went a long way to help her win over uncommitted listeners to her point of view.

Patrick wrote more than 200 scientific papers and several books. She and coauthor Charles W. Reimer made a seminal contribution to diatom taxonomy with the publication of their two-volume *The Diatoms of the United States*. Her five-volume series entitled *Rivers of the United States* brought together the important characteristics of the major river systems of the United States. Her book *Groundwater Contamination* addressed an important aspect of that valuable aquatic resource. Among her many awards and honors, Patrick became the 12th woman elected to the National Academy of Sciences in 1970, received the Eminent Ecologist Award from the Ecological Society of America in 1972, and was awarded the National Medal of Science from President Bill Clinton in 1996. She died at age 105.

Author and environmentalist Rachel Carson (1907–1964) was born in Springdale, Pennsylvania.[88] She developed passions for reading and being outdoors and walked the large family property with her mother. She loved writing and regularly submitted short works to magazines such as *St. Nicholas*, which catered to children and was open to young contributors.[89]

After high school, Carson attended Pennsylvania College for Women.[90] She majored in English until she encountered the dynamic and demanding biology professor Mary Scott Skinker.[91] Carson was transformed by Skinker, and in late February

[86] Today this station is run by Morgan State University, in Baltimore, and is known as the Patuxent Environmental and Aquatic Research Laboratory.
[87] Bott and Sweeney (2014).
[88] Souder (2012), p. 24.
[89] Souder (2012), pp. 24–25.
[90] Souder (2012), p. 28.
[91] Souder (2012), pp. 31–34.

1928, Carson's junior year, she began thinking about merging her interests in biology and English: "I have always wanted to write, but I don't have much imagination. Biology has given me something to write about."[92] In particular, the biology Carson was drawn to was in the sea, even though she had never seen the ocean. During an English assignment, she read the closing lines of Tennyson's *Locksley Hall*, "Let it fall on Locksley Hall. With rain or hail, or fire or snow. For the mighty wind arises, roaring seaward, and I go" and felt the sea was her destiny.[93]

Skinker encouraged Carson to consider a summer research fellowship at the Marine Biological Laboratory. It didn't take much to convince Carson.[94] She entered graduate school at Johns Hopkins University and received a full scholarship. She found Woods Hole to be a "delightful place to biologize." She became fond of the library, which never locked its doors, and enjoyed exploring the tide pools, where the anemones and urchins fascinated her.[95]

Not only did Woods Hole affirm Carson's passion for the sea, but it also taught her that neither her demeanor nor training predisposed her to be a scientist. For the first time in her life, she struggled.[96] She completed her MS degree at Hopkins a year behind schedule. Her professors admired her teaching skills but not her research abilities.[97] Nevertheless, in the fall of 1932, she enrolled in a PhD program at Hopkins. She did not finish.

In 1935, she took and passed civil service examinations in parasitology, wildlife biology, and aquatic biology, and in the fall, she was hired by the U.S. Bureau of Fisheries. She was to assemble "information for public distribution on the natural history and conservation of the fishes of the Atlantic Ocean."[98] In addition, Carson began writing popular articles covering fish and wildlife topics such as upcoming fishing seasons, oyster farming, and the recovery of waterfowl numbers.[99]

Carson was deeply influenced by the pseudoscientific literature of the day ("nature-fakers"). She was particularly taken by Henry Williamson's books.[100] These—the "wellspring of all Carson's work"[101]—represented legitimate adult reading then but today would be packaged and published as children's books. Indeed, Carson initially envisioned her first book *Under a Sea Wind* as a children's book.[102] In 1937, Carson published "Undersea" in the *Atlantic Monthly*. She used the byline "R. L. Carson," so readers would assume she was a man. It caught the attention of the literary community and resulted in her first book *Under a Sea Wind*, published in November 1941, shortly before Pearl Harbor was bombed. She gave names to the animals she featured and mixed scientific fact with fiction. Nevertheless, she received good reviews from scientists such as William Beebe and organizations such as the Scientific Book Club.

---

[92] Souder (2012), p. 35.
[93] Souder (2012), p. 38.
[94] Souder (2012), pp. 41–42.
[95] Souder (2012), p. 43.
[96] Souder (2012), pp. 43–44.
[97] Souder (2012), pp. 48–49.
[98] Souder (2012), p. 50.
[99] Souder (2012), p. 53.
[100] *Tarka the Otter* (1927) and *Salar the Salmon* (1935).
[101] Souder (2012), p. 315.
[102] Souder (2012), p. 90.

In 1944, Carson was promoted to information specialist.[103] A year later, she wrote the first of three press releases on the hazards of DDT. Carson warned fish processors of the potential dangers of using this new pesticide. Two weeks later, Fish and Wildlife Service (FWS) issued a warning that even in low doses, DDT killed nontargeted species. This topic niggled as she took on other projects.

By 1951, Carson had been working several years on her new book, tentatively titled "*Return to the Sea*."[104] She had enough on paper that in 1951 she took a leave of absence from the FWS to finish and promote the book. During the first half of June, the *New Yorker* published ten chapters under the title "The Sea: Unforgotten World."[105] Oxford University Press published the entire book later that year, entitled *The Sea Around Us*. It was a sensation. By early September 1951, it was number one on *The New York Times* bestseller list, where it stayed for the next 7 months. Francesca La Monte, Associate Curator of Fishes at the American Museum of Natural History, wrote "it [is] one of the most beautiful books of our time."[106] Rachel Carson became a phenomenon, [107] although many reviewers were surprised that a woman could write a book like this.[108] In June 1952, Carson quit FWS to devote herself to writing.[109] In 1955, Houghton Mifflin published Carson's *The Edge of the Sea* to similar popular success.[110]

Since her 1945 and 1946 articles on DDT, Carson had been saving newspaper clippings covering pesticide issues. By 1958, she had been convinced for years that chemical pesticides were "a threat to the entire balance of nature and even more immediately to the welfare of the human population."[111] That spring she began considering a book addressing the pesticide problem. It began with the working title "The Control of Nature"[112] and became *Silent Spring*.

As Carson began writing, she was plagued with health issues, including a duodenal ulcer, pneumonia, chronic sinus infections, and a staph infection that developed into septic arthritis. Most seriously, she developed breast cancer. In March 1960, she had two masses biopsied. One tumor was benign, the other was "suspicious enough" that she had a radical mastectomy. Her surgeon implied the surgery was precautionary and required no follow up.[113] Then Carson found a swelling in her sternum.[114] It turned out that the "suspicious" tumor in her breast was indeed malignant and had metastasized. Carson had asked her surgeon about this possibility and was dismissed.

Carson's health issues delayed the publication of *Silent Spring* until the fall of 1962. It became a runaway bestseller and, due to industry backlash, one of the most controversial books in the history of publishing. Nevertheless, *Silent Spring* triggered a wave of environmental protection legislation including the Clean Air Acts

---

[103] Souder (2012), p. 110.
[104] Souder (2012), p. 140.
[105] Souder (2012), p. 146.
[106] Souder (2012), p. 153.
[107] Souder (2012), p.156.
[108] Souder (2012), pp. 152–153.
[109] Souder (2012), p. 164.
[110] Souder (2012), pp. 217–218.
[111] Souder (2012), p. 275.
[112] Souder (2012), pp. 279–280.
[113] Souder (2012), p. 309.
[114] Souder (2012), p. 312.

FIGURE 5.5   Rachel Carson; image in the public domain.

of 1963 and 1972, Wilderness Act of 1964, Endangered Species Preservation Act of 1966, Endangered Species Conservation Act of 1969, National Environmental Policy Act of 1970, Environmental Pesticide Control Act of 1972, Insecticide, Fungicide, and Rodenticide Act of 1972, Endangered Species Act of 1973, and the National Forest Management Act of 1976.

Finally succumbing to breast cancer, Rachel Carson died less than a year-and-a-half after *Silent Spring* was published. She lived long enough to know that she had accomplished what she had set out to do (Figure 5.5).

Environmentalist and author Margaret Murie (1902–2003) was born in Seattle, Washington. When she was nine, her family moved to Fairbanks, Alaska, where her father, who was an assistant U.S. attorney, had been assigned.[115] When she was 17, she attended Reed College, in Portland, Oregon. Following her sophomore year, she returned to Fairbanks, where a friend introduced her to the Bureau of the Biological Survey biologist Olaus Murie. She studied at Simmons College in Boston, before returning to Alaska to enroll in the newly established Alaska Agricultural College and School of Mines,[116] where she graduated with a degree in business administration in 1924. Later that year she and Olaus married.

---

[115] Murie (1978).
[116] Now the University of Alaska, Fairbanks.

FIGURE 5.6   Margaret Murie; image in the public domain.

In 1946, Olaus resigned his survey appointment to become director of The Wilderness Society. She assisted where she could, and when the society's signature piece of legislation, The Wilderness Act, became law a year after Olaus died, she famously represented him at President Lyndon Johnson's signing ceremony. Later, John Denver befriended Margaret Murie and wrote *A Song for All Lovers*[117] to commemorate her relationship with Olaus.

Murie is best known for her simple but lyrical stories of life in the field with Olaus and his collaborators. She published several of these stories separately while Olaus was alive, and then in 1978, long after he had died, as a compilation she titled *Two in the Far North*, classified as an "outdoor adventure book." It was both an influential and inspirational account of field life with a 1920s vintage Survey biologist. The Muries also collaborated on *Wapiti Wilderness*,[118] published in 1966, 3 years after Olaus died. The book recalls their life in Jackson Hole, Wyoming (Figure 5.6).

Of all the women profiled here, Valerie Jane Morris-Goodall (1934–) needs the least introduction.[119] Goodall was born in London, England. Her father was an engineer and an elite race car driver for Aston Martin, and her mother worked in business

[117] Denver (1995).
[118] Murie and Murie (1966).
[119] Goodall (1971).

but her true ambition was writing.[120] Goodall was raised by a nanny, who described her as "...a lovely child, very patient and happy, never needing scolding because she never did anything wrong!"[121] When Goodall was a year old, her father gave her a child-sized toy chimpanzee, created and sold to commemorate the first baby chimp born at the London Zoo.[122] Goodall called it Jubilee, and it became her most cherished possession; she carried it with her everywhere and still has it. She never much liked school when she was young. Instead, she thought about nature, animals, and distant lands.[123] Her favorite books were Lofting's Doctor Dolittle series, Kipling's *The Jungle Book*, and Burroughs' *Tarzan of the Apes* series. When Goodall was eight, she decided that after she grew up, she would go to Africa and live with wild animals.[124]

Goodall attended Queen's Secretarial College, where she learned to type and took shorthand at a professional level. Her evaluations noted she was "a clever girl, but rather smug ... quite immature and not really ready for responsibility ... will eventually make a good secretary."[125] What seemed to impress her instructor most were her cartoons of fishes.[126]

When she finished school, a classmate invited her to stay at her parents' farm outside Nairobi, Kenya. Goodall raised the money and on 3 April 1957, her 23rd birthday, arrived in Nairobi. She wanted to work with wild animals and was advised that she should meet Louis Leakey at what is now the Nairobi National Museum of Kenya. She called the museum and told the person on the other end of the line that she'd like to make an appointment to meet Dr Leakey. The person responded "I'm Dr Leakey, what do you want?"[127] Leakey gave her a job on the spot as an assistant secretary. Soon after, Leakey offered Goodall and another woman the chance to accompany him and his wife Mary to Olduvai Gorge. Both accepted the invitation.

Toward the end of the expedition, Leakey began talking about his ambitious plan to study the great apes of Africa—chimpanzees, pygmy chimpanzees, and gorillas—to illuminate the origins of human behaviors. After they returned to Nairobi, Leakey again brought up the subject of studying apes in the wild and an exasperated Goodall said "Louis, I wish you wouldn't keep talking about it, because that's just what I want to do."[128] Leakey told Goodall that he was interested in the Lake Tanganyika chimp population because the remains of prehistoric man were often found on lakeshores and it was possible that chimp behavior in these environments might shed light on the behavior of early hominids.[129] Reversing the traditional prejudices against women scientists, Leakey felt that because of their patience, women might be better observers of animal behavior than men.[130] Leakey also devalued

---

[120] Peterson (2006).
[121] Peterson (2006), p. xx.
[122] Peterson (2006).
[123] Peterson (2006), p. 44.
[124] Peterson (2006), p. 46.
[125] Peterson (2006), p. 71.
[126] Indeed, few accounts mention Goodall's artistic ability, but her field sketches of chimp behaviors (which can be seen on pages 38 and 47 in the 2017 *National Geographic* article) are first rate.
[127] Peterson (2006), p. 101.
[128] Peterson (2006), p. 117.
[129] Peterson (2006), p. 180.
[130] Peterson (2006), p. 120.

formal education and academic credentials; he wanted someone with an uncluttered mind, who would conduct the study because of their quest for knowledge. Goodall agreed, and Leakey set about obtaining funding.

Leakey may have had a more than a professional relationship with Goodall in mind when he hired her, and we relate this story because it's one many women today can relate to (Chapter 9). Within months of working as Leakey's personal secretary, Leakey told Goodall he was in love with her. She wrote in letters that she was "horrified" by his advances and love letters—he was 30 years older, and she had no interest. He still sent letters. She wrote "He really does behave like a child over this, and I begin to see why Mary [Leakey] has taken to the brandy."[131] They finally "thrashed it out," Louis promising to be "merely a father to me," and Goodall promising to "trust him with everything as he valued my friendship more than anything else in the world." [132]After their altercation, she reported, "… it is so pleasant now."[133]

Goodall began her work at The Gombe Stream Chimpanzee Reserve in 1960.[134] The chimps' response to her began as fear, then shifted to anger and aggression, and then, thanks to a bold male chimp Goodall called David Greybeard, to acceptance.[135] She soon began to identify individuals by their distinctive features; when she was sure she could recognize an animal, she named it. At the end of 1961, Goodall was admitted into a PhD program working with Robert Hinde at Cambridge University.[136]

During Goodall's second field season, the *National Geographic* photographer/cinematographer Hugo van Lawick showed up. He had been making a film of the Leakeys' work at Olduvai when Leakey realized he was both an excellent photographer and had a deep compassion for animals. Leakey wrote to Goodall about his abilities and mentioned to a mutual friend that he had found the right person to be Goodall's husband.[137]

Goodall became widely known after her article "My life among wild chimpanzees" appeared along with photographs by van Lawick in the August 1963 issue of *National Geographic*. However, it was the 1965 film produced by National Geographic and aired by CBS called "Miss Goodall and the Wild Chimpanzees,"[138] and a second *National Geographic* article entitled "New discoveries among the chimpanzees" published the same month with Goodall's picture on the cover, that made her famous. But it also created a duality. Was Jane Goodall the courageous scientist who discovered that chimpanzees can be carnivorous and use and make tools or was she the National Geographic's latest cover girl used to sell magazines—"A

---

[131] Peterson (2006), p. 124.
[132] Peterson (2006), p. 130. In a later interview with Virginia Morrell, author of a book on the Leakeys, Goodall said "what I was most afraid of was what my rejection of him might mean for my study of the chimpanzees." But Leakey's love was unconditional, and in fact, in 1962 sent Hugo van Lawick to Goodall thinking they would be compatible.
[133] Peterson (2006), p. 130.
[134] Peterson (2006), pp. 153–154.
[135] Peterson (2006), p. 211.
[136] Peterson (2006), pp. 261–263.
[137] This turned out to be true; the two were married on 3 April 1964, and 3 years later their son Hugo Eric Louis van Lawick [Grub] was born.
[138] Gerber (2017).

willowy blonde with more time for monkeys than men?"[139] In fact, she was both, as she first demonstrated in a series of lectures given in February 1966 at the DAR Constitution Hall in Washington, D.C., Goodall painted a picture of Gombe as a beautiful and peaceful tropical forest, home to chimps that she described as having human personalities ("feels he's a little bit superior," "beginning to find her feet").[140] Her conservation message was even stronger: "Surely it is up to us to do something to ensure that at least some of these fantastic, almost human creatures continue to live undisturbed in their natural habitat."[141]

Goodall learned to use her duality—part scientist, part male fantasy—to her advantage. Because of this, she

had to contend with a primarily male science establishment that didn't take her seriously; with media executives whose support hinged on her willingness to be glamorized; with men who said they'd be her partner or patron but also sought control, concessions, or relationships that she did not want. Through it all, Goodall's philosophy seemed the same: She would endure slights, accommodate demands, tolerate fools, make sacrifices—if it served to sustain her work.[142]

As she tells it:

There's this glamorous young girl out in the jungle with potentially dangerous animals. People like romanticizing, and people were looking at me as though I was that myth that they had created in their mind. And the [National] Geographic helped create it too. ... There was nothing I could do about it because as far as they knew, it was me. And there was no way I could be portrayed differently. It wasn't inaccurate. ... at some point, I realized that if people were going to think this way, then they would listen to me, which is true. And this would conserve the chimps and do all the other things I need to do.[143]

Following the awarding of her PhD in February of 1966, Leakey transferred responsibility for the station to Goodall. By 1970, the National Geographic Society had decided that Goodall and van Lawick's projects were reaching the point of diminishing returns—"The 'curve' is flattening out."[144] Later that year, David Hamburg, at Stanford, and Goodall began serious talks about a Gombe–Stanford partnership. The partnership was finalized in May 1971, when the William T. Grant Foundation agreed to support the Stanford Outdoor Primate Facility, which Goodall called "Gombe West."[145] The University of Dar es Salaam, in Tanzania, was included in the partnership. Goodall agreed to become a faculty member at Stanford, be there during portions of the spring and fall semesters and contribute to the Primate Facility.[146] Beginning in 1972, both Stanford and the University of Dar es Salaam began sending students to Gombe.[147] That year, Goodall and van Lawick became estranged. In

---

[139] Gerber (2017), p. 43.
[140] Gerber (2017), p. 51.
[141] Gerber (2017), p. 51.
[142] Gerber (2017), p. 39.
[143] Gerber (2017), pp. 48–49.
[144] Peterson (2006), p. 467.
[145] Peterson (2006), pp. 478–479.
[146] Peterson (2006), p. 483.
[147] Peterson (2006), pp. 484–485.

1974, Goodall divorced van Lawick and in early 1975 married Derek Bryceson, the Director of Tanzania National Parks.[148]

On the night of 19 May 1975, 40 Marxist guerillas entered the camp at Gombe and kidnapped four Stanford students. Bryceson became the spokesperson for Tanzania, and his behavior was "difficult," "irascible," and "anti-American."[149] Hamburg thought Goodall could help, but Bryceson worked to keep Goodall and the Americans out of the picture and away from each other. Finally, after paying a ransom and releasing two political prisoners, all of the students were released.[150]

At Stanford, Goodall's perceived behavior—caring more about the chimps than the students and unwillingness to do more to release the students and contribute to the ransom—generated considerable animosity. In 1976, she resigned her appointment and contributed $14,000 to retire the bank loan used to pay the ransom. She then established the Jane Goodall Institute for Research, Conservation, and Education, modeled on the L.S.B. Leakey Foundation.[151]

In 1992, Conoco bought 50 acres of woodland, marsh, and savanna in the Tchimpounga Valley to host the Jane Goodall Institute chimpanzee sanctuary. The sanctuary was expensive to run, and a year later her institute had burned through its endowment.

Goodall began fundraising in earnest, bringing with her a stuffed toy monkey mascot she called Mr. H. She was raising about $2 million per year. As Goodall traveled, she was bombarded by philosophical questions—the source of her serenity, did she pray, was she religious, did she meditate, what was the secret ingredient for her optimism, what was her philosophy of life? In 1995, the spiritual philosopher Phillip Berman asked Goodall to work with him on a book that would answer these questions. Goodall said yes. In 1999, *Reason for Hope: A Spiritual Journey* was published.[152] Following the success of this book, Goodall began incorporating elements of it into her lectures. She gives four reasons for hope: the power of the human brain to innovate, the resilience of nature (when given a chance), the creative energy of young people, and the indomitable human spirit. More than any other woman of her generation, Goodall inspired, indeed empowered, women to work with animals in the field. She is still inspiring them (see comments in Chapter 8).

REFERENCES

Bott, T. L., and B. W. Sweeney. 2014. *Ruth Patrick 1907–2013. A biographical memoir.* 16 pages. Washington, D.C: National Academy of Sciences.

Brown, W. S., and A. R. Breisch. 2005. Margaret McBride Stewart. *Copeia* 2005:701–708.

Campbell River Mirror. 2016. Chris Pielou. https://web.archive.org/web/20160918064100/ http://www.campbellrivermirror.com/mobile/obits/?id=10047784; accessed 3 February 2021.

---

[148] Peterson (2006), p. 541.
[149] Peterson (2006), p. 557.
[150] Peterson (2006), pp. 546–561.
[151] Peterson (2006), p. 569.
[152] Peterson (2006), pp. 640–642.

Colman, P. 1995. *Rosie the riveter: women working the home front in World War II*. New York: Crown Publishers.

Debakcsy, D. 2019. The professor and the frogs: the ecology and herpetology of Margaret Stewart. October 22, 2019. https://womenyoushouldknow.net/frogs-ecology-herpetology-margaret-stewart/; accessed 29 September 2021.

Denver, J. 1995. A song for all lovers. The wildlife concert. London: Sony Music.

Flader, S. 2010. Biographical portrait: Estella Bergere Leopold, paleoecologist and conservationist. *Forest History Today* Spring/Fall:55–57.

Flint, P. P. 2013. A brother's perspective. In *Hamerstrom stories: recollections from the life of Hammy and Fran Hamerstrom*, ed. E. H. Paulson, pp. 21–22. Stevens Point, WI: R. Schneider Publishers.

Furlong, K. 2014. Elsie Quarterman, who rediscovered Tennessee coneflower, dies at 103. Vanderbilt News. Vanderbilt University; accessed 1 February 2021.

Gerber, T. 2017. Becoming Jane. *National Geographic* 232 (October):30–51.

Goodall, J. V.-L. 1971. *In the shadow of man*. Boston, MA: Houghton Mifflin Company.

Hamerstrom, F. 1980. *Strictly for the chickens*. Ames, IA: Iowa State University Press.

Hamerstrom, F. 1988. Touchstone. In *Aldo Leopold: mentor*, ed. R. E. McCabe, pp. 28–31. Madison, Wisconsin: Proceedings of an Aldo Leopold Centennial Symposium.

Lacey, S. 2005. Brazil: the lady who loved the river. https://www.telegraph.co.uk/travel/destinations/southamerica/brazil/732927/Brazil-The-lady-who-loved-the-river.html; accessed 9 June 2018.

Langenheim, J. H. 1996. Early history and progress of women ecologists: emphasis upon research contributions. *Annual Review of Ecology and Systematics* 27:1–53.

Langenheim, J. H. 2010. *The odyssey of a woman field scientist: a story of passion, persistence, and patience*. Bloomington, IN: Xlibris Corporation.

Lannoo, M. J. 2018. *This land is your land: the story of field biology in America*. Chicago: University of Chicago Press.

Leakey, M. 1984. *Mary Leakey: disclosing the past: an autobiography*. New York: McGraw-Hill Book Company.

Lewis-Jones, H., and K. Herbert. 2016. *Explorers' sketchbooks: the art of discovery & adventure*. San Francisco, CA: Chronicle Books.

Martin, D. 2005. Miriam Rothschild, high-spirited naturalist, dies at 96. *The New York Times*, January 25, 2005. www.nytimes.com/2005/01/25/science/miriam-rothschild-highspirited-naturalist-dies-at-96.html; accessed 27 February 2018.

McCabe, R. A. 1987. *Aldo Leopold: the professor*. Madison, WI: Published privately by R.A. McCabe.

McCormick, J., and R. B. Platt. 1980. Recovery of an Appalachian Forest following the Chestnut blight or Catherine Keever—you were right! *American Midland Naturalist* 104:264–273. doi:10.2307/2424865.

Meine, C. 1988. *Aldo Leopold: his life and work*. Madison, WI: University of Wisconsin Press.

Murie, M. 1978. *Two in the far north*. New York: Alfred A. Knopf.

Murie M., and O. Murie. 1966. *Wapiti wilderness*. New York: Alfred A. Knopf.

Peterson, D. 2006. *Jane Goodall: the woman who redefined man*. Boston, MA: Houghton Mifflin Company.

Ryle, J. 1988. Margaret Mee and the moonflower. Sunday Times Magazine, London, 1 August 1988. Posted 2016. https://johnryle.com/?article=margaret-mee-and-the-moonflower; accessed 9 June 2018.

Simberloff, D., N. Sanders, and P. Peres-Neto. 2017. A homage to EC Pielou: on of the 20th century's most accomplished scientists. British Ecological Society. https://methodsblog.com/2017/03/10/ec-pielou/#more-5655; accessed 3 February 2021.

Souder, W. 2012. *On a farther shore: the life and legacy of Rachel Carson*. New York: Crown Press.

Stevens, T. 2021. Pioneering plant scientist Jean Langenheim dies at 95. https://news.ucsc.edu/2021/04/langenheim-in-memoriam.html; accessed 28 September 2021.

Ware, S. 2015. Resolution of respect: Elsie Quarterman, 1910–2014. *Bulletin of the Ecological Society* 96:74–76. https://doi.org/10.1890/0012-9623-96.1.74.

6 1972 to Present (Civil Rights Era)

It was during and following the 1970s that women began truly being assimilated into field biology-related programs and institutions. For example, the Animal Ecology Department at Iowa State University was established in 1975 and housed the prototypical Fish and Wildlife Cooperative Unit in the country. In 1979, Marilyn Bachmann became the first tenured female faculty member, after serving 10 years as a nontenure-track instructor. She remained the only female faculty member (there were either 14 or 15 male members during this time) until her retirement in 1993. Her replacement, Diane Debinski, then served as the only female faculty member until 2003. A history of the department reveals that gender disparity was not considered in the first two self-studies.[1] It is a different story today. The now-combined Animal Ecology and Natural Resource Ecology and Management Department is comprised of 32 faculty members, 11 of whom are women. A great improvement, but the percentage of female faculty still lags behind the percentage of female undergraduates. While the percentage of female faculty rose from about 7% in 1979 to about 34% today, over roughly the same period the percentage of women undergraduates grew from about 20% to over 50%.[2]

This time period also marks a transition in our narrative, from accounts of mostly historical women to women we knew or know. There are a large number of women who devoted their lives to field biology during this time; unfortunately, we have space to profile only a few. It is also true that there were so many women working and training in so many fields that most discipline-based patterns break down.

Our first two profiles highlight wetland experts. Ecologist and conservationist Joy Zedler (1943–) grew up on a farm and attended a one-room school through grade 8. She earned a PhD in botany (1968) from the University of Wisconsin, where her study of drained wetlands led to a career aimed at solving land-care problems. She followed her husband Paul to San Diego State University and later joined the SDSU faculty. Her salt marsh research with students and postdoctoral researchers pioneered adaptive restoration (large field experiments to "learn while restoring") and helped activists Mike and Patricia McCoy protect Tijuana Estuary from urban development. She created SDSU's Pacific Estuarine Research Lab, which trained researchers and restoration practitioners to fulfill the needs of the region and beyond. In 1998, Zedler returned to UW as the Aldo Leopold Professor of Restoration Ecology at the UW Arboretum, where the field of restoration ecology had taken root.[3] She led her students to solve problems in wetland ecology to find ways to favor native plants over

---

[1] Atchison et al. (2018).
[2] Atchison et al. (2018).
[3] Lannoo (2018).

DOI: 10.1201/9781003311508-7

FIGURE 6.1   Joy Zedler. (Photo courtesy of Joy Zedler.)

invasive species and to restore degraded wetlands using adaptive and watershed approaches. Zedler's field research showed how and why wetland restoration often falls short of full recovery (Figure 6.1).[4]

Zedler is notable among academics for putting her ideas into action, which is reflected in her 280+ publications, including the *Handbook for Restoring Tidal Wetlands* (2000). She helped the National Research Council define wetland boundaries and improve the mitigation of damages to wetlands, as required under the Clean Water Act. She continues to bring science to bear on issues confronting The Nature Conservancy and other conservation organizations. Meanwhile, her family tackles science-based management of wildfires (Paul) and coastal waters (their twin daughters), and the grandchildren inspire her to write books for school kids.

Rebecca Sharitz (1944–2018) was born in Wytheville, Virginia.[5] She received her BS in biology from Roanoke College in 1966 and her PhD in botany/ecology from the University of North Carolina in 1970. After a brief appointment at Saginaw Valley State University, in 1972 she became an adjunct assistant professor in Botany at the University of Georgia. She was centered at the Savannah River Ecology Laboratory (SREL) and rose through the ranks, becoming a Senior Research Ecologist at SREL in 1986 and a full professor at UG in 1989 (Figure 6.2).

---

[4] Zedler (2000).
[5] https://esa.org/history/deep-ecology-remembering-becky-sharitz; https://news.uga.edu/rebecca-sharitz-srel-fellowship; both accessed 9 November 2021.

FIGURE 6.2 Patricia Werner (left) and Rebecca Sharitz; photographer unknown. (Photo courtesy of Patricia Werner, Personal Collection.)

Sharitz is renowned for her groundbreaking research on southeastern wetlands, including bottomland forests, coastal plains floodplains, and Carolina Bays—ecosystems she fought hard to conserve and restore. Sharitz authored more than 160 peer-reviewed publications and was celebrated as a teacher and mentor. In 2010, the Environmental Law Institute presented her with its National Wetlands Award in Science Research.

The next four women we profile explored plant-based, ecosystem-level phenomena, including pollination biology and successional patterns. Community ecologist Beverly Rathcke (1945–2011) was born in Wadena, Minnesota.[6] She received her BS in biology at Gustavus Adolphus College in St. Peter, Minnesota (1967), her MS in applied entomology at Imperial College London, England (1968), and her PhD in environmental sciences (Ecology Program) at the University of Illinois Urbana-Champaign (1973). In an interview, she shared:

> At Illinois I had a woman role model for the first time: Dr. Mary Willson [see below]. She was on my PhD committee, and we did a research project together on *Asclepius* (milkweed) pollination that turned out to be one of the first studies on sexual selection in plants. Although at the time, I did not think of [Willson] as a role model, I am sure it helped me remain in ecology knowing that it was possible, albeit difficult, for women.[7]

---

[6] Greiling (2011).

[7] https://esa.org/history/beverly-j-rathcke-challenging-dogma-training-students; accessed 2 October 2021.

Rathcke did postdoctoral work at Cornell University with ecologist Richard Root, where she was introduced to community ecology and the importance of empirical testing of ecological theory.

After her postdoctoral work, Rathcke began her career as an Investigator and Research Professor in the Section of Population Biology and Genetics at Brown University, where her husband, Robert W. Poole, was a faculty member. In 1978, Rathcke joined the faculty in the Department of Ecology and Evolutionary Biology at the University of Michigan, where she remained until her retirement in 2010. Her research revolved around plant–animal interactions, including herbivory, competition, and pollination ecology. She also studied how environmental changes, such as introduced species, habitat fragmentation, and hurricane disturbances, affect species' reproductive success. Rathcke conducted field experiments on the impacts of introduced plants in Michigan and on the pollination biology of mangroves in the Bahamas, Mexico, and Florida. Her work was noted for "challenging current dogma."[8]

Rathcke strongly supported women in science.

> I became aware of myself being a role model at Brown University when women students told me that I was the first woman scientist they had teach a course. I also became more aware of the sexism that women face. I organized a seminar for women in science at the University of Michigan and I tried to ensure equal opportunities and support for women faculty and students.[9]

Rathcke served as advisor or co-advisor for 29 PhD students. For this commitment, in 2008 she received University of Michigan's Rackham Distinguished Graduate Mentoring award.

Rathcke's advisor was Mary Willson (1938–), an ecologist with broad interests that include both plants and animals in the areas of conservation biology, animal behavior, and evolutionary ecology. Willson earned her PhD in 1964 at the University of Washington, working with ecologist and ornithologist Gordon Orians.[10] She has published over 200 peer-reviewed papers and numerous reports.[11] She taught at the University of Illinois for 25 years before moving to Juneau, Alaska, in 1989, to take a position as an ecological researcher with the U.S. Forest Service's Forestry Sciences Laboratory.[12] In 1999, Willson retired but has continued to work as an ecological consultant and independent researcher both in Alaska and in Chile. In 1996, she and Chilean botanist Juan Armesto cofounded Senda Darwin Biological Station on Chiloé Island, Chile, where researchers have produced cutting-edge studies, including conservation of biodiversity and ecosystem processes. Her research interests have included such diverse topics as the potential for seed dispersal by the banana slug, community ecology of birds in Chile, nesting success and polygyny in Winter Wrens, nestling provisioning by American Dippers, pinniped foraging ecology, pollination in milkweeds, and the effect of loss of habitat connectivity on pair formation and juvenile dispersal of Chucao Tapaculos.

---

[8] https://esa.org/history/beverly-j-rathcke-challenging-dogma-training-students; accessed 2 October 2021.

[9] https://esa.org/history/beverly-j-rathcke-challenging-dogma-training-students; accessed 2 October 2021.

[10] Langenheim (1996).

[11] https://academictree.org/evolution/publications.php?pid=53690; accessed 2 February 2021; https://www.researchgate.net/scientific-contributions/Mary-F-Willson-72697601.

[12] https://onthetrailsjuneau.wordpress.com/about; accessed 2 February 2021.

Patricia Werner (1941–) was born in Flint, Michigan.[13] She received both of her graduate degrees at Michigan State University (MS in 1968 and PhD in 1972). In an online interview, she revealed "Ultimately, I went into ecology because I took a class and it was big, messy science and I loved it, trying to find a pattern among a lot of complexity in biology."[14] After graduation, Werner accepted a faculty position at Michigan State and began conducting research at the W. K. Kellogg Biological Field Station. There, she collaborated with Hal Caswell at the Woods Hole Oceanographic Institute to develop models of population growth based on data from her work with the European invasive teasel (*Dipsacus fullonum*). She also developed a research program based on old-field succession that laid the groundwork for long-term research on plant communities at Kellogg.[15]

In the early 1980s, Werner took a research position at the Tropical Ecosystems Research Center in Australia, continuing her work in successional or disturbed plant communities. From 1985 to 1990, Werner was the director of the large Commonwealth Science and Industrial Research Organisation laboratory in Darwin, where she ran two field stations: one in the town of Katherine, where the focus was on tropical crops and pastures, and one called Kapalga in Kakadu National Park (a World-Heritage Site) that focused on wildlife. In Kapalga, she continued her research on the role of fire and feral buffalo on savanna canopy tree populations. She has recently published a population dynamics model of these trees using her 25–30 years of field data (Figure 6.3).[16]

In 1992, Werner moved to the University of Florida and later served as Director of the division of Environmental Biology at the National Science Foundation (United States).[17] In 2002, she moved back to Australia to take up an appointment in the Fenner School of Environment and Society at the Australian National University in Canberra.

During her career, Werner became adept (to the tune of about $6 million dollars in funding) at asking "for anything using one page of paper—and getting it."[18] In 2013, Werner established the "Patricia A. Werner Scholarship for Ecological Field Studies," which annually supports students at the Kellogg Biological Station. Werner is Emeritus Professor (University of Florida) and Honorary Professor at the Fenner School of Environment and Society at the Australian National University.

Ecologist Deborah Rabinowitz (1947–1987) was born in Willimantic, Connecticut. She received her BS in biology from New College of Florida and her PhD (1975) in theoretical population biology from the University of Chicago.[19] She was the first female faculty member in the Department of Ecology and Evolutionary Biology at the University of Michigan. In 1982, she moved to the Section of Ecology and Systematics at Cornell. She published a classic paper in 1981 describing seven

[13] https://esa.org/history/werner-p-a/; accessed 4 February 2021; https://ohms.libs.uga.edu/viewer.php?cachefile=russell/RBRL416ESA-007.xml; accessed 4 February 2021.

[14] https://hi-in.facebook.com/MSUISP/videos/alumni-feature-patricia-werner-phd/389279244974817; accessed 4 February 2021.

[15] Zoellner and Gross (2019).

[16] Werner and Peacock (2019).

[17] https://fennerschool.anu.edu.au/news-events/events/lessons-learned-during-career-research; accessed 4 February 2021.

[18] https://fennerschool.anu.edu.au/news-events/events/lessons-learned-during-career-research; accessed 4 February 2021.

[19] Simberloff (1988).

FIGURE 6.3 Patricia Werner. (Photographer unknown, photo courtesy of Patricia Werner, Personal Collection.)

different meanings for the concept of "rarity."[20] Her most notable discovery was "frequent type rarity"—characterized by sparse numbers scattered over a wide geographic range—a category most ecologists would not have previously recognized.[21] Tragically, Rabinowitz developed cancer and died at the age of 39. Those who knew her pointed out that she was an important role model for women scientists both at

---

[20] Rabinowitz (1981).
[21] https://hi-in.facebook.com/MSUISP/videos/alumni-feature-patricia-werner-phd/389279244974817; accessed 4 February 2021.

Michigan and at Cornell. She was broad-minded and generous and maintained "the strength of character and ... grace ... to the end."[22]

Continuing the tradition begun by Jane Goodall, our next two women profiled built their careers studying primates. Sarah Blaffer Hrdy (1946–) was born in Dallas, Texas. She attended Wellesley College as a philosophy major, where she became interested in Mayan culture and then transferred to Radcliffe College and majored in anthropology (BA in 1969). She then studied filmmaking at Stanford, but while there, she fell under the influence of Paul Ehrlich, who piqued her interest in primate social systems.[23] In 1970, she entered Harvard, where she met her husband, David, and worked with Irven DeVore in India on Hanuman langurs. Hrdy received her PhD in 1975.[24] She wrote *The Woman that Never Evolved* (1981),[25] which was chosen by *The New York Times Book Review* as one of the Notable Books of the Year. In 1984, she joined the University of California at Davis as a professor of anthropology. Her studies on primate social systems were groundbreaking, and she is considered a pioneer in the evolutionary basis of female behavior in both human and nonhuman primates.[26] In 2002, *Discover* magazine listed Hrdy as one of the 50 most important women in science. She retired in 1996, and now lives with her husband in northern California, where they operate a walnut plantation.[27]

The life of Dian Fossey (1932–1985) has been portrayed in both books and movies.[28] Like Jane Goodall, she was mentored by Louis Leakey. Unlike Goodall, Fossey's upbringing was not nurturing. As an adult, Fossey was socially awkward and had difficulty with interpersonal relations. Working with Leakey, though, she found herself when she fulfilled a long-held dream and began studying Mountain Gorillas in the Virunga Mountains of East Africa.

(As he did with Jane Goodall, Leakey claimed to be in love with Fossey. Fossey felt the same way toward Leakey as Goodall did [Chapter 5], handled the situation similarly, and managed to remain on good terms with Leakey for the rest of his life. Again, we mention this only because a subset of professional women today have been subjected to the same sort of uncomfortable and unwelcome advances that Goodall and Fossey received from Leakey, and in the context of this book we would be remiss if we did not call attention to this history.)

Fossey began her work in 1967. In early 1970, a young blackback she had named Peanuts left his tree, approached, and stood next to her. She slowly extended her hand. Peanuts extended his hand and they touched.[29] Photographer/cinematographer Bob Campbell recorded the event.[30] Not long afterward, Fossey was in Cambridge to begin her PhD program (completed in 1976) with Goodall's former advisor Robert

---

[22] https://ecommons.cornell.edu/bitstream/handle/1813/19292/Rabinowitz_Deborah_1987.pdf;jsessioni
d=D73A591E5968D4F5772FE3F5C6718403?sequence=2; accessed 17 November 2021.

[23] Sridhar (2018); https://anthropology.ucdavis.edu/people/sbhrdy; accessed 16 February 2021.

[24] Sridhar (2018).

[25] Harvard University Press, Cambridge, Massachusetts.

[26] https://thisviewoflife.com/on-the-origin-of-hbes-sarah-hrdy; accessed 16 February 2021.

[27] https://thisviewoflife.com/on-the-origin-of-hbes-sarah-hrdy; accessed 16 February 2021.

[28] For example, Farley Mowat's book *Women in the Mists* (1987) and the Universal Pictures' film *Gorillas in the Mist* (1987).

[29] Mowat (1987), pp. 84–85.

[30] Mowat (1987), pp. 84–85.

Hinde, when the January 1970 issue of *National Geographic* with Fossey on the cover appeared on the bookstands. She was instantly famous and, as happened with Goodall, was inundated with candidates wanting to work with her.[31] In 1972, a second *National Geographic* article appeared showing Campbell's pictures of her with Peanuts.

In 1972, Fossey had a similar experience with a young male gorilla she called Digit. One day, as she settled in to watch Digit's group, the adolescent came over to her, put his arm around her shoulders, patted her head, and sat down. She laid her head in his lap, and a special bond began between the two.[32] As Digit matured, he became his group's sentry. He was the first to face intruders, whether from another gorilla group or threatening humans (typically poachers, who at that time deliberately killed gorillas to make souvenirs, such as ashtrays from dried hands, cleaned skulls and mounted head to sell to tourists; poachers also caught baby gorillas for sale to zoos[33]). Digit's loyalty ultimately led to his death. On 2 January 1978, one of Fossey's trackers found Digit's mutilated body. His head and hands were missing, and his abdomen had been sliced open. The story, as later reconstructed, went as follows. During the final days of December, Digit's group had been pursued by six poachers and their dogs and were finally cornered. Digit charged the men to cover the retreat of the rest his group, killing one of their dogs, before being speared to death.[34] Fossey, of course, was devastated. Once she publicized the news of Digit's death, there was a huge public outcry. Seeking to retrieve some good from this tragedy, Fossey then established the Digit Fund to solicit and accept donations, which became a crucial source of funding.

On 3 August 1980, Fossey left Rwanda to take an appointment as a visiting professor at Cornell and to write her bestseller *Gorillas in the Mist*. She was gone for almost 3 years and then returned to the Virungas in late June 1983.[35] Two years later, on the night of 7 October 1985, she found a wooden image of a puff adder on her doorstep—the curse of death.[36] Not long after, on the morning of 27 December 1985, one of her African attendants went to Fossey's cabin and discovered her body. Her skull had been split open from a panga machete wound across her face.[37] There were other wounds from blows on the top and back of her head.[38]

Fossey was buried near her cabin, in the graveyard she had created for the gorillas killed by poachers. In his eulogy of her, Reverend Elton Wallace said:

> She will lie now among those with whom she lived, and among whom she died. And if you think that the distance Christ had to come to take the likeness of Man is not so great as that from man to gorilla, then you don't know men. Or gorillas. Or God.[39]

Fossey's legacy lives on through the gorillas she helped conserve. Following her murder, the Digit Fund was renamed the Dian Fossey Gorilla Fund International,

---

[31] Mowat (1987), p. 88.
[32] Mowat (1987), pp. 101–102.
[33] Mowat (1987), p. 59.
[34] Mowat (1987), pp. 160–161.
[35] Mowat (1987), p. 284.
[36] Mowat (1987), pp. 355–356.
[37] Mowat (1987), pp. 365–366.
[38] Mowat (1987), p. 368.
[39] Mowat (1987), pp. 366–367.

and the number of gorillas around Karisoke, where Fossey worked, has more than quadrupled.[40]

One discernable trend during this time was the increasing numbers of women choosing to work in the relatively new discipline of marine biology. We are reminded of the founding of the discipline of ecology a century before, when women were given unprecedented opportunities to work in this once new field. New disciplines appear to be founded based on the standards of the time, freed from a confining history based on a "this is the way we do things" approach. Here we profile seven women who built remarkable careers in the field of marine science.

Eugenie Clark (1922–2015), known as "The Shark Lady," was born and raised in New York City. Her father died before her second birthday. Clark's interest in marine life began during elementary school. On Saturdays, while her mother worked at the cigar and newspaper stand in the lobby of the Downtown Athletic Club, Clark spent delightful hours watching fishes at the New York Aquarium at Battery Park.[41]

> … I couldn't get enough of peering into the fish tanks, especially the big tank that held the sharks (sand tiger sharks) in water slightly murky green, where I could press my face against the glass by hanging over the railing. I couldn't see the back or sides of the tank and I pretended I was on the sea bottom with sharks swimming around me. I thought the sharks were beautiful, graceful, and magnificent. It was my dream to learn more about them and all the other beautiful and wondrous smaller fish.[42]

As a child, Clark's hero was William Beebe.

> I told my family I would like to go down [in the bathysphere] and be like William Beebe. They said maybe you can take up typing and get to be a secretary to William Beebe or somebody like him. I said, I don't want to be anybody's secretary! I want to be like William Beebe going down …[43]

Clark realized her dream, and, like Beebe, became a celebrated naturalist, writer, and ocean explorer. She also became a world authority on sharks and other fishes.

Clark received her BA in zoology from Hunter College in 1942. Then,

> When I applied to graduate school at Columbia University, the Chairman of the Zoology Department (a famous geneticist) told me, "Well, I guess we could take you but to be honest, I can tell by looking at you, if you do finish you will probably get married, have a bunch of kids, and never do anything in science after we have invested our time and money in you."[44]

New York University accepted her, and she received her MA (1946) and PhD (1950) from NYU. She was a faculty member at the University of Maryland from 1968 to 1992. Over the years, Clark carried out research in association with the American Museum of Natural History in New York City, Woods Hole Marine Biological Laboratory in Massachusetts, the Lerner Marine Laboratory in Bimini, Bahamas, and the Scripps Institute of Oceanography in La Jolla, California.

---

[40] Sneed (2017); see also gorillafund.org.
[41] Balon (1994b).
[42] Balon (1994a).
[43] Balon (1994b).
[44] Balon (1994a).

Clark's research included studies of whale sharks, deep sea sharks, and spotted oceanic triggerfish. She studied reproductive behavior, territoriality, and ecology of tropical marine sand-dwelling fishes. Her research involved snorkeling, SCUBA diving, and submersibles. Famously, she discovered that sharks were not the dimwitted creatures some had assumed but rather that they could be trained to learn visual tasks as quickly as some mammals; she published this research in the journal *Science*. She made it her mission to dispel the public's fear of sharks, through education.

In 1955, Clark started the one-room Cape Haze Laboratory in Placida, Florida, with philanthropic support from the Vanderbilt family, as a place where she could do her research. The laboratory was moved to Sarasota in 1967 and renamed Mote Marine Laboratory; it now hosts two dozen marine research and conservation programs, offers educational programs for the public, and has a major public aquarium.[45] After leaving the University of Maryland in 1992, Clark returned to Mote Marine Laboratory as Senior Scientist and Director Emerita. (Figure 6.4)

Rather than leaving science to raise a family, as feared by that department chair at Columbia, she combined the two. She took her four children along with her on research expeditions around the world. All four became enthusiastic divers.[46]

Clark was an avid promoter of marine conservation and spokeswoman for marine biology. She engaged the public by sharing her love of the ocean in public lectures and appeared in dozens of documentaries and television specials. She wrote three books—*Lady with a Spear* (1953), *The Lady and the Sharks* (1969), and *The Desert Beneath the Sea* (1991; coauthor with Ann McGovern)—and over 160 scientific papers and 12 popular articles in *National Geographic* magazine. She was chief scientist for 72 submersible dives as deep as 12,000 feet to study deep sea sharks and led more than 200 field research expeditions to more than 20 countries, including the Red Sea and Gulf of Aqaba, Caribbean, Mexico, Japan, Palau, Papua New Guinea, the Solomon Islands, Thailand, Indonesia, and Borneo.[47] In addition to her many awards and honors, she received The Explorers Club Medal of Excellence from the American Society of Oceanographers, a Gold Medal from the Society of Women Geographers, and the Distinguished Fellow Award from the American Elasmobranch Society.[48]

Clark was still actively engaged in research when she died in 2015 of complications from nonsmokers' lung cancer at the age of 92. She had made her last dive just the year before. Her family placed her ashes into the Gulf of Mexico from aboard the Mote Marine Laboratory's ship, the R/V Eugenie Clark.[49] Through her passion for science, innovative research, ability to connect with the public, and devotion to marine conservation, she inspired countless others to become marine biologists.

---

[45] https://mote.org/news/article/remembering-the-shark-lady-the-life-and-legacy-of-dr.-eugenie-clark; accessed 4 October 2021.

[46] https://mote.org/news/article/remembering-the-shark-lady-the-life-and-legacy-of-dr.-eugenie-clark; accessed 4 October 2021.

[47] https://mote.org/news/article/remembering-the-shark-lady-the-life-and-legacy-of-dr.-eugenie-clark; accessed 4 October 2021.

[48] https://mote.org/news/article/remembering-the-shark-lady-the-life-and-legacy-of-dr.-eugenie-clark; accessed 4 October 2021.

[49] https://mote.org/news/article/remembering-the-shark-lady-the-life-and-legacy-of-dr.-eugenie-clark; accessed 4 October 2021.

FIGURE 6.4   Eugenie Clark, measuring a shark. (Courtesy of Mote Marine Laboratory.)

Sylvia Earle (1935–) was born in Gibbstown, New Jersey.[50] She was raised on a small farm near Camden, where she explored the nearby woods. Neither of her parents had a college education, but both loved nature and encouraged their daughter's interests. When Earle was 13, her family moved to Clearwater, Florida—and the ocean—where she beachcombed and explored salt marshes and seagrass beds. She received scholarships to Florida State University, where she learned SCUBA diving. She majored in botany, believing that understanding vegetation is the first step to understanding any ecosystem.[51] Earle received her BS from Florida State University (1955) and her MS from Duke University (1956). She took a few years off from formal education to marry and start a family but continued her explorations, including participating in a 6-week National Science Foundation expedition in the Indian Ocean. She earned her PhD from Duke in 1966. Her dissertation study of brown algae of the Eastern Gulf of Mexico was the first such detailed study of its kind.[52] Earle became a research fellow at Harvard, and then served as resident director of the Cape Haze Marine Laboratory in Florida that Eugenie Clark had started. This must have been rewarding. When Earle was 18, she had read Eugenie Clark's book *Lady with a Spear* and the following year had met Clark at her newly established laboratory.

---

[50] https://achievement.org/achiever/sylvia-earle/#interview; accessed 5 October 2021.
[51] https://achievement.org/achiever/sylvia-earle/#interview; accessed 5 October 2021.
[52] https://achievement.org/achiever/sylvia-earle/#interview; accessed 5 October 2021.

The two became close friends. Earle saw Clark as "an intrepid and beloved leader, explorer, scientist, witness, teacher, communicator, and friend."[53]

In 1969, Earle applied to participate in the "Tektite project," the nation's first underwater laboratory, sponsored jointly by the U.S. Navy, the Department of the Interior, and NASA. Based off the coast of St. John in the U.S. Virgin Islands, the project enabled teams of scientists to live in an enclosed area on the ocean floor, 50 feet below the surface. She was rejected. By the time of her application, Earle had spent more than 1,000 research hours underwater—more than any other applicant. But, she said, "The people in charge just couldn't cope with the idea of men and women living together underwater."[54] Tektite I was an all-male crew.

The following year, Earle led Tektite II, Mission 6, an all-female research expedition. She and four others (four scientists and one engineer) spent nearly 2 weeks together in their small four-room habitat capsule, 50 feet below the surface, carrying out ecological studies. The adventure made Earle and the other researchers celebrities, complete with a ticker-tape parade and a reception at the White House. More importantly, the experience gave Earle a public platform. She became an outspoken advocate for undersea research; raised public awareness of the damage caused by pollution and environmental degradation; was in high demand as a public speaker; and began to produce books and films and write for *National Geographic* magazine. Her research has focused on the ecology and conservation of marine ecosystems. She has also been actively involved in developing technology to facilitate study of the ocean depths.

Earle has contributed to marine science in many different capacities. From 1979 to 1986, she served as Curator of Phycology at the California Academy of Sciences. In the early 1990s, she was appointed Chief Scientist of the National Oceanographic and Atmospheric Administration (NOAA)—the first woman so appointed. In 1992, she founded Deep Ocean Exploration and Research, a company devoted to advancing marine engineering. From 1998 to 2002, she led the Sustainable Seas Expeditions, a program sponsored by the National Geographic Society to study the U.S. National Marine Sanctuary. She led several research investigative teams following the Exxon Valdez (1989) and Mega Borg (1990) oil spills and consulted after the Deepwater Horizon disaster (2010).

After receiving the million-dollar TED Prize in 2009, Earle launched Mission Blue, also known as the Sylvia Earle Alliance, a nonprofit global coalition with the goal of establishing a worldwide network of marine preserves ("Hope Spots") to save "the blue heart of the planet." Earle is an Explorer in Residence at the National Geographic Society. She has published more than 225 papers and has led more than 100 expeditions. Among her many awards and honors, she was named *Time Magazine*'s first Hero for the Planet (1998) and a Living Legend by the Library of Congress (2014); she received the Explorers Club Medal (1996), the Walter Cronkite Award for Excellence in Journalism (2014), and the National Geographic 2013 Hubbard Medal, the Society's highest honor for distinction in exploration and discovery.[55]

---

[53] https://mote.org/pages/for-eugenie-clark-a-love-letter-by-sylvia-earle; accessed 5 October 2021.

[54] https://achievement.org/achiever/sylvia-earle/#interview; accessed 5 October 2021.

[55] https://tektite2020.com/sylvia-earle.html; accessed 5 October 2021.

FIGURE 6.5   Miriam Kastner. (Photo courtesy of Special Collections and Archives, UC San Diego.)

Miriam Kastner (1935–) was born in Bratislava, Czechoslovakia. She received a master's degree in geology, with a minor in chemistry from Hebrew University in Jerusalem, Israel, in 1964. She then went to Harvard University, where she was introduced to oceanography. She received her PhD in geosciences from Harvard in 1970. Kastner found that Harvard was not an encouraging environment for women (Figure 6.5).

> At that time at Harvard, women students were told from time to time that Harvard's mission is to educate the future leaders of the country and women will not be such leaders, therefore why bother to educate them. Most of the faculty members did not expect much from women students; the perception was that the few women present would be there until they found a husband.[56]

Kastner went to Scripps Institute of Oceanography in 1972, becoming the second female faculty member and the first woman professor in the Geosciences Research Division. Her research in geochemistry has focused on diverse aspects, from the effects of fluids at subduction zones on seawater chemistry and other processes to the abundance and distribution of methane hydrates below the sea floor.[57] Kastner was interviewed

---

[56] https://scripps.ucsd.edu/news/scientists-life-miriam-kastner; accessed 6 October 2021.
[57] https://scripps.ucsd.edu/news/scientists-life-miriam-kastner; accessed 6 October 2021.

in her office on 23 May 2006 by Laura Harkewicz as part of the Scripps Institution of Oceanography's Oral History Project. Following are some insights from her interview.[58]

> I was the first woman on the faculty at Scripps … what younger women don't realize today is there were no special perks for women, women were an oddity at that stage, especially in the top institutions. I didn't get startup money. I got an empty laboratory and I was happy with that, at least I got a laboratory. I grew up in Israel and Israel is a very different society where women had early on a different status … When I grew up the notion that women cannot achieve certain positions did not exist. When I was an undergraduate, I majored in geology with a minor in physical chemistry. I had great women professors in almost every topic: for example, in physical chemistry, in biology, and in statistics. … At Harvard, I very much enjoyed the great education but from a social point of view it seemed to me like the U.S. was in the Middle Ages compared with Israel. … I ended up staying here because once I was offered the opportunity at Scripps, which is the best oceanographic institution in the world, I accepted it.
>
> I'm trying to tell the women students of today how it was and that they're very spoiled today. Women have special grants at NSF, they receive a considerable amount of startup money, but some still complain. I think the women of today receive a golden carpet, which is really helpful to their career, but they should realize it wasn't always like that.

Deborah Dexter (1941–) was born into an academic family. "If I had told my parents, when I was 17, that I wanted to get married and have kids they would have had a fit. But I never even thought of that as a possibility," she says. "There was no question, when I was that age, that I would go to the university." [59] Dexter's curiosity about marine life started when she was a child. "My grandmother lived in Hermosa Beach and I got to go down there as a kid. We would walk down the beach every morning and dig in the sand." She earned her BA in biology and MA in education at Stanford, and then received her PhD in zoology from the University of North Carolina-Chapel Hill in 1962. From 1964 to 1967, she was a researcher at the UNC Institute of Marine Sciences. In 1964, she joined a research cruise sampling the Atacama trench, 100 miles off the coast of Peru. There were 30 researchers on board; Dexter was the only woman.

> At the time, I just thought I was so lucky to be on the ship and to be doing those things. … It was a really unbelievable thing—to have one female scientist on board. It just didn't happen that way back then.

Dexter then joined the faculty at San Diego State University, where she was the only woman in the Department of Zoology.[60] She taught marine invertebrate biology, biological oceanography, marine ecology, and the popular course Life in the Sea for nonscience majors.[61] Dexter took her students into the field and introduced them to the intertidal and nearshore fauna, including sponges, sea urchins, star fish, and sea cucumbers.

---

[58] We thank Dr Paul Dayton for bringing this work to our attention and Peter Brueggeman, Director of Scripps Institution of Oceanography Library and Archives (retired), for making this archive available to us.

[59] https://endeavors.unc.edu/?s=Dexter; accessed 11 February 2021.

[60] https://plannedgiving.sdsu.edu/gifts-in-action/deborah-dexter; accessed 6 March 2021.

[61] Parker (2016).

We would go down to the beach and see seaweed all over the rocks. So, you have to slip your fingers into a crevasse or turn over a rock so that they begin to see where these organisms are. Learning about their biology in a lab setting doesn't do any good if you haven't got a clue where or how to find them 10 miles away.

"If you do the right things, it doesn't take that much to get your students excited," she says. "And if you get them exited, they're much more likely to work harder."[62]

Dexter felt she had to prove herself time and time again, but it got easier. "You have to go with the flow," she says. "If you can do that, you will get more and more opportunities because people will start to say, 'This is a woman who can do this.'"[63]

Elizabeth Louise "Pooh" Venrick (1941–) was born in Chicago, Illinois. She received a BA from Pomona College in Claremont, California (1962). The following fall she entered the doctoral program of Scripps Institution of Oceanography (University of California, San Diego) in biological oceanography. Two years later, she went on her first scientific cruise, *Ursa Major*, studying the plankton along a mid-Pacific transect from Alaska to Hawaii; this was the research on oceanic phytoplankton that she continued throughout her career. In 1969, she received her PhD and joined the tri-agency California Cooperative Oceanic Fisheries Investigations (CalCOFI) where she spent her career. In 1976, she was the first PhD, the first woman, and the first expert in marine science to be appointed by Governor Brown to serve on the California Fish and Game Commission, and she served on several ocean- and fishery-related government committees throughout her career. She retired in 2009, at which time she was the Scripps representative on the CalCOFI Committee, and a founder and codirector of the Integrative Oceanic Division at Scripps (Figure 6.6).

Venrick was interviewed in her office on 15 December 2005 by Laura Harkewicz as part of the Scripps Institution of Oceanography's Oral History Project. The following quotes from Venrick's interview offer some insights of what it was like to be a woman in oceanography during the 1960s.[64]

When I first came in, I was the first woman admitted to biological oceanography. And I learned later that my major professor ... who was one of the fairest people I have ever met, was concerned enough about admitting a woman that he asked the guys did they think it would be okay. I gather that this started off the usual bit about, "Well, what's she look like?" and one of them started a rumor that I was six foot two, and they all believed it until the day I walked in the door. (I was 5 foot four.)

I'm sure there were a lot of things that happened then that now ... would be interpreted as sexism. I was just totally oblivious. I've heard other stories from colleagues of mine who had more pointed experiences. A good friend told the story that when she was, I think, applying for admissions, one of the male professors said, "Well, do you realize you're going to be taking food out of some family's mouth," reflecting the then-prevalent dogma of Man-as-the-Breadwinner. Going to sea, it turned out I was also the first woman to go to sea alone. There had been a few women who had gone to sea in

---

[62] https://endeavors.unc.edu/?s=Dexter; accessed 11 February 2021.

[63] https://endeavors.unc.edu/?s=Dexter; accessed 11 February 2021.

[64] We thank Dr Paul Dayton for bringing this work to our attention and Peter Brueggeman, Director of Scripps Institution of Oceanography Library and Archives (retired), for making this archive available to us.

FIGURE 6.6   Elizabeth Louise Venrick; photo by Susan Hamilton. (Photo courtesy of Elizabeth Venrick, Personal Collection.)

groups or in pairs, or with husbands, but my very first cruise, I was slated to be the only woman. ... I knew the guys. I mean, these were my graduate school friends and so I had no problem with being the only woman.

In more recent years, I've seen women join the crew. I've seen women be chief scientists. But back in the mid-sixties women at sea were viewed with skepticism. I remember once I was on the first leg of a cruise (San Diego to Kodiak, Alaska) and I wanted to continue on the cruise across the subarctic. This was in January–February. It was a pretty rough cruise. The chief scientist at that time said, "No. It's too rough for a woman."

Another woman friend was on a cruise where the chief scientist told the women "Don't wear a bathing suit" even though the other scientists often did. These were experiences that accompanied the early acceptance of women at sea. Now [in 2005] I think our graduate student incoming class is slightly more than half women. And, people don't think anything of both sexes at sea. I've been on cruises where I've bunked with a guy, because the distribution of bunks didn't match the distribution of sexes. You know, you try to avoid the situation but if it happens, it just happens. That's been a big change, just the way we're treated.

It might have been [harder being a woman], but I never felt it. In some ways, being a woman was an advantage. I've talked to people who went through graduate school at the same time and they often have very different perceptions. I think that so much is just your own particular outlook, or whatever you want to call it. But I simply was not looking for discrimination and I never found it.

One thing that I do remember of those early days is that so few of the women actually went on to work in the field. Most of them either got married as soon as they got their Ph.D., and started raising families, or they got married during their career and quit research. To that extent, the concern about women in the field was being justified, by women. And now, of course, women do both. I don't know how they do it. Nobody thinks anything of it.

In retrospect, I see now that my early career caught the first swell of the women's movement, and I rode that swell throughout my early career. My timing, though accidental, was excellent. I received my Ph.D. in a very young, heavily male-dominated field. Were my opportunities all because I was a convenient "token woman"? Some, probably. Was that why I hired by CalCOFI? Appointed by the Fish and Game Commission? The California Condor Advisory Committee? The Scientific and Statistical Committee of the Pacific Marine Fisheries Commission? The Marine Fisheries Advisory Committee? etc. I reassure myself that I couldn't have been too bad, or the searches would have passed me by, but I will never know for sure. And it doesn't matter—it's been a great career, with wonderful colleagues and interesting adventures.

Kathy Frost (1949–) was born in Topeka, Kansas.[65] Her parents took her camping, hunting, and fishing, and she grew to love the outdoors. Her grandfather lived by the ocean and perhaps because of this exposure she attended the University of California Santa Cruz, where she received her master's degree. She became a marine biologist through serendipity. She met her partner, Lloyd Lowry, while taking a course at Hopkins Marine Station in 1972. Lloyd wrote a "Dear Sir" letter to the Alaska Department of Fish and Game asking for a job. He was offered a position and brought Frost along. She began volunteering at Fish and Game, eventually started getting paid, and was on soft money for 25 years. Some insights from her interview (Figure 6.7):

FIGURE 6.7   Kathy Frost. (Photographed by and used with permission, Denali Whiting.)

---

[65] Interviewed along with her partner Lloyd Lowry by The Wildlife Society Student Chapter on 9 October 2017. Available at https://www.youtube.com/watch?v=O0ZPV9DWkiM; accessed 16 January 2020.

There have been big changes in marine mammal science in the last 40 years. When I started there were very few women working in the field. In Alaska, I think that was even more exaggerated. Fish and Game had one other woman biologist when I was hired and she basically didn't go in the field. I was lucky to have a couple of field partners, Lloyd one of them … who really didn't care if I was a girl or a guy, they just cared that I did my job. And I think they helped me realize too, that everybody doesn't have to do everything everyone else does—nobody is the same. People are people but they're not the same. The quicker you realize that you have your own set of skills and someone else has theirs, the easier it is to figure out how you can contribute to a field project. I was never going to pick up something as heavy as Lloyd and my other field partners could, but my hands were smaller and fit in places their hands didn't. I could fit in spaces they couldn't. I often looked at things from a different point of view. If we had to fix something or do something, I had something different to contribute. Lloyd was a stickler for being competent. He insisted that if I was going to be in the field operating small boats and working around nets, snow machines and other equipment, I not only do it right but I had to do it well. That was huge. I guess that was something I tried to pass on to other young women as I went along—that it's important to be competent and to be good. You can't just go out and say "I want I want I want I want – I want to be equal, I want to be the same" and then turn around and I say "I can't, I don't know how, I don't want to." You can't have it both ways. "I don't know how" is something you can fix. In the early days, I was out there as me, but I was also out there as a woman in the field. A lot of guys, especially those who had been doing it a different way for a very long time, were watching and saying they didn't want a woman in the field and they just didn't want women around. So, you weren't just acting for yourself, you were paving the way for people who were coming behind you. There was a responsibility to do well and be good so that the women who came after me and others like me would have an easier way forward. When I look around me now, the science world is very different. At a marine mammal conference, 60 or 70% of the people in the room are women now. But guy or gal, it's important to be good at what you do and to figure out what you're good at. You need to specialize in something. If you want to go in the field and run small boats then take the time to learn how to do that well. One of the differences when I grew up was that boys pulled apart car engines and repaired boats and most girls didn't. If you're going to be a field biologist in the Arctic you've got to be able to do your part and contribute. I wasn't nearly as good at engines as Lloyd and Bob were but I learned how to replace water pumps in outboard engines and we went through a lot of water pumps when we worked in Bristol Bay in silty water. You need to make a real contribution—not just be a hanger on.

We end our consideration of marine scientists in this chapter with Jane Lubchenco (1947–), born in Denver, Colorado. She received her BA in biology from Colorado College in 1969.[66] As an undergraduate, she took a summer course in invertebrate zoology at the Marine Biological Laboratory in Woods Hole and became fascinated with marine biology and research. She received her MS in zoology from the University of Washington (1971) and her PhD from Harvard in marine ecology (1975). She remained at Harvard, where she was hired as an assistant professor. In 1977, she and her husband moved to Oregon State University, where they shared a tenure-track position.[67] Her early research focused on rocky seashore ecology and biodiversity. Later she became interested in human impacts on the oceans (Figure 6.8).

[66] https://gordon.science.oregonstate.edu/lubchenco/sites/default/files/general/dr_lub-chenco_20_17_08_bio_700_wd.pdf; accessed 28 February 2020.
[67] Lubchenco and Menge (1993).

FIGURE 6.8   Jane Lubchenco; image in the public domain.

Beginning in 2009, she served as part of President Barack Obama's Science Team, first as Under Secretary of Commerce for Oceans and Atmosphere and Administrator of the National Oceanic and Atmospheric Administration (NOAA) and then as the first U.S. Science Envoy for the Ocean. She founded three programs—the Aldo Leopold Leadership Program, COMPASS, and Climate Central—designed to train scientists to communicate their research more effectively to nonscientists. A colleague pointed out that Lubchenco's approach to problems has always been "we gotta do something about it."[68] Lubchenco's career is legendary. She is one of the world's most highly cited ecologists and is a member of the National Academy of Sciences, the American Academy of Arts and Sciences, the American Philosophical Society, and the Royal Society.[69] She received the Eminent Ecologist Award from the Ecological Society of America in 2014.

We conclude this chapter by coming full circle to the tradition of populariz-ing natural history, a field where women have always excelled. During this time period, from the early 1970s to the present, nature writers such as Alison Hawthorne Deming, Annie Dillard, Camille T. Dungy, Gretel Ehrlich, Robin Kimmerer, Leslie

---

[68] Finkbeiner, A. https://www.hakaimagazine.com/features/an-ecologist-organizes-the-world; accessed 28 February 2020.

[69] https://gordon.science.oregonstate.edu/lubchenco/sites/default/files/general/dr_lub-chenco_20_17_08_bio_700_wd.pdf; accessed 28 February 2020.

Marmon Silko, Terry Tempest Williams, Andrea Wulf, Ann Zwinger, and many others have shared their love of nature through writing, in the hopes of inspiring readers to understand, appreciate, and protect their natural surroundings.

One nature writer extraordinaire was Ann Haymond Zwinger (1925–2014), who continued the theme Susan Fennimore Cooper (profiled in Chapter 2) began a century earlier—a plea for preservation of forests and natural resources. Zwinger was born in Muncie, Indiana, the daughter of a lawyer father and artist mother. She studied art history in college and graduated from Wellesley College with a BA in 1946. Four years later, she graduated from Indiana University with a master's degree in Fine Arts. She was working toward a doctorate degree at Harvard when she met her future husband, an Air Force pilot. She became a military wife and the couple moved around the country while raising three daughters. In 1960, her husband was transferred to Colorado Springs, where they stayed until his death in 2012.

Zwinger and her husband bought 40 acres of land and built a cabin in the mountains west of Colorado Springs. Her goal was to become intimate with her surroundings. To do that, she began by cataloging and illustrating the plants on their land.[70] Zwinger got her break into the publishing world in 1969 when she met Rachel Carson's literary agent Marie Rodell, through a mutual friend. When Rodell asked Zwinger what she would most like to do, she answered: "Write about our land."[71] With Rodell's encouragement, Zwinger first wrote a chapter that was accepted by Random House for *Beyond the Aspen Grove* (1970). In her book, Zwinger explores life in the Rockies between 7,000 and 9,000 feet elevation, in streams and lakes, aspen groves and ponderosa forests, marshes, and meadows.

During the next several decades, Zwinger published another 20 books, ranging from desert to alpine tundra environments. She emphasized preservation, environmental stewardship, and individual accountability and filled her books with beautiful pen-and-ink illustrations. In 1973, she was a finalist for the National Book Award for The Sciences for *Land Above the Trees* (1972), coauthored with Beatrice Willard. She received the John Burroughs Memorial Association Gold Medal for nature writing in 1976 for *Run, River, Run* (1975) and the Western Arts Federation Award for nonfiction in 1995. Zwinger also wrote pieces for *Audubon*, *Natural History*, and *The Smithsonian*. She taught at Carleton College, Smith College, and Colorado College, serving as a mentor and role model for aspiring nature writers. Zwinger emphasized the value of careful observation to understand nature and of sharing those observations with others.

This ends our discussion of the *history* of women field biologists. Early European and North American women field biologists—motivated, determined women largely working in isolation—came from scattered locations. They, in turn, motivated subsequent generations of women who had the advantages of women's colleges in the East, land grant colleges in the Midwest, and a frontier to explore in the West. In the following two chapters, comprising the next section of this book, we will meet a subset of today's women field biologists; each has generously shared her path, experiences, and perspectives.

---

[70] Norwood (1993), p. 52.

[71] https://obits.gazette.com/us/obituaries/gazette/name/ann-zwinger-obituary?pid=172384209; accessed 29 June 2018.

REFERENCES

Atchison, G. J., J.E. Morris, and S. J. Dinsmore. 2018. A history of the Fisheries and Wildlife Programs, the Animal Ecology Department, and the Natural Resource Ecology and Management Department at Iowa State University. Unpublished final draft. https://www.nrem.iastate.edu/files/page/files/history_of_animal_ecology_dept_final_draft_2018.pdf; accessed 29 December 2020.

Balon, E. K. 1994a. An interview with Eugenie Clark. *Environmental Biology of Fishes* 41:121–125.

Balon, E. K. 1994b. The life and work of Eugenie Clark: devoted to diving and science. *Environmental Biology of Fishes* 41:89–114.

Greiling, D. 2011. *Bulletin of the Ecological Society of America*. https://www.esa.org/history/obits/Rathcke_B.pdf; accessed 8 February 2021.

Langenheim, J. H. 1996. Early history and progress of women ecologists: emphasis upon research contributions. *Annual Review of Ecology and Systematics* 27:1–53.

Lannoo, M. J. 2018. *This land is your land: the story of field biology in America.* Chicago: University of Chicago Press.

Lubchenco, J. and B. Menge. 1993. Split positions can provide a sane career track: a personal account. *BioScience* 43:243–248 doi:10.2307/1312127. JSTOR 1312127.

Mowat, F. 1987. *Woman in the mists: the story of Dian Fossey and the mountain gorillas of Africa.* New York: Warner Books.

Norwood, V. 1993. *Made from this Earth: American women and nature.* Raleigh, NC: University of North Carolina Press.

Parker, M. L. 2016. Deborah Dexter interview. 14 March. https://endeavors.unc.edu/woman_on_board; accessed 15 February 2021.

Rabinowitz, D. 1981. Seven forms of rarity. In *The biological aspects of rare plant conservation,* ed. H. Synge, pp. 205–217. New York: John Wiley & Sons.

Simberloff, D. 1988. Deborah Rabinowitz: In Memoriam. *Conservation Biology* 2:119–120.

Sneed, A. 2017. 3 Decades after Dian Fossey, gorillas still face extinction. https://www.scientificamerican.com/article/3-decades-after-dian-fossey-gorillas-still-face-extinction; accessed 6 July 2021.

Sridhar, H. 2018. Revisiting Hrdy 1974. Reflections on Papers Past. https://reflectionsonpaperspast.wordpress.com/2018/01/02/revisiting-hrdy-1974; accessed 16 February 2021.

Werner, P. A. and S. J. Peacock. 2019. Savanna canopy trees under fire: long-term persistence and transient dynamics from a stage-based matrix population model. *Ecosphere.* doi:10.1002/ecs2.2706.

Zedler, J. B. 2000. Progress in wetland restoration ecology. *Trends in Ecology & Evolution* 15:402–407.

Zoellner, D. C. and K. L. Gross. 2019. *In the founder's footprints: a history of Michigan State University's W.K. Kellogg Biological Station.* Lansing, MI: Michigan State University.

Section 2

Current Perspectives

7 Backgrounds, Paths, and Careers

To gain a broader understanding of what it is like now to be a woman in field biology, we interviewed 75 women field biologists. They focus on systematics, animal behavior, ecology, conservation, and land restoration. They work with plants, invertebrates, and vertebrates. They are employed by federal and state agencies, colleges and universities, natural history museums, conservation NGOs, and consulting firms. For many, a childhood love of nature and being outside led to their chosen careers. Others had serendipitous conversions. Some followed linear paths to their careers; others took circuitous routes. Many women profiled here have served as department chairs, deans, directors, and society and organization presidents. They have received teaching, research, service, and lifetime achievement awards. Some are elected fellows of scientific societies and members of the U.S. National Academy of Sciences. All have had unique experiences, backgrounds, and opportunities that have shaped their careers and lives as field biologists. Following are their stories, arranged historically, in chronological order by birth year.

Frances C. James (b. 1930; PhD 1970; avian ecology; fieldwork primarily in the United States and Mexico; Professor Emerita, Florida State University; President, American Ornithologists' Union; President, American Institute of Biological Sciences; Eminent Ecologist Award, Ecological Society of America, 1997; Margaret Morse Nice Medal, Wilson Ornithological Society, 1999). I grew up in suburban Philadelphia, where I was a birdwatcher as a teenager. After earning an AB degree from Mt. Holyoke College, I earned my master's degree from Louisiana State University. While raising three daughters, I served as a part-time instructor at the University of Arkansas and eventually earned a PhD there. My dissertation focused on intraspecific size variation among birds in the central and eastern United States. It suggested that climatic variables like wet-bulb temperature that combined various avenues of heat loss and gain are more important than temperature alone in accounting for size variation. They are also more aligned with Bergmann's intent. From 1973 to 1977 I served as assistant and associate program director in Ecology, and later in Population Biology at the National Science Foundation. Then I joined the faculty at Florida State University. With lots of help from students, I conducted field studies of geographic variation in the development of nestling red-winged blackbirds in the United States and Mexico to discover whether their regional variation in size and shape had a partly nongenetic basis. Other research involved the ecology of the endangered red-cockaded woodpecker in relation to the history of prescribed fire in the Apalachicola National Forest. I retired from FSU in 2003 but am continuing my research (Figure 7.1).

DOI: 10.1201/9781003311508-9

FIGURE 7.1   Frances James. (Photo courtesy of and photographed by Helen Roth.)

Rosemary Grant (b. 1936; PhD 1985; evolutionary biology; fieldwork in the Galápagos Islands; Senior Research Biologist, Emeritus, Princeton University; Darwin Medal, The Royal Society of London, 2002; Grinnell Medal, UC Berkeley, 2003; Balzan Prize, 2005; Darwin-Wallace Medal, 2008; Kyoto Prize, 2009; Royal Medal of Biology, Royal Society of London, 2017; BBVA Foundation Frontiers of Knowledge Award, 2018; Fellow, Royal Society of London, 2007; Member of National Academy of Science, elected 2008). I grew up in a small village, Arnside, in the English Lake District, where as a child I roamed the surrounding fells and fossil-rich carboniferous limestone cliffs, finding wildflowers, butterflies, and fossils. I became intrigued with variation in nature, and when I was 12 years old my father (a medical doctor) suggested I should read Darwin's *Origin of Species*, which I did. In high school, I realized that genetics would provide a fundamental approach to understanding the variety of organisms, and I was fortunate to study genetics at Edinburgh University in Professor Conrad Waddington's department. I was interested in the question: how do populations maintain such a large amount of genetic and morphological variation and how do they diverge to the point of becoming different species? During a Research Associate position at the University of British Columbia, I met my husband, Peter (also from England), who was interested in similar questions but more from an ecologist's point of view. Trying to shed light on these questions led us to our joint research in the Galápagos where we have been studying Darwin's finches since 1973. After Peter and I married, I held several Research Associate positions and raised two daughters. At one point, I was invited to give a talk at Uppsala University in Sweden about our Darwin's finches work. The renowned ecologist Dr Staffan Ulfstrand introduced me as Dr Grant. I thanked him, but told the audience that I did not have a PhD. That night at the social, Dr Ulfstrand suggested I come to Uppsala and earn my PhD with him. The next day I asked him if he had been drunk when he invited me to come and be his student. He said no! Thus, I earned my PhD degree at age 49. Peter and I accepted positions at Princeton University in 1986. Our work combines ecology, evolution, and behavior with genetics and more recently with genomics (Figure 7.2).

FIGURE 7.2 (Peter and) Rosemary Grant. (Photo courtesy of and photographed by K. T. Grant.)

Michelle Pellissier Scott (b. 1939; PhD 1984; behavioral ecology; fieldwork in Australia and New Hampshire; Professor Emerita, University of New Hampshire; Excellence in Teaching Award, University of New Hampshire, 2011). I was not particularly interested in nature as a child. I grew up in the suburbs and spent all my free time reading. At Wellesley College I was a liberal arts major and took few science courses. I didn't have a career plan. My philosophy while job-hunting was, "I'm smart, I can do what you want me to do." I married an architect and adopted two children in the 1970s. By the time the kids were 3 years and 1.5 years, I was bored out of my mind. I happened to see a fund-raising advertisement: "Would you give a lion a home?" I thought, not only will I give one a home, I will go and visit you. Within 5 minutes, my husband and I decided to move to Nairobi, Kenya, for an adventure. My husband got a job in Nairobi, and I was a docent at the Nairobi Museum of Natural History. I took courses from anthropologist Richard Leakey at the museum and became interested in human evolution. During my time in Kenya, I fell in love with wild animals. A year and a half later, when we returned to Cambridge, Massachusetts, I started auditing courses at Harvard, first in anthropology and then in biology. I became intrigued with animal behavior. Eventually I made up the necessary science courses, applied to graduate school, and was accepted at Harvard. I moved my family (now three adopted children) to Australia for two and a half years while I did my fieldwork on reproductive tactics of male marsupials.

FIGURE 7.3   Michelle Pellissier Scott. (Photo courtesy of and photographed by Michelle Pellissier Scott.)

Six years after earning my PhD, I got a job as an Assistant Professor at the University of New Hampshire, where I worked for 21 years. Much of my research and fieldwork in New Hampshire focused on the ecology and behavior, including male parental behavior, of burying beetles (Figure 7.3).

Marvalee Wake (b. 1939; PhD 1968; morphology, development, and reproductive biology of caecilians; fieldwork primarily in Mexico, Central America, and Vietnam; Professor of the Graduate School, University of California, Berkeley; Henry S. Fitch Award for Excellence in Herpetology, American Society of Ichthyologists and Herpetologists (ASIH), 2014; Fellow, California Academy of Sciences, elected 1978; Fellow, AAAS, elected 1983; Fellow American Academy of Arts and Sciences, elected 2003; President, American Institute of Biological Sciences, ASIH, Society for Integrative and Comparative Biology, International Union of Biological Sciences, and International Society of Vertebrate Morphology). My fourth-grade teacher was passionate about field biology, and we explored tide pools, the chaparral, and other habitats. I thought I might be able to be like her someday—spry and wonderfully knowledgeable about the environment. I was pre-med as an undergraduate at the University of Southern California and had delayed taking most of my zoology major requirements until my senior year.

FIGURE 7.4 Marvalee Wake. (Photo courtesy of and photographed by Charles R. Crumly.)

I had applied to medical schools and was shadowing physicians, but I was becoming disenchanted with what I saw. At the same time, I was finding my zoology courses fascinating. The professor of my evolution course, Jay Savage, suggested I do a senior research project. He had a set of caecilians from Costa Rica to be identified and suggested that would be a good project. I was hooked. Jay suggested I apply to graduate school at USC and work with him. I did (I was Jay's first female graduate student) and I have been continuing my senior thesis research—work with caecilians—ever since. After 2 years of graduate work at USC, I moved to Chicago with my husband David Wake (he had accepted a faculty position at the University of Chicago); I did much of my dissertation research in Chicago. I was an instructor, then assistant professor at the University of Illinois at Chicago. In 1969, Dave and I moved to Berkeley where I have been on the faculty ever since. I nominally retired in 2003 but remain active with research, teaching, and extramural scientific commitments. My field research has involved collecting and observing the behavior of caecilians, especially feeding and locomotion. Through my fieldwork, the animals themselves have shown me how to approach my research investigations that have focused on developmental biology, reproduction, anatomy, and evolution of amphibians (Figure 7.4).

Jeanne Altmann (b. 1940; PhD 1979; behavioral ecology of baboons; fieldwork in Kenya; Professor Emerita, Princeton University; Fellow, Animal Behavior

Society (ABS), elected 1989; Exemplar Award, ABS, 1996; Member, National Academy of Sciences, elected 2003; Distinguished Animal Behaviorists Award, ABS, 2012; Sewall Wright Award, American Society of Naturalists, 2013; Lifetime Achievement Award, International Society of Primatology, 2014; President, ABS). I grew up in suburban Maryland, just outside Washington, D.C., in a sequence of small apartments or flats where I had minimal exposure to animals. My early life centered on reading, math, science, and puzzles. My parents valued education, but in the nineteenth century model; I should go to college to be a better wife and mother and to have an employment skill in case something happened to my future husband. After my freshman year as a math major at UCLA, I had a summer job at the National Institutes of Health in Bethesda, Maryland. There I met Stuart Altmann, who was finishing his PhD at Harvard, having studied rhesus monkey behavior. Stuart and I married the next year. The UCLA math department was not the right place for me, as the faculty made it clear they considered women math majors a waste of time. Stuart accepted a faculty position at the University of Alberta, where I finished my undergraduate math degree and became a mother. In 1963, Stuart, toddler Michael, and I made our first trip to Amboseli, in southern Kenya, to study baboons—my immersive introduction to fieldwork, animal behavior, and baboons. That initial project led to one of the world's longest-term field studies of primates. In 1965, Stuart accepted a position at the Yerkes Primate Research Center at Emory University. A few months later, our daughter Rachel was born. During this time in Atlanta, I developed and taught a remedial math program for the local school system and earned my MA in teaching (math). In 1970 Stuart accepted a faculty position at the University of Chicago. Once Rachel was in school full-day, I began my PhD in behavioral ecology, focusing on female baboons and their infants. We continued studying baboons at Amboseli, with my focus being demography, ontogeny, and female life histories. In 1984, I joined the new Conservation and Research Department at the Brookfield Zoo—my first full-time employment. By 1998, Stuart had transitioned to emeritus status and I accepted a faculty position at Princeton University. Researchers from Kenya, the United States, and elsewhere have now worked on the Amboseli Baboon Research Project, with studies ranging from hormones and genetics to nutrition, behavior, and ecology. I am one of four women who currently direct the project (Figure 7.5).

Mary Jane West-Eberhard (b. 1941; PhD 1967; behavior and evolution of social wasps; fieldwork mainly in Colombia and Costa Rica; Senior Scientist Emerita, Smithsonian Tropical Research Institute; Member, U.S. National Academy of Sciences, elected 1988; Fellow, Animal Behavior Society (ABS), elected 2009; Sewall Wright Award, American Society of Naturalists, 2003; Quest Award for Lifetime Achievement, ABS, 2012; Hamilton Award, International Union for the Study of Social Insects, 2014; Rudolph Raff Pioneers Award, Pan-American Society of Evolutionary Developmental Biology, 2019; selected as one of the 21 "Leaders in Animal Behavior;" President, Society for the Study of Evolution). When I was born in 1941, it was still unusual for women to become scientists. My mother was a teacher, and I attribute to her my early confidence in school. My parents encouraged my curiosity about the world. I was an avid reader of pioneer stories, and I still have my coonskin cap from childhood hanging in my study—encouraging

FIGURE 7.5   Jeanne Altmann. (Photo courtesy of and photographed by Susan C. Alberts.)

exploration of new frontiers and a monument to understanding parents. As a young child, I loved to watch turtles, snakes, and ducklings near my home on Maceday Lake in southern Michigan, perhaps setting the stage for a lifelong career as an animal behaviorist. As an undergraduate at the University of Michigan, I worked as an assistant in the Insect Division in the Museum of Zoology. I loved sleuthing in the university library and liked the excitement of exploring ideas beyond textbooks. I earned my master's and PhD degrees from the University of Michigan and did a postdoc at Harvard University. I spent 10 years as an Associate in Biology at the Universidad del Valle in Cali, Colombia. After that my family and I moved to Costa Rica where the Smithsonian Tropical Research Institute (STRI) showed great flexibility in giving me and my husband part-time positions while living in Costa Rica. My husband taught half time at the Universidad de Costa Rica while maintaining his half-time STRI position. My half-time STRI appointment eventually became full time. As an animal behaviorist, the ability to study organisms in my own backyard in Colombia and in Costa Rica, without the need for grant funds, contributed greatly to my ability to focus on research. My work with social insects led to my interest in sexual selection and the evolutionary role of social competition, and in phenotypic plasticity and alternative phenotypes. As part of that, I reexamined Darwin's ideas about sexual selection and focused on the importance of the environment in development ("developmental plasticity") for understanding speciation and other major themes of evolutionary biology (Figure 7.6).

FIGURE 7.6   Mary Jane West-Eberhard. (Photo courtesy of and photographed by Marcelo Casacuberta.)

Julie Denslow (b. 1942; PhD 1977; forest ecology; USDA Forest Service (retired), Adjunct Professor, Tulane University; Honorary Fellow, Association for Tropical Biology and Conservation (ATBC), 2009; President, Organization for Tropical Studies; President and Executive Director, Association of Tropical Biology (now ATBC)). I grew up in Miami where our family enjoyed boating, fishing, snorkeling, gardening, and summer vacations in the Appalachians. I credit those early years for my love of nature. We lived in a rural area with woods, canals, and fields, where we were free to wander, build forts, climb trees, and explore. At Oberlin College I majored in zoology where we read Rachel Carson's *Silent Spring*, a book that helped shape my view of the world. I married just out of college. Although my marriage took me out of a career track, I maintained a strong interest in ecology and biology, leading eventually to a master's degree at the University of Miami and part-time jobs as lab technician, lab instructor, and park naturalist. Eventually I started graduate school in botany at Rutgers University, planning to pursue field research in the tropics. I took the Organization for Tropical Studies (OTS) course in tropical ecology in 1971, which convinced me that I could contribute to science and that I had the right stuff for field-work. It also gave me life-long connections with colleagues and mentors. On returning from the course, I transferred to the University of Wisconsin to complete a PhD in plant ecology. My dissertation project took me to the Cordillera Central of Colombia to follow early succession in fields cleared for shifting agriculture. After I earned my PhD,

FIGURE 7.7 Julie Denslow. (Photo courtesy of Julie Denslow, photographer unknown.)

I taught as a lecturer at UW, and worked with my then husband, Timothy Moermond, on the ecology of frugivorous birds and plants whose seeds they dispersed. We set up our project at La Selva Biological Station in Costa Rica, which brought me back into contact with OTS. I joined the faculty of Tulane University in 1986. In 1995, I moved to Louisiana State University where I hoped to offer more support to graduate students in tropical ecology. At Tulane and at LSU I began work on the effects of sea level rise on bottomland hardwood forests in Louisiana. From LSU, I moved to Hilo, Hawaii, to join the USDA Forest Service as a research scientist and team leader for the invasive species unit. I retired from the Institute of Pacific Island Forestry in 2007. In 2015 I was honored when the Association for Tropical Biology and Conservation named their annual award the "Julie S. Denslow Prize for Best Paper in *Biotropica*." Since retirement I have continued an interest in watercolor painting. I enjoy the close observation required, a trait shared with science (Figure 7.7).

Mercedes Foster (b. 1942; PhD 1974; ecology and social behavior of birds; fieldwork primarily in Mexico, Central America, and South America; U.S. Geological Survey Curator of Birds at the National Museum of Natural History, retired; Founding Director, American Bird Conservancy; Fellow, American Ornithologists' Union, 1980; Fellow, American Association for the Advancement of Science, elected 1988; Washington Biologists' Field Club, 1995; Alexander Skutch Medal for Excellence in Tropical Ornithology, Association of Field Ornithologists, 2006; President, Cooper Ornithological Society). I grew up in Oakland, California. My family was not outdoorsy, and I didn't have much opportunity to explore the out-of-doors other than during summer camp. As an undergraduate in zoology at the University of California Berkeley, I intended to go to medical school. During my senior year, however, I took a required course in Vertebrate Natural History, which included weekly field trips. I loved it and immediately decided to go to graduate school instead. I got a master's at Berkeley in zoology, with a minor in botany. When I took an Organization for Tropical Studies field class in Costa Rica, I had

FIGURE 7.8   Mercedes Foster. (Photo courtesy of and photographed by Fiona A. Wilkinson.)

my second epiphany and have been doing research in the tropics ever since. I married another field biologist and followed him around the country as he pursued his career. I attended the University of Chicago for a while and later got my PhD at the University of South Florida in Tampa. I continued to do research in the tropics, got divorced, and got a non-tenure lecturer position at UC Berkeley and was a Curatorial Associate in Ornithology at the Museum of Vertebrate Zoology, Berkeley. In 1980 I was hired as Curator of Birds for the National Ecology Research Center of the U.S. Fish and Wildlife Service at the National Museum of Natural History in Washington, D.C. My research has always focused on avian ecology and social behavior, particularly lek behavior. However, because the lek birds that I study are frugivorous, I became increasingly interested in frugivory—in particular, why birds choose the fruits (species and individuals) they do. My fieldwork involves making long-term behavioral and ecological observations of lek species and their populations, collecting information on behavior, relationships, longevity, habitat use, diet, etc. In my spare time (at night), I visit frog ponds to survey who is breeding and to record their behavior, calls, habitat, etc. This is for fun as much as for science (Figure 7.8).

Patricia (Pat) Wright (b. 1944; PhD 1985; conservation and primatology; fieldwork in Peru, Paraguay, Borneo, East Malaysia, the Philippines, and Madagascar; Distinguished Professor, Stony Brook University; MacArthur Fellow (MacArthur "Genius" Grant), 1989; Fellow, AAAS, elected 2004;

Distinguished Primatologist Award, American Society of Primatologists, 2008; Member, American Philosophical Society, elected 2013; Indianapolis Prize for Conservation, 2014; Finalist, St. Andrews Prize for the Environment, 2018; Cincinnati Zoo Wildlife Conservation Award, 2021). My dad taught his six children to love nature. I was raised in rural western New York and loved animals. I was unable to get a reasonable-paying job after receiving my BA in biology from Hood College. Instead, I became a social worker; my job was to help women get off welfare in NYC during the LBJ Great Society days. It wasn't until I fell in love with an owl monkey in a pet store before a Jimi Hendrix rock concert that my life had a focus. The first time I saw the rain forest, my husband and I had gone to Amazonian Colombia to get that monkey a mate. After I had a daughter in 1973 and my owl monkeys, Herbie and Kendra, had a daughter 2 weeks later, I realized our parenting styles were different. I was now a stay-at-home housewife, while my husband worked as an artist to support us. Herbie was taking care of his infant daughter (carrying her 93% of the time, teaching her new foods, playing with her as she got older), while Kendra just ate, rested, and nursed the baby every few hours. The 1970s were the beginning of the "sexual revolution," and I knew this father care in our distant cousins was important. How and why had this system evolved? I had to find out in the wild where father-care "worked." National Geographic Society rejected a grant to fund this research, because I didn't have a PhD. Everywhere I turned to for funds, I was rejected and dismissed. Funding came from a kind 81-year-old woman in my hometown who had provided financial support for George Eastman (Kodak) at his beginning. I did the study in Peru, accompanied by my 3-year-old daughter and husband, and published my research. After that I started graduate school at City University of New York and worked with Dr Warren Kinzey, an expert on primate evolution. By this time, I was divorced with a 4-year-old daughter and no money. Amanda attended my classes because I couldn't afford child care. I did a postdoc at Duke University with Dr Elwyn Simons, who introduced me to Madagascar and lemurs. For nearly 30 years I have studied wild lemurs in Madagascar. I discovered a new species of lemur, rediscovered one thought extinct, spearheaded the formation of Ranomafana National Park, and established the Centre ValBio research station at the edge of the park. An IMAX 3D film, "Island of Lemurs: Madagascar," narrated by Morgan Freeman, featured me and my conservation project (Figure 7.9).

Dee Boersma (b. 1946; PhD 1974; ecology and behavior of penguins; fieldwork mainly in the Barren Islands, Alaska, Galápagos Islands, Falkland Islands, and Argentina; Professor and Director, Center for Ecosystem Sentinels, University of Washington; Distinguished Teaching Award, University of Washington, 1993; Fellow, American Ornithologists' Union, elected 1994; Fellow, AAAS, elected 2000; Member, American Academy of Arts and Sciences, elected 2021; American Ornithologists' Union Elliott Coues Award, 2008; Heinz Environmental Award, 2009). I've always loved wildlife and being in nature. I collected butterflies as a kid, caught turtles, and spent lots of time outdoors. I thought I would study insects and particularly silk moths. I collected their cocoons and hatched them. When I got to college, I took an entomology course and found studying insects was really about DDT and killing pests. Fortunately, I took ornithology, which was steeped in natural history. As an undergraduate I collected birds for Central Michigan University

FIGURE 7.9   Patricia Wright. (Photo courtesy of and photographed by David J. Lowe.)

during spring breaks in Florida. I wanted to go to the Galápagos, and Paul Colinvaux said if I came to The Ohio State University for graduate work, I could do research in the Galápagos. I did my dissertation on Galápagos penguins because I thought it was odd that a penguin lived on the equator and 50 years later am still studying those penguins. Since 2010 I have been going twice a year to Galápagos to check nests, measure the penguins, and determine how well they are doing. Penguins are using our constructed nests made from lava and we hope to provide more nesting sites. The creation of nests has increased reproduction in the Galápagos penguins, the world's rarest species of penguin. For nearly 40 years I have been working at Punta Tombo, Argentina, on Magellanic penguins studying their natural history, migration routes, growth rates, foraging locations, population changes, and impacts of oil pollution. The study started because a Japanese company wanted to harvest breeding penguins for oil, meat, and their skins. Even the size of the colony at Punta Tombo was unknown. With funding from the Wildlife Conservation Society and interested volunteers we banded over 60,000 penguins, and we follow reproductive success in several hundred known-age individuals. Our data, locals, and NGOs helped convince the Argentine government to move oil tanker lanes farther offshore and that has decreased the occurrence of oiled birds. I have proven a link between climate change (increased storms and rain) and mortality of chicks. In the 1980s and 1990s, I also worked with Fork-tailed storm-petrels in Alaska showing how unusual their natural history is. These robin-size seabirds can reflect oil spills. They also can neglect their eggs, resulting in incubation periods that range from under 40 days to more than

FIGURE 7.10   Dee Boersma. (Photo courtesy of and photographed by Sue E. Moore, Center for Ecosystem Sentinels, University of Washington.)

70 days, and the chicks still hatch. I established the Center for Ecosystem Sentinels at the University of Washington and serve as its Director (Figure 7.10).

Martha (Marty) Crump (b. 1946; PhD 1974; behavioral ecology and conservation of amphibians; fieldwork primarily in Costa Rica, Brazil, Ecuador, Argentina, and Chile; Professor, University of Florida, retired; Adjunct Professor, Utah State University and Northern Arizona University; Distinguished Herpetologist Award, The Herpetologists' League, 1997; Henry S. Fitch Award for Excellence in Herpetology, American Society of Ichthyologists and Herpetologists, 2020; President, Society for the Study of Amphibians and Reptiles). I began my fascination with nature as a child exploring the Adirondack woods with my geologist father. I was particularly intrigued with frog metamorphosis and raised hundreds of tadpoles. I entered the University of Kansas in 1964 planning to become a wildlife biologist. After I was told that women did not go into that line of work, I sought a part-time job at the KU Museum of Natural History to learn what museum biologists do. William Duellman, curator of herpetology, needed a student to catalogue and tie tags onto preserved specimens. I got the job and the rest is history. Early during my senior year, Bill invited me to join a summer expedition to survey amphibians and reptiles in eastern Ecuador. That summer I became hooked on the tropics, amphibians, and fieldwork. I stayed at KU and earned my master's degree working in Belém, Brazil, on the ecological distribution of amphibians and reptiles in varzea, igapó, and terra firme forests. An Organization for Tropical Studies course in Costa Rica expanded my tropical experience beyond

FIGURE 7.11   Martha Crump. (Photo courtesy of and photographed by Alan H. Savitzky.)

amphibians and reptiles and cemented my love of natural history. I earned my PhD working in Santa Cecilia, Ecuador, on the reproductive behavioral ecology of a community of 81 species of frogs. After a postdoc at Brooklyn College with Stanley Salthe, I joined the faculty at the University of Florida, where I continued my research with amphibians in Central and South America. Much of my research has focused on amphibian modes of reproduction. My involvement in amphibian conservation began with the disappearance of the iconic golden toads I was studying in Monteverde, Costa Rica; the species soon became a "poster frog" for worldwide amphibian declines. During a year sabbatical in Argentina, my husband and I found that we really loved working with Latin American students. We returned to UF for 2 years and then moved to Flagstaff, where we joined Northern Arizona University as adjunct professors. Since then, I have been offering workshops to graduate students in Latin America on amphibian biology and field survey methods; continuing fieldwork in Chile, working on behavioral ecology and conservation of Darwin's frogs; and writing natural history books to make biology accessible for a general audience and for children (Figure 7.11).

Susan Riechert (b. 1946; PhD 1973; behavioral ecology of spiders; fieldwork primarily in the southwest USA, Tennessee, and West Africa (Gabon); Professor Emeritus, University of Tennessee; President, American Arachnological Society; President Animal Behavior Society; Fellow, Animal Behavior Society (ABS), elected 1993; Fellow, AAAS, elected 2008; UTK Distinguished Service Professorship; UTK Chancellors Professorship; UTK Chancellors Teaching Award; UTK Macebearer, 2011; Notable UT Woman Award, UT, 2014; Penny Bernstein Distinguished Teaching Award, ABS, 2018). I have always loved music. As a child, I played the French horn. But I battled with scarlet fever, which cost me most of the hearing in my left ear and affected my balance. I regained my balance but never my full hearing. As an undergraduate at the University of Wisconsin, I thought I would major in English because I love to read, but I found the first offering I had in that department to be boring. Then I thought maybe I'd become a veterinarian, but chance led me to enroll in a field zoology course in which we spent a lot of time collecting fish. I was one of only two women in the class. My thigh-high wading boots were many sizes too large and kept filling with water. I went underwater in a creek

FIGURE 7.12   Susan Riechert. (Photo courtesy of and photographed by W. M. Post.)

we were sampling and had to be rescued. That did it. I made collections along the shorelines as two fellows managed the seine. Spiders were abundant in these habitats and out of boredom, I collected them while waiting for my teammates to bring the seine to shore. I excelled at identifying the spiders and decided that to pass the class I would make a spider collection rather than a fish collection. I subsequently was invited to join the field trips the biology faculty took (Galápagos Islands, Central America, and the desert southwest); I was the spider-collector. I stayed at UW for my PhD. I worked at a recent lava bed in New Mexico, surrounded by desert grassland habitats, where I found that the spiders living on the lava flow were territorial and competed for nest sites. At the time, biologists thought that only vertebrates exhibited this kind of territorial behavior. The central focus of my research has been the extent to which spider populations are at adaptive equilibria with respect to their physical and biotic environments. I have also devoted considerable energy to STEM outreach projects. I retired in 2020 after 47 years on the faculty at the University of Tennessee, but I'm still doing research, working with students, and involved in science outreach (Figure 7.12).

Jane Brockmann (b. 1947; PhD 1976; animal behavior; fieldwork mainly in the United States; Professor Emerita, University of Florida; Program Director, Animal Behavior Program, National Science Foundation; President, Animal Behavior Society (ABS) 1992; Distinguished Teaching Award, ABS, 1995; Fellow ABS, elected 1995; Secretary General, International Ethological Conference, 1999–2003; Fellow AAAS, elected 2008; Distinguished Animal Behaviorist Award, ABS, 2016). I spent my early years in Louisville, Kentucky, and Indianapolis, Indiana, before moving to Chicago when I was nine. During high school, I was an active participant in bird-watching field trips led by the Chicago Ornithological Society (the youngest person by 30 years). I enjoyed reading books about birds, which led me to Tinbergen and Lorenz and the field of animal behavior. The summer after my freshman year I worked at the Cornell Lab of Ornithology, where I helped with various field and lab projects, and spent a lot of time talking with graduate students and faculty. As an undergraduate at Tufts University, I took field courses and participated in spring field trips to Jamaica. I studied territorial behavior of damselfish for my master's

FIGURE 7.13   Jane Brockmann. (Photo courtesy of Jane Brockmann, photographed by Mary K. Hart.)

degree at the University of Wisconsin, and then nesting behavior of great golden digger wasps for my PhD also at UW. In 1977–1978, I joined the Animal Behaviour Research Group at Oxford University, UK, on an NSF NATO Postdoctoral Fellowship working with Richard Dawkins on the economics of decision-making in great golden digger wasps. After being hired at the University of Florida in 1976 and returning from the postdoc in 1978, I initiated a new project on the reproductive behavior of pipe-organ mud-daubing wasps (male behavior, sex ratios, life history); my field sites were the undersides of bridges in Alachua County. I conducted one project on a related species at Monteverde, Costa Rica. I pursued the wasp research until about 1990, when I decided to work on alternative reproductive tactics of horseshoe crabs. The horseshoe crab project got its start by taking undergraduates on weekend field trips to our island field station at Seahorse Key, where horseshoe crabs nest in considerable numbers. It was a great opportunity for students to try out field research. I have continued this field study adding in new components on the genetics, development, population biology, and conservation of horseshoe crabs. Since retirement I have added a citizen scientist component in collaboration with State and University marine biologists (Figure 7.13).

Ellie Prepas (b. 1947; PhD 1980; freshwater ecology; fieldwork in Ontario and Alberta, Canada; Professor Emeritus, Lakehead University). I had the privilege of growing up in the country—first on the Niagara Escarpment, outside Hamilton, then on the Frontenac Arch outside of Kingston, ON, Canada. I spent all my time outdoors,

FIGURE 7.14   Ellie Prepas. (Photo courtesy of and photographed by Jonathan Russell R.P.F.)

making grass huts, finding secret paths through the woods, befriending the wildlife (especially bugs), and learning to canoe in the Thousand Islands. I attended country schools (some one- and two-room schools, little biology) where I showed an aptitude for math. My academic path was nonlinear (a bout with business; then degrees in math, urban planning, and biology). I decided on biology after a stint in the windowless computing centers of the 1960s–1970s and acknowledging my obsession with visiting lakes and woods. I got my math undergraduate degree at the University of Waterloo and my master's degree in environmental studies at York University, studying the history of the Canadian petroleum industry. I received my PhD from the University of Toronto in limnology; my thesis focused on predicting short-term changes in the phosphorus concentration in lake water. I realized that processes and change in natural systems was my calling. I worked at the University of Alberta from 1979 to 2001 and at Lakehead University from 2001 to 2013. My research focused on lakes and streams and their watersheds: water quality and quantity and biogeochemical characteristics therein. In Ontario I worked in the limestone escarpment near Toronto to Lake of the Woods on the western boundary, and in Alberta I worked on lakes and streams throughout the province. My early work focused on within-lake factors; my later work focused on the linkage between watershed disturbance and surface water quality, including drinking water supplies. In retirement I continue to enjoy the outdoors living on a headwater lake in the Frontenac Arch. I keep in touch with some colleagues from my working days, contribute to discussions on local water quality, garden, and dream of travelling to places I never got to appreciate during my career (Figure 7.14).

Sandra Vehrencamp (b. 1948; PhD 1976; behavioral ecology, mainly birds; fieldwork primarily in Central America, Trinidad, East Africa, and West Africa; Professor Emerita, Cornell University; Faculty Research Mentor Award, Cornell University, 2004; William Brewster Memorial Award, American Ornithologists' Union, 2011; Member, American Academy of Arts and Sciences, elected 2013). Biology became my passion in junior high school. I began college at the University of California, Riverside, where I largely focused on learning German so I could study abroad in that country and learn about my cultural heritage. While in Germany my junior year, I recall sitting on a bench watching the pigeons, curious about their behaviors and wondering if one could study that. I learned from my fellow students that UC Berkeley had a much greater variety of upper division classes in biology, including one called animal behavior. I transferred to Berkeley and took advantage of the famous courses (animal behavior, endocrinology, vertebrate natural history, marine biology) and research opportunities in many different labs. During my first year of graduate work at Cornell University, I took the Organization for Tropical Studies course in Costa Rica. I fell in love with the tropics and the incredible diversity of plants, animals, life strategies, and complex biological interactions there. My advisor had a pet smooth-billed ani and told me about their unusual communal nesting system. After the OTS course, I stayed in Costa Rica and conducted pilot research. I discovered several formerly unknown things about them, most notably that the eggs typically found under their nests were not caused by sloppy nesting but by intentional egg tossing. Although multiple pairs of birds were cooperating to incubate and feed their communal brood, they were also competing with each other to get a few more of their own eggs into the communal clutch. After I completed my PhD, I joined the faculty at the University of California, San Diego. I continued to study anis, and later worked on bats, sage grouse, and gazelles with my spouse, Jack Bradbury. I became interested in bird song and wondered why the males of many avian species sang a repertoire of song types. With students, Jack and I pioneered interactive song playback studies where we could switch song types in response to the wild birds' songs. In 1998, we were offered two positions at the Lab of Ornithology at Cornell University (Figure 7.15).

Anne Pusey (b. 1949; PhD 1978; social behavior of mammals, primarily chimpanzees and lions; fieldwork in Tanzania; Professor Emerita, Duke University; Member, American Academy of Arts and Sciences, elected 2005; Fellow, Animal Behavior Society, elected 2013). As a child growing up in England, I was interested in animals and animal behavior. I read all the books about people studying animals, watched all the documentaries, raised tadpoles, and watched birds. I discovered early on that animal behavior was a science that could be studied as a career. My father was a zoologist at Oxford University and knew Niko Tinbergen. I worked as a "slave" at Ravenglass, Tinbergen's field station (black headed gull colony), for two seasons while still in high school. After I received my BA at Oxford in 1970, I was put in touch with Jane Goodall by Robert Hinde (Professor at Cambridge) while researching graduate school opportunities. Jane was looking for an assistant to study chimpanzees. I interviewed, got the job, and went to Gombe in 1970, straight from my final exams. I met David Hamburg (psychiatrist at Stanford, collaborating with Jane) in 1971, who shared my interest in behavioral changes during adolescence, and went to Stanford as his graduate student. I met fellow student Craig Packer at Gombe. We

FIGURE 7.15 Sandy Vehrencamp. (Photo courtesy of Sandy Vehrencamp, photographed by John Burt.)

carried on a long-distance relationship while I did my dissertation work at Stanford, and he at University of Sussex. We married in 1977 and went to Japan for 3 months to study Japanese macaques, then we switched to the Serengeti lion project as our friends David Bygott and Jeannette Hanby were finishing their 4-year study. Because of the extensive genealogical information, we were able to tackle some burning questions in the developing field of sociobiology. I kept in touch with Jane Goodall and eventually, after Craig and I landed a shared assistant professorship at the University of Minnesota (initially spending half our time each year in Tanzania), I began archiving and digitizing all the field data collected on the Gombe chimps. By that time Jane had become an activist and was no longer spending field time in Gombe. After my kids reached school age, I stopped spending long periods in the field, but I continued to study the Gombe chimps for the rest of my career, curating the archive first at Minnesota and then at Duke, and advising many graduate students and postdocs. I am retired, but still collaborating with colleagues on Gombe research (Figure 7.16).

Joan Berish (b. 1950; MS 1981; wildlife biology, especially reptiles; fieldwork in eastern United States; Research Biologist, Florida Fish and Wildlife Conservation Commission, retired; Lifetime Service Award, Gopher Tortoise Council, 2001; Louise Ireland Humphrey Achievement Award, Wildlife Foundation of Florida, 2013; Influential Woman in Wildlife Award, Florida Chapter of The Wildlife Society, 2021). My childhood fascination with animals (including secretly saving my school lunch money to buy a horse) led me on a somewhat convoluted path toward

FIGURE 7.16   Anne Pusey. (Photo courtesy of Anne Pusey, photographed by Anthony Collins.)

becoming a "wild biologist." Working as a summer substitute mail carrier outside Washington, D.C., was my first taste of arduous outdoor conditions. Later, my stint as a helitack firefighter in New Mexico proved that I could withstand the rigors of the field. Working for a volatile vet gave me clinical experience, and my time as a research technician in Nuclear Medicine at the V.A. Hospital in Albuquerque led to my first publication. Without the firefighting background, I'm not sure I could have gotten into Auburn University. Women were rare in wildlife biology back then. And without the medical publication, I would not have qualified as a Senior Biologist for the Florida Game and Fresh Water Fish Commission. I received my master's degree in wildlife biology from Auburn University. My thesis work involved the distribution of the eastern indigo snake in Georgia. I began working in Florida in 1980 at the Commission's Wildlife Research Lab in Gainesville where I conducted research on a keystone species, the gopher tortoise. Initially, my emphasis was on gopher tortoise status, distribution, harvest levels, population dynamics, and movements. Later, I conducted studies on the distribution and effects of upper respiratory tract disease on gopher tortoise populations. I also investigated the harvest of rattlesnakes and softshell turtles. My memoir, *Fire and Fauna*, was published in 2019 (Figure 7.17).

Margaret ("Canopy Meg") Lowman (b. 1953; PhD 1983; canopy biology (aka, "Arbornaut"); fieldwork worldwide, including Australia, Malaysia, the Amazon, Ethiopia, and India; Executive Director, TREE Foundation; Adjunct Professor, Arizona State University; Research Professor, Universiti Sains

FIGURE 7.17 Joan Berish. (Courtesy of the Florida Fish and Wildlife Conservation Commission.)

Malaysia; Eugene Odum Prize for Excellence in Ecology Education, Ecological Society of America, 2002; Kilby Laureate Medalist, 2002; Lowell Thomas Medal, Explorers Club, 2006; Mendel Medal, Villanova University, 2007). My childhood was centered in upstate New York, where I was a geek-child who played in nature and made tree forts. I collected every treasure of nature imaginable, from butterflies and shells to twigs, collections that fed my scientific curiosity. In fifth grade, I won second prize in the New York State Science Fair. I felt shy, but very proud, standing in the science fair hall with my wildflower collection, surrounded mostly by boys with electronic experiments and chemistry displays. I got my BA degree (biology) from Williams College and my MSc (ecology) from Aberdeen University. To thaw from the chill of the Scottish Highlands, I decided to study tropical forests at the University of Sydney, Australia. My dissertation research focused on leaf growth and herbivory in the rainforest canopy. I did a postdoc studying a malady striking gum trees (*Eucalyptus*), rendering their canopies leafless and ultimately dead ("eucalypt dieback"). After living in Australia for 12 years, I returned to the United States in 1990 to seek intellectual freedom and resume science. I had been living in rural outback Australia, married to an Aussie who could not allow his wife the freedom of her career (he and his family actually ridiculed it), and immersed in a society in which being a professional woman was not socially acceptable. I flew to New England with my two boys, ages 3 and 5, and began life as a single mother and professor at Williams College. Two years later I became Director of Research, then CEO, at the Marie Selby Botanical Gardens in Sarasota, Florida. Other positions I have held include Professor at New College of Florida, Director of the Nature Research Center and Professor at North Carolina State University, and Director of Global Initiatives/Lindsay Chair of Botany and Chief of Science and Sustainability

FIGURE 7.18   Margaret Lowman. (Photo courtesy of and photographed by Meg Lowman.)

at the California Academy of Sciences. My fieldwork has taken me to 46 countries, tackling questions about forest canopies/biodiversity/insect pests. In addition to canopy biology, my other passion is mentoring girls in science. The rumor is true—I bring Oreo cookies on all my tree-climbing expeditions (Figure 7.18).

Lucinda McDade (b. 1953; PhD 1980; tropical botany; fieldwork in Central and South America, South Africa, Madagascar, and Namibia; Director of Research and Executive Director, California Botanic Garden and Professor of Botany, Claremont Graduate University; President, Association for Tropical Biology; President, American Society of Plant Taxonomists (ASPT); Merit Award honoree, Botanical Society of America, 2013; Asa Gray Award, ASPT, 2019). I grew up in south Florida, in a working-class family. I spent as much time outside as I could, planting vegetables, collecting land crabs, and exploring. I was the first in my family to finish college, but I had no idea about career options. I was good in math and science, so everyone assumed I would become a doctor. During my sophomore year at Tulane University, I took a course in plant morphology. After the first exam, the professor suggested to me that because I was "good with plants," maybe I should focus on plants for a career, but to do that I would need to go to graduate school. I had no idea what that was, but I found out. I took many more botany courses and especially loved the fieldwork. I went to graduate school at Duke University and then did a postdoc at STRI, Panama. Personal life intervened when I married a Duke zoology professor. I figured I was stuck there as an adjunct for life, about which I grew increasingly dissatisfied, but my husband swore we were always going to move—emphasis on my career—when the kids finished college. And we did. I got a job as Assistant Professor at the University of Arizona. He had a decent job

FIGURE 7.19 Lucinda McDade. (Photo courtesy of Lucinda McDade, photographer unknown.)

there too, but he was recruited to the Academy of Natural Sciences in Philadelphia in 2000, and I was offered a job as Curator of Botany. It was a nightmare—an herbarium with 1.5 million specimens absolutely full of bugs and now *my* problem. I am really a westerner. I missed being able to grow plants and I missed having graduate students, so I accepted a job offer at California Botanic Garden. That led to a very long bicoastal episode, but it worked. My husband retired and eventually joined me. My research focuses on diversity of the plant family Acanthaceae using phylogenetic reconstruction and plant reproductive biology, and I've dabbled in hybridization as a mode in plant evolutionary history (Figure 7.19).

Melanie Stiassny (b. 1953; PhD 1980; systematics, biodiversity, and evolutionary morphology of tropical freshwater fishes; fieldwork in Africa; Research Curator, Department of Ichthyology, American Museum of Natural History; Robert H. Gibbs, Jr. Memorial Award for Excellence in Systematic Ichthyology, 2016, American Society of Ichthyologists and Herpetologists). When I was a child growing up outside of London, my nature-loving mother used to take me to the British Museum of Natural History. I loved all animals, but chance drew me to fish! Their amazing diversity cemented my interest in evolutionary biology and steered

FIGURE 7.20   Melanie Stiassny. (Photo courtesy of Melanie Stiassny, photo credit AMNH.)

my career path away from becoming a veterinarian. I did my undergraduate work at the University of London. During my first year in graduate school at the University of London, Dr Ethelwynn Trewavas, retired curator at the British Museum of Natural History, told me I needed to go to Africa—and she handed me a check to pay for my first African fieldwork. I went to Malawi and studied cichlids. I continued my research with cichlids on a Royal Society Fellowship at the University of Leiden and then accepted a position as Assistant Professor at Harvard University, where I taught courses in systematics and ichthyology. After 5 years at Harvard, I got my dream job as Curator at the American Museum of Natural History in 1987. I enjoy working with students through my two adjunct appointments at Columbia University and the City University of New York, and now in the museum's own graduate program. I am currently working with an international team of researchers studying freshwater biodiversity in the Congo River and trying to increase awareness of conservation issues in central Africa. We are helping to build in-country capacity for the study of fishes so that local students and scientists can develop solutions to local problems. I also participate as a scientific advisor to World Resources Institute, MacArthur Foundation, International Foundation for Science, World Wide Fund for Nature, and Conservation International (Figure 7.20).

Maureen (Mo) Donnelly (b. 1954; PhD 1987; tropical amphibian biology; fieldwork in Mexico, Central, and South America; Professor and Associate Dean for Graduate Studies in the College of Arts, Science, and Education,

FIGURE 7.21   Maureen Donnelly. (Photo courtesy of and photographed by Lilly Linden.)

Florida International University; University Mentorship Award, FIU, 2008; Robert K. Johnson Excellence in Service Award, American Society of Ichthyologists and Herpetologists (ASIH), 2017; President, ASIH). As a child I was fascinated by metamorphosis of tadpoles. I vividly remember catching frogs and tadpoles from the small wetland "borrow pit" created when the railroad tracks were installed. I also remember getting whipped for going down into the drainage ditch behind our house to catch those frogs. After high school I thought about becoming a dental hygienist as a way to support myself. But when I asked a question to my professor in the anatomy and physiology course and he said he didn't know, I realized there were many unanswered questions in science. I started going to southern Sonora state, Mexico, to watch birds when I was 20 years old. That first summer I traveled with several friends through Mexico and became fascinated with tropical frogs. Jay Savage, who did herpetology research in Costa Rica, invited me to apply to work with him for my PhD. I jumped at the chance because my goal was to get to the tropical belt, and Costa Rica was within that belt. My dissertation work at La Selva Biological Station in Costa Rica focused on reproductive ecology and resource use of strawberry poison frogs. After I finished my PhD, my husband got a job at Florida International University, but my only job offer was a postdoc position with Chuck Myers at the American Museum of Natural History. I wanted to use my PhD and continue with tropical fieldwork, so I went to New York City and my husband went to Florida. After my postdoc, I got a job at FIU in 1994 ("spousal hire"—they wanted to keep my husband). When my department wanted to evict me from my lab to make way for medical school personnel, I leapt to the deanery in return for being allowed to keep my lab, ensuring my students a safe home. Guiding graduate students is honorable and rewarding work that has never felt like work; my students are my proudest achievements. My research interests shifted from "pure" to "applied" science as I watched deforestation and other forms of human disturbance take place in the tropics. My focus now (and that of my students) is to understand and conserve biodiversity—to learn what amphibians and reptiles can teach us about our changing world (Figure 7.21).

Sue Moore (b. 1954; PhD 1997; visual and acoustic surveys of marine mammals; fieldwork in Pacific Arctic, Antarctic Peninsula, and California; Affiliate Professor, Center for Ecosystem Sentinels, University of Washington; IASC Medal, International Arctic Science Committee, 2020). I grew up in Detroit, Michigan, where my father managed the Wonder Bread bakery. I always enjoyed being outdoors as a child. I roamed with my two older brothers and their friends, bicycling along the Rouge River collecting tadpoles and once a snapping turtle. My high school was not the greatest in terms of science classes, but it did have a strong English faculty. I entered the University of Michigan as a journalism major, but by my sophomore year I knew journalism was not for me. So, I followed my strong interest in the ocean (and dolphins) to change my coursework to science. I was dumbfounded. With no science in high school, I sat in some of my first biology lectures wondering what xylem and phloem were—a completely new and foreign language! I decided to move to San Diego and complete my education at the University of California, San Diego. I figured that with Scripps Institution of Oceanography, Sea World, and the Navy, I could someday work with dolphins and in an ocean setting. In the end, I worked for all three entities. After completing my PhD, I joined NOAA's National Marine Mammal Laboratory (NMML), Alaska Fisheries Science Center, as Cetacean Program Leader. In 2002, I became the Director of NMML, but quickly realized that I missed field research. In 2004, I was invited to join an engineering team attempting to add passive acoustic sampling capability to sea gliders at the Applied Physics Laboratory (APL), University of Washington. I remained at APL until 2008, combining my work there with participation in NOAA-related field studies in the Pacific Arctic. I returned to NOAA to join oceanographers at the Pacific Marine Environmental Laboratory conducting research on how climate change was impacting marine ecosystems offshore Alaska. Working with atmospheric, sea ice, and ecosystem modelers, I led efforts to routinely include visual and acoustic surveys of marine mammals in large ocean research programs funded by NOAA, NSF, and BOEM. In 2010, I partnered with academic scientists to form the Distributed Biological Observatory (DBO) and continue to participate in annual cruises to conduct surveys for marine mammals in conjunction with biophysical sampling in productivity "hotspots" (Figure 7.22).

Nalini Nadkarni (b. 1954; PhD 1983; ecosystem roles of canopy-dwelling plants; fieldwork primarily in Costa Rica and Olympic National Park, Washington State; Professor, University of Utah; Guggenheim Award, 2001; AAAS Public Engagement with Science Award, 2011; Monito del Giardino Prize for Environmental Action, 2012; Archie F. Carr Medal for Conservation, 2013; Fellow, Ecological Society of America, elected 2016; Union of Concerned Scientists "Inspiring Scientist," 2019; International Canopy Network Cofounder and President; President, Association for Tropical Biology). I am the child of immigrant parents; my dad was from India (became a pharmacologist) and my stay-at-home mom is an Orthodox Jew of Russian parentage. I grew up in Bethesda, Maryland, and was a tree-climbing child. I have always wanted to protect trees and forests. I started at Brown University on a pre-med course because that was the primary path into science, but I hated being in the lab for long hours and the competitive atmosphere of pre-meds. After I listened to lectures given by ecologist Dr Jonathan

FIGURE 7.22 Sue Moore. (Photo courtesy of and photographed by Andrew Trites, University of British Columbia.)

Waage and went on fieldtrips, I realized people could earn a living as field biologists and that ecology was a way I could both understand and protect trees. But since the age of four, I had taken modern dance lessons and loved to express myself through dance. Did I want to pursue field biology or dance: the forest or the stage? I decided to give each a try. I worked for a year as a field assistant for a biologist studying leaf-feeding beetles in Papua New Guinea. Then I lived as a full-time dancer at a modern dance studio in Paris. I preferred being in the forest to living in the city, and I felt I could contribute more as a biologist. While I was a graduate student at the University of Washington, I took the Organization for Tropical Studies field course in Costa Rica and learned that the organisms and interactions of the forest canopy were virtually unknown. I learned mountain-climbing techniques and did research on how canopy plants interact with whole-forest systems. I served on the faculty at the University of California, Santa Barbara, from 1984 to 1989 and then was Director of Research at the Marie Selby Botanical Gardens in Sarasota, Florida. From 1991 to 2011, I split a single faculty position at The Evergreen State College with my entomologist husband, Jack Longino. I am now Professor at the University of Utah. In the 1990s, I shifted my research to investigate the negative effects of human activities on forests. I am deeply interested in public engagement of science, particularly with underserved groups and those with little access to science in traditional educational venues. I bring science lectures, conservation projects, and nature imagery to faith-based groups, urban youth, and incarcerated adults and youth, much of which is facilitated through collaborations with artists, humanists, and corporate partners (Figure 7.23).

Lynne Parenti (b. 1954; PhD 1980; ichthyology; fieldwork primarily in the western and central Pacific; Curator and Research Scientist, National Museum of Natural History, Smithsonian Institution; Fellow, California Academy of Sciences, elected 1989; Fellow AAAS, elected 2001; President, American Society

FIGURE 7.23   Nalini Nadkarni. (Photo courtesy of and photographed by Sybil G. Gotsch.)

of Ichthyologists and Herpetologists (ASIH); Robert H. Gibbs, Jr., Memorial Award for Excellence in Systematic Ichthyology, ASIH, 2013). As a child, I loved the natural world and being outside, especially exploring the salt marshes on Staten Island where I grew up. I was always interested in biology, and I had a deep appreciation for art and the relationship between art and science. My father worked as a commercial artist, one of my sisters is a quilter/systems analyst, and another sister is a painter. If I was not an ichthyologist, I would be a botanist. I was fortunate to have benefitted from the New York City and State school system. I got into Cornell but had to decline because my family could not afford it. I went to Stony Brook instead and became interested in comparative anatomy and, ultimately, fishes. Mary Mickevich, a graduate student who mentored me at Stony Brook, told me that with my interests I should work with Donn Rosen at the American Museum of Natural History. She introduced me to the research community at the AMNH, and the rest is history. I received my PhD from the City University of New York in a joint program with the AMNH; Rosen was my major advisor. At first, I supported my research through grants and postdoctoral fellowships. I was a postdoc at the Smithsonian, then a NATO-NSF postdoctoral fellow at the Natural History Museum in London, and lastly a Research Scientist at the California Academy of Sciences. In 1990, I was offered my dream job—Curator of Fishes and Research Scientist at the National Museum of Natural History, Smithsonian Institution. My main research interests include systematics; phylogeny and biogeography of tropical freshwater and coastal marine fishes; and comparative anatomy, development, and reproduction of bony fishes. My fieldwork has taken me to such exotic places as Papua New Guinea, Borneo, Sulawesi, the Malay Peninsula, Singapore, Taiwan, China, Hawaii, Tasmania, New Zealand, and Cuba (Figure 7.24).

Kathy Winnett-Murray (b. 1954; PhD 1986; behavioral ecology, primarily birds; fieldwork mainly in Costa Rica and Michigan; Professor, Hope College, retired; Michigan College Science Teacher of the Year Award, 1998; Sigma Xi Hope College Chapter Research Award, with husband Greg Murray, 2001; Janet L. Anderson Provost's Excellence in Teaching Award, Hope College, 2006). As a child, I had an innate curiosity about the natural world, encouraged and indulged by many family camping trips. I grew up in southern California, so the camping trips usually involved visits to some of the most amazing places on earth—the western

FIGURE 7.24   Lynne Parenti. (Photo courtesy of and photographed by Zeehan Jaafar.)

National Parks. I majored in biology at the University of California, Irvine, think-ing I might become a veterinarian, but it didn't take long for me to realize that the clinical aspect of working with animals was not what I wanted. A huge turning point came when I connected with George Hunt, who was putting together teams to study marine bird populations in the southern California Bight region in advance of proposed offshore oil drilling. I was painfully shy, but while riding up an eleva-tor together one day, I asked about his seabird research. He already had the island team assembled for that season, but he needed someone to measure ants for a side project. I spent a semester measuring hundreds of ants and got to hob-nob with the graduate students at the weekly Hunt lab lunches. Not the glamorous fieldwork I had envisioned, but I gave it my best shot, cultivated a meticulous data collection, and was there when another team member was needed the following year. My master's work (California State University, Northridge) focused on marine birds on Santa Barbara Island, California, and my PhD work (University of Florida) was a compara-tive behavioral/ecological study of four species of wrens in Monteverde, Costa Rica. My participation in the Organization for Tropical Studies field course in Costa Rica was a pivotal experience, which included exposure to many ecological systems and researchers. After completing our PhDs, my husband, Greg Murray, and I accepted a joint tenure-track position at Hope College in Michigan. We have continued to work in Costa Rica, primarily on projects relating to life and death of pioneer trees. I have also carried out many local Michigan field research projects concerning Eastern

FIGURE 7.25   Kathy Winnett-Murray. (Photo courtesy of and photographed by K. Greg Murray.)

Bluebirds and painted turtles in the context of human-altered environments, projects that offer undergraduate research opportunities and fieldwork (Figure 7.25).

Janis Dickinson (b. 1955; PhD 1987; behavioral, evolutionary, and molecular ecology of insects and birds; fieldwork primarily in California, New York, and Arizona; Professor Emerita, Department of Natural Resources, Cornell University; elected Fellow of AAAS, 2009, California Academy of Sciences, 2012, and Animal Behavior Society, 2014; Loye and Alden Miller Research Award, American Ornithological Society, 2018). Growing up on the north shore of Long Island, I lived less than two blocks from the beach. After school I would rush to the ocean to beachcomb. I loved insects, spiders, and other arachnids. At ten, I started a club for the neighborhood girls. To join, the initiate had to allow a daddy longlegs to climb up her arm. I didn't want any squeamish club members! My father died when I was 20. Still, I think being raised as one of three girls in a family that valued education is a form of privilege in itself, and this gave me the grit to work two jobs and put myself through Binghamton University. In college, I took a natural resources course and ended up focusing on environmental studies and biology. After taking animal behavior and studying digger wasps at Flathead Lake, I was inspired to do graduate work in entomology at Cornell University. Early on, I studied sexual selection and sperm competition in insects. When I wanted to study paternity in the field, I switched to birds because at the time there was no way to study paternity in insects without curtailing their survival. My primary research has been a long-term field study (28 years) of the evolutionary behavioral ecology of Western Bluebirds (cooperative breeding, winter sociality, sexual selection, parental care, sex ratio, incest avoidance, and vocal communication). My approach was to use field experiments and demographic data to test hypotheses for the fitness advantages of behaviors. I lived year-round and did my research at the Hastings Natural History Reserve in California, first with NSF and NIH postdocs, then as a research associate on soft money (NSF grants), then as half-time Assistant and Associate Research Zoologist, Museum of Vertebrate Zoology, Berkeley. In 2005 I accepted a position as Associate Professor of Natural Resources and Arthur A. Allen Director of Citizen Science at

FIGURE 7.26 Janis Dickinson. (Photo courtesy of and photographed by Walt Koenig.)

the Cornell Lab of Ornithology and Cornell University. I retired in 2018 and moved back to California to focus on painting and playing the fiddle (Figure 7.26).

Debra Moskovits (b. 1955; PhD 1985; conservation; fieldwork mainly in South America; Vice President for Environment, Culture & Conservation, and later for Science and Education, The Field Museum, Chicago, retired; 2020 Conservation Leadership Award, Openlands). Growing up in São Paulo, Brazil, I loved to watch animals in the forests and coastal beaches not far from home but had no idea field biology could be a profession. I started undergraduate work at Barnard College in New York City, where David Ehrenfeld introduced me to fieldwork. The city felt overwhelming, so I transferred to Princeton University after my freshman year. There, Henry Horn, John Terborgh, and Bob May were exceptionally encouraging. As an undergraduate, I joined Terborgh and a few students to study monkeys in the forests of Manu National Park, Peru. For my senior thesis I studied woodpeckers at Archbold Biological Station in Florida, and then rejoined Terborgh's team studying monkeys in Manu for almost a year. After my PhD work at the University of Chicago with tortoises (Ilha de Maracá, Brazil), I knew I wanted to focus on conservation in the tropics. I accepted a job at the Field Museum in Chicago, developing exhibits about nature and conservation. I soon asked the question "Can museums be taken seriously in conservation action?" The response to "Try it out" allowed me to

FIGURE 7.27 Debra Moskovits. (Photo courtesy of and photographed by Flor Angela Peña A.)

launch an experiment channeling the Museum's resources—global collections and scientific expertise—to achieve real-time conservation results on the ground. After initial work with Conservation International, we developed the Museum's rapid biological and social inventories in western Amazonia and the Andean foothills, with the goal of helping local governments establish vast conservation landscapes. The conservation team has grown into the Keller Science Action Center (KSAC), and conservation and quality of life are now integral to the Field Museum's mission. KSAC works in the remote wilderness of the Andes/Amazon, where the Team's rapid inventories and work with local partners provide information and initiate collaborations that support governments in establishing conservation landscapes benefitting local residents. KSAC's other geographic focus is at home in Chicago, where the Museum is a leader in Chicago Wilderness (250 organizations) and its efforts to green the 8 million acres in the region. The Museum is an ideal venue to put science to work for conservation because it has the collections, the expertise, and the rapport and credibility with the public. In retirement I volunteer at the Museum, facilitating conservation efforts (Figure 7.27).

Gwen Kolb (b. 1956; BS 2001; private lands biology; fieldwork in the mid-west United States and New Mexico; Private Lands Biologist, U.S. Fish and Wildlife

FIGURE 7.28   Gwen Kolb. (Photo courtesy of Gwendolyn Kolb, Credit/USFWS.)

Service; Lifetime Achievement Award, Pheasants Forever-Illinois, 2016). I grew up in Cleveland, Ohio. My parents were from Tennessee, so every summer we went down south and I was turned loose in the Tennessee woods. I had the best of both worlds. I am an avid reader, and anything about the natural world was on my reading list. My mother encouraged me to take it all on, that no hurdle was too high. Biology caught my attention in middle school. Finally, I had a name for all of my interests. You have to figure that was in the early 1970s, so not something that women should have been interested in—still, no hurdle too high. I received my education at The Ohio State University, where I majored in wildlife management with wetlands as my specialty. My advisor suggested attending career day and talking to the U.S. Fish and Wildlife Service representatives. I did, and they hired me almost on the spot, despite the fact I hadn't yet gotten my degree. I was 45 years old when I began working for USFWS. I began as a student intern and got hands-on experience as a Wildlife Tech. While working at National Wildlife Refuges, I was exposed to Private Lands Biologists and decided that was what I wanted to do. Being a woman of color sometimes made the path rough, but as my mother would say—no hurdle too high. I have never shied away from controversy or who I was. I created my identity based on being different, of being a 45-year-old Black woman in a White man's FWS world. I got things done, but I did things differently. I am now a Private Lands Biologist in New Mexico, where I am the State Coordinator for the Partners for Fish and Wildlife Program, New Mexico. I love educating land owners. I use test plots to disprove nonsense ideas and to "smack egos." My test plots show landowners "this is how this works, and this is how that works." I am eligible now to retire at the GS13 level, but I have no plan to do so. People inspire me. I especially love mentoring young people and working with the Tribal Nations of New Mexico (Figure 7.28).

Judith (Judie) Bronstein (b. 1957; PhD 1986; plant/animal interactions; field-work mainly in Costa Rica, south Florida, Colorado, and Arizona; Professor, University of Arizona; Distinguished Career Teaching Award, UA, 2007; Distinguished Service Award, NSF, 2008; Fellow Ecological Society of America, elected 2016; President, The American Society of Naturalists, 2022). As a child

I was a crazy reader and academic superstar. My family had no tradition of doing outside activities such as camping. In high school, I became inspired by the environmental movement and the first Earth Day. After a couple of weeks on my undergraduate campus at Brown University, I wanted to stay forever. It's the academic/lifelong education part that I latched onto. I now realize that I could have been equally happy being a professor of comparative literature or history. It has never really been biology, per se, that drives me. Fieldwork is fine, and I enjoy it when I'm doing it, but unlike a lot of my colleagues, it isn't my passion. Learning stuff and making connections across disparate material is my passion. My research integrates field biology, conceptual synthesis, and theory. At Brown, I designed a personalized curriculum called "The use of the science of ecology in environmental policymaking." I figured policy was a good career direction and very socially useful. Ultimately, it became clear that environmental policy wasn't "me." I became progressively more interested in basic ecology during this period and took courses from a very inspiring instructor, Beverly Rathcke. She was a PhD ecologist who was married to a professor, but was treated as merely a faculty wife who was allowed occasionally to teach. When I was a junior, Beverly divorced her husband, left Brown, and obtained a professorship at the University of Michigan. I became her first graduate student. I struggled to find a project until I took an Organization for Tropical Studies course and hit the jackpot. I ended up spending 2 wonderful years in Costa Rica working on figs and fig wasps, which launched me on a career studying mutualism. I did a year of postdoctoral research in southern France, and next went to the University of Alberta for a second postdoc. I got my job at the University of Arizona soon afterward. I still gravitate to the learning process (Figure 7.29).

Robin Chazdon (b. 1957; PhD 1984; forest ecology and forest and landscape restoration; fieldwork mainly in Costa Rica, Brazil, and Australia; Professor, University of Connecticut, retired; currently (part time) Professor, University of the Sunshine Coast, Queensland, Australia and lead consultant with Forestation International; President's Medal, British Ecological Society, 2003; Provost's Award for Excellence in Research, 2004; Woman of Innovation Award in Research Innovation and Leadership, Connecticut Technology Council, 2014; Finalist, Prince Bernhard Chair, Utrecht University, 2014; Executive Director, Association for Tropical Biology and Conservation). I grew up in inner-city Chicago and was a tomboy, climbing trees in our neighborhood park. I loved the natural world and was fortunate to have impressionable experiences in forests at camp and during family camping trips. In high school I developed a strong interest in ecology and the environment. Barry Commoner, E. F. Schumacher, and Rachel Carson were my heroes. I went to Grinnell College in Iowa and immersed myself in liberal arts and biology, with an emphasis on plant ecology and physiology. The highlight of my college experience was a semester-long field study program in Costa Rica during my sophomore year. I fell in love with the tropics and Costa Rica. I returned to Costa Rica for my doctoral research while a student at Cornell University, and just kept going back. When I got married to Robert Colwell, 1 year after completing my PhD, I struggled to gain recognition as my husband was older and more established and wanted to remain at Berkeley. But he encouraged me to apply for jobs after I completed three postdocs in the Bay Area and exhausted the opportunities there. I was

FIGURE 7.29 Judith Bronstein. (Photo courtesy of and photographed by Marco A. R. Mello (https://marcomellolab.wordpress.com).)

offered a faculty position at the University of Connecticut. The department was also interested in hiring him, and we moved to Connecticut with our 6-month-old daughter. My long-term research in Costa Rica stimulated my current interest in forest and landscape restoration and in using the power of natural regeneration to let forests restore themselves under suitable conditions. I began working in Brazil's Atlantic Forest in 2014 with a Science without Borders Fellowship with Dr Pedro Brancalion at the University of São Paulo. Our project focused on landscape effects on natural regeneration in several areas of Atlantic Forest in São Paulo State and comparisons between planted forests and naturally regenerating forests. Since retiring from UConn in 2016, after 28 years, I have been consulting for several organizations with work in the Philippines, in Queensland, Australia, and with the International Institute of Sustainability in Brazil and Australia (Figure 7.30).

Rosemary Gillespie (b. 1957; PhD 1986; community ecology and evolution of spiders; fieldwork primarily in Hawaii; Professor, University of California Berkeley; President, American Arachnological Society, The American Genetics Association, and The International Biogeography Society; IBS Alfred Russel Wallace Award, 2019). I grew up in rural southwest Scotland and have always felt my calling in biology. My education through early high school, however, had a strong emphasis on deportment and "domestic" science, lacking any obvious route to academia. I spent my last 2 years of high school in the north of Scotland, where this

FIGURE 7.30   Robin Chazdon. (Photo courtesy of Robin Chazdon, photographer unknown.)

trajectory changed and allowed me to go to Edinburgh University to study ecology. I graduated in 1980, having experienced the excitement of research: I had written a paper on spider feeding behavior, working with my undergraduate advisor, Philip Ashmole. After I completed my BSc, Ashmole suggested an internship in the United States. During that trip, I met Susan Riechert, a leader in the field of behavioral ecology, specifically working with spiders. Susan opened the prospect of graduate school at the University of Tennessee. I left Scotland and did my PhD in Knoxville with Susan, working on the behavioral ecology of spiders in North Carolina. I had intended to return to Scotland after my PhD, but instead I ended up in Hawaii, taking over a project on feeding behavior in happy face spiders. There I discovered a previously unknown adaptive radiation of Hawaiian long jawed spiders. The finding of such an extraordinary and totally unknown phenomenon was intoxicating, and ever since I have been developing that system. My current research focuses on community assembly over ecological and evolutionary time, using geological chronosequences provided by island archipelagoes, in particular the Hawaiian Islands. Using the island age gradient, coupled with various molecular genetic approaches, my research integrates macroecological and microevolutionary approaches to provide insights into the temporal dynamics of species diversification. I am basically asking, how does biodiversity develop through time, and what are the recent anthropogenic stressors? (Figure 7.31)

FIGURE 7.31 Rosemary Gillespie. (Photo courtesy of and photographed by George Roderick.)

Donna Shaver (b. 1959; PhD 2000; sea turtle research and conservation; Chief Sea Turtle Science and Recovery and Supervisory Wildlife Biologist, National Park Service, Padre Island National Seashore; National Park Service Director's Award for Natural Resource Research, 2012; U.S. Fish and Wildlife Service Endangered Species Recovery Champion Agency Partner Award, 2013; Harte's Heroes Legends of Gulf Award, 2014; International Sea Turtle Society Lifetime Achievement Award, 2018; Distinguished Alumni Award, Texas A&M University-Kingsville, 2020; Texas Academy of Science 2021 Distinguished Texas Scientist Award). While growing up in upstate New York, I was influenced by my nature-loving grandfather to love science and animals. At 20 years old, while attending Cornell University, I decided to devote my life to helping save endangered species. But I didn't yet know what species I would focus on. Through the Student Conservation Association, I learned of a position at Padre Island National Seashore where the efforts to save the Critically Endangered Kemp's ridley sea turtle were just beginning. I worked with the Kemp's ridley program there during the summer of 1980 and returned after graduation to work seasonal and temporary jobs at the park that would keep me near the Kemp's ridley project. I had decided that Kemp's ridleys, the world's most endangered sea turtles, would be my life's focus. I completed my master's at Texas A&M University, Kingsville, while doing seasonal work with the National Park Service and my PhD under Dr David Owens at Texas A&M University, College Station, while working with the U.S. Geological Survey. A few years after completing my PhD, I returned to work with the National Park Service and have been with them ever since. I have spent over 40 years working with sea turtles, on

FIGURE 7.32   Donna Shaver. (Photo courtesy of Donna Shaver, photo credit National Park Service.)

stranding and nesting projects, public outreach, and more, focusing on both research and conservation. My staff members and I collaborate with other researchers in the United States and Mexico and provide training and leadership to hundreds of biologists and volunteers working with local sea turtles (Figure 7.32).

Judy Stone (b. 1959; PhD 1994; botany and evolutionary biology; fieldwork in Central and South America and Maine; Professor, Colby College). I grew up in the country, at a time when parents did not always need to know where their children were. I spent many hours wandering by myself, occasionally getting temporarily lost, in the fields and forests surrounding my home in southern Michigan. I studied forestry as an undergraduate at the University of Michigan, because it seemed like a field that would give me a job. I realized, however, that I had no interest in treating the forest as an agricultural system. That background, however, provided me a strong foundation for field biology. In addition to field-intensive coursework, I spent three summers working for the Forest Service or Bureau of Land Management in the western United States. After working as an environmental educator for several years, I pursued a master's degree in Forestry at Yale, hoping to become a tropical dendrologist. While I was at Yale, for reasons both positive and negative, I realized that I would rather hold a traditional academic position. By the time I started my PhD program at Stony Brook, I already had more than a year of tropical field experience under my belt, and I chose a dissertation topic that would allow me to work at La Selva Biological Station in Costa Rica. For my postdoc at Duke, I learned molecular

FIGURE 7.33   Judy Stone. (Photo courtesy of and photographed by Christel Ramos.)

genetic skills and carried out manipulative field experiments in gardens on campus. I joined the faculty at Colby College, a small liberal arts college in Maine in 1999. At Colby I have continued to do field research, both locally and internationally. In most of my projects, I focus on the genetic and ecological processes that govern variation within plant species. In particular, I have looked to see how natural selection acting on the male function of flowers creates biochemical and morphological features that dictate plants' capacity for self-fertilization (Figure 7.33).

Lora Smith (b. 1960; PhD 1999; wildlife biology/herpetology; fieldwork primarily in Madagascar and southeastern United States; Scientific Researcher, Jones Center at Ichauway). My parents were tolerant of my inclination to be outdoors catching frogs, lizards, and snakes as a child, but the only career they imagined for me was a veterinarian. For many years I thought I might do that, but once I realized how competitive vet school was and what a veterinarian does day to day, I realized it was not for me. I received a BS in biology from Eckerd College, a small liberal arts college in Florida. The program was great, with small classes and young enthusiastic professors, but it did not prepare me for a job in biology or natural resources. I worked for a year selling shoes and eventually got a job as a lab technician and then as a field technician for a biological consulting company. Consulting gave me the opportunity to work in wetlands and do endangered species surveys. I learned a lot but realized I did not want to be a consultant, paid by developers. I left consulting and got a master's degree in Wildlife Ecology and Conservation from the University of Florida working on gopher tortoise reproduction.

FIGURE 7.34   Lora Smith. (Photo courtesy of and photographed by Miguel Pedrono.)

I did my PhD, also at UF, on ploughshare tortoise ecology in western Madagascar working under Dr Ken Dodd. I then worked with the U.S. Geological Survey on an amphibian
monitoring program in Okefenokee Swamp. In 2001, I accepted a position at the Jones
Center at Ichauway in southwestern Georgia. The Center has a permanent research staff
of seven scientists working in wildlife, aquatic ecology, and forestry. We all have adjunct
faculty status at southeastern universities that allows us to co-advise graduate students
who do fieldwork here at the Center. I've been able to mentor dozens of students and field
technicians and have witnessed a transition to female-dominated programs in Wildlife
Ecology. I am currently the only female scientist at the Jones Center (1 of only 2 in our
history). My research program is centered on the ecology of amphibians and reptiles of
the Southeastern Coastal Plain, and, in particular, on herpetofaunal linkages between
aquatic and terrestrial systems and keystone species in these systems. I also study the
interactions between natural and human disturbance legacies in longleaf pine forests
and herpetofaunal communities (Figure 7.34).

**Patricia (Pacha) Burrowes (b. 1961; PhD 1997; amphibian ecology, amphibian
population declines; fieldwork mainly in Colombia, Puerto Rico and other islands
in the Caribbean, Bolivia, and the Philippines; Professor, University of Puerto Rico;
Research in Biology Education Award, National Association of Biology Teachers,
2007).** I grew up in the mountains of Colombia, with cloud forest as my backyard. There
I became fascinated with frogs and snakes, and it was obvious to my parents that I would
become a biologist. I did my undergraduate work at Iowa State University; Michael

FIGURE 7.35 Patricia Burrowes. (Photo courtesy of and photographed by I. De la Riva.)

Lannoo, coauthor of this book, was the TA in my herpetology course. That course convinced me that amphibians were the coolest animals in the world. My first herpetological research (master's thesis University of Kansas) was in Colombia, where I conducted an inventory of the amphibians and reptiles of an unknown cloud forest reserve in the Andes. Because at the time, few Colombians had a chance to go abroad for graduate school, I was determined to do my research in my country as a way to give back. My PhD research focused on the reproductive biology and population genetics of a cave-dwelling frog in Puerto Rico. After receiving my PhD from the University of Kansas, I joined the faculty of the University of Puerto Rico, Río Piedras Campus, where I am Principal Investigator of the Amphibian Disease Ecology Lab. In Puerto Rico, population monitoring of *Eleutherodactylus* populations in the late 1980s reflected drastic declines of montane species, and this sad finding led me to study the pathogenic chytrid fungus (*Bd*), its ecological interactions and potential synergies with local climate. I started working in Bolivia in 2012, when Ignacio De la Riva, worried by the amphibian declines in the Andean region, invited me to investigate the status of *Bd* in the Andean cloud forests where he had noticed drastic amphibian declines. Our work there led to unraveling the historical, taxonomic, and geographic distribution of *Bd* among Bolivian host species, discovering that aquatic birds can carry *Bd* across complex landscapes, and studying the genomics of the Bolivian *Bd* strain. In the Philippines Ignacio and I are studying species that are very similar ecologically to those in the New World, to evaluate the ecological versus phylogenetic signature in the response of amphibians to modern stressors. In Colombia, we are looking at changes in a once mega-diverse community of amphibians, 34 years after it was originally studied (Figure 7.35).

FIGURE 7.36   Erin Muths. (Photo courtesy of and photographed by E. Muths.)

Erin Muths (b. 1961; PhD 1997; amphibian disease, demography, and conservation; fieldwork mainly in Rocky Mountains, United States; Zoologist, U.S. Geological Survey Fort Collins Science Center; Charlie W. Painter Award, Southwest Partners in Amphibian and Reptile Conservation, 2019). I grew up in Wyoming, mostly in Jackson, where I hiked nearby hills, played on the banks of small creeks, and cross-country skied out the back of our house. My parents exuded a sense of responsibility, appreciation, and awe of the natural world that I didn't recognize as such until much later. In high school, I focused on speech and debate and student government, but my biology teacher was one of my favorites. Along with the valley itself, the teachers and opportunities at Jackson Hole High School set the stage for my interest in conservation and biology. While in high school, I attended "Nature in Literature," a winter program focused on writing, taught by Ted Majors and Terry Tempest Williams, with guest lectures by Mardy Murie. There I learned to write, think, and deeply appreciate my surroundings. I didn't go to college "for biology." I think I was aiming for a life spent saving the wilderness through political science. It took a while before my major at the University of Wisconsin morphed into wildlife ecology, but by the time my lengthy undergraduate sojourn was over, the course was set. After UW, I had a couple of internships including one at National Wildlife Federation in D.C. I went to Kansas State University for my master's degree and then did an internship at Archbold Biological Station in Florida, working with Florida scrub jays. After that, I ran off to Australia and completed my PhD on red kangaroo reproductive ecology at the University of Queensland. Upon returning to the United States, I landed a field tech position with the U.S. Geological Survey in Fort Collins, Colorado. After 6 months, I marched into the director's office, put my CV down, and asked what he was going to do about it (my supervisor was moving). I'm not sure that would fly these days, but it worked then. They hired me as a permanent federal employee. My current research projects include reintroductions of boreal toads in Rocky Mountain National Park, and demography and disease in chorus frog and boreal toad populations in Colorado and Wyoming (Figure 7.36).

Michele (Nish) Nishiguchi (b. 1962; PhD 1994; evolutionary ecology, microbiology; fieldwork in the Indo-West Pacific and the Mediterranean Seas; Professor, University of California, Merced; National Academies Education Fellow in the Life Sciences, 2011; Regents Professor and Dennis W. Darnall Faculty Award, New Mexico State University, 2012, 2015; American Society of Microbiology Committee Recognition on Microbiological Issues Impacting Minorities, 2013). I have always loved the sea. I grew up in Northern California near Monterey, where I was fascinated with marine life. As a child in the early 1970s, I watched "The Undersea Adventures of Jacques Cousteau" every Friday night. I would dive off the living room sofa into the deep sea of shag carpet and imagine I was swimming with the undersea critters. I went to the University of California, Davis, as a premed major, but then I took an invertebrate zoology course. I discovered that my passion was marine invertebrate biology, not medicine. I spent 10 weeks at the Bodega Marine Station working on my undergraduate senior thesis with intertidal sea anemones. After receiving my BS in Biochemistry/Theatre Arts, I spent 2 years working at Genentech, in S. San Francisco, where I honed techniques in protein biochemistry; I also took courses in marine ecology and invertebrate physiology at San Francisco State University and UC Berkeley. I earned my master's degree in Marine Science at Scripps Institution of Oceanography, University of California, San Diego. I received my PhD at the University of California, Santa Cruz, working on epiphytic bacteria and macroalgae. My NSF postdoctoral work at the University of Southern California and the University of Hawaii focused on beneficial symbioses between squids and their luminous *Vibrio* symbionts, investigating species specificity, competitive exclusion, and co-speciation. I next did a postdoc at UCLA to strengthen my skills in phylogenetics and systematic biology. Finally ready to start my own laboratory, in 1999 I accepted a faculty position at New Mexico State University. In 2020 I moved back to California and joined the faculty at the University of California, Merced. I continue to study the evolutionary ecology between marine invertebrates and their symbiotic bacteria, specifically the coevolution between bobtail squids and their bioluminescent *Vibrio* bacterial symbionts (Figure 7.37).

Melissa (Missy) Sparrow-Lien (b. 1965; MS 1992; population management of wildlife, wildlife surveys, habitat management; fieldwork in Wisconsin; Wildlife Biologist, Wisconsin Department of Natural Resources). I grew up in a small town in western Minnesota but spent considerable time on my grandparents' farm in eastern North Dakota. A small river ran through the farm, where I explored, fished, and skated. My family did a lot of camping, hiking, and canoeing when I was young, which sparked my interest in a career that involved the outdoors. It took me a few years in college to figure out what I wanted to do, but once I discovered what a wildlife biologist did, I was hooked. I received my undergraduate degree at the University of Minnesota-Duluth. I then attended Northern Michigan University for my master's degree, where I studied American woodcock in an aspen clearcut. Following graduate school, I worked seasonally for a couple of years. I first worked with the U.S. Forest Service capturing and banding spotted owls in northern California, and then with the U.S. Fish and Wildlife Service on a wetland research project in Jamestown, North Dakota. I was offered a full-time job with the Wisconsin DNR in 1994 as a wildlife biologist, private lands specialist. I worked in southeast Wisconsin at the Plymouth Service Center for 18 years, working with many conservation partner groups and

FIGURE 7.37 Michele Nishiguchi. (Photo courtesy of Michele Nishiguchi, photographed by Steffen Binke.)

landowners to restore wetlands and native grasslands on their properties. I married another DNR wildlife guy, which made it tricky finding two jobs in the same field in the same area. Luckily, he did more administrative work and I did fieldwork. When he got an offer with Minnesota DNR, I transferred to a wildlife biologist position in St. Croix County, on the MN/WI border. Currently I am the wildlife biologist for Wisconsin DNR for two counties in western Wisconsin. I am involved with property acquisition and management, population management, wildlife surveys and monitoring of chronic wasting disease; however, most of my time involves habitat management on our diverse properties, including woodlands, wetlands, prairies, and oak savannas. I make sure we have enough funds for the acquisitions and habitat projects through grant writing. I also work with conservation partners and landowners on habitat projects throughout the area and with our county deer advisory council to determine quota levels for upcoming hunting seasons. I also band waterfowl, carry out prescribed burns, and give presentations to school groups, university classes, etc. (Figure 7.38).

Karen Lips (b. 1966; PhD 1995; effects of global change on biodiversity of amphibians and reptiles; fieldwork primarily in Central America and the United States; Professor, University of Maryland, College Park; Aldo Leopold Leadership Program Fellow, 2005; Fellow, AAAS, elected 2011; Fellow, Ecological Society of America, elected 2016; President's Award, Chicago

FIGURE 7.38 Melissa Sparrow-Lien. (Photo courtesy of and photographed by Missy Sparrow-Lien.)

Zoological Society, 1998; Sabin Award for Amphibian Conservation, 2012). I always knew I wanted to work with animals, but until I was in college I had no idea what kinds of jobs that meant beyond veterinarian or zoo keeper. As a zoology major at the University of South Florida, I wanted to take a bunch of "ology" courses. They were all upper level and scheduled for the same semester, so I could only take one. I took herpetology with Henry Mushinsky, after which he hired me to work on their gopher tortoise project. I learned that scientist was a job, and you could get paid more than the minimum wage I was earning at the local mall. Henry encouraged me to consider graduate school, which until then I had never heard of and had no idea what that meant. As a result, I missed all the application deadlines and spent a "gap year" as an intern at Archbold Biological Station. There I learned how to be a field ecologist, did my first independent project, and met other interns and graduate students who helped me apply for graduate school. I ended up at the University of Miami working with Jay Savage. Jay had an invitation to send students to a new field station on the Costa Rica-Panama border, and so I spent my second summer collecting and surveying herps. One of those sites, Las Tablas, became my dissertation field site where I focused on the population ecology of a tree frog. After receiving my PhD, I served on the faculty at St. Laurence University and then at Southern Illinois University. I joined the faculty at the University of Maryland, College Park, in 2009. I study how global change affects biodiversity of amphibians and reptiles in Latin America and the United States. A primary focus of my research is the ecological

FIGURE 7.39   Karen Lips. (Photo courtesy of and photographed by Carly Muletz Wolz.)

and environmental factors that influence amphibian species' response to disease, and how that information might be used in conservation and recovery plans; how the loss of biodiversity affects communities and ecosystems; and how human activities contribute to the spread of disease and loss of biodiversity. I also work on increasing engagement on environmental issues, promoting scientific leadership, and fostering international scientific collaborations (Figure 7.39).

Amy Boyd (b. 1967; PhD 2001; ecology and evolutionary biology; fieldwork in Arizona and North Carolina; Professor, Warren Wilson College; Faculty Teaching Excellence Award, Warren Wilson College, 2006; Faculty Service Learning Award, Warren Wilson College, 2006). As a child, I loved being outdoors, even though I was raised in a city (Louisville, Kentucky). My mother knew I would be a biologist from the moment I could walk. I loved watching nature shows and reading about natural history. When I went to Earlham College, though, I was interested in so many things that I wasn't sure which path to take. I got a job at the natural history museum on campus and helped one of my professors with bird-banding projects, and I TA'd a field botany course for 3 years. I worked with my botany professor and other students on ecological field research projects during my junior year, which resulted in several publications. All of this set me up to move into graduate school and into a career as a field biologist … but I still wasn't sure. I spent a few years teaching high school, then running a youth hostel and teaching English to migrant workers before I went to graduate school; my plan was to get certified to teach high school. But while starting my master's program, I was also teaching

FIGURE 7.40   Amy Boyd. (Photo courtesy of Warren Wilson College, photographed by Mary Bates.)

college courses and finding that I loved teaching at that level. So, I switched gears and changed programs. I moved to the University of Arizona for my PhD, with a very specific purpose: to teach at a teaching-focused small liberal arts college. Lucinda McDade was a fantastic mentor who helped me navigate the world of academia and taught me to be a good scientist without expecting me to work *all* the time like I saw many of my fellow graduate students doing. I took an OTS Tropical Biology course in Costa Rica, which helped shape who I would become as a field biologist and gave me the foundation to take my own students to Costa Rica to study ecology and conservation. Although I wasn't looking for jobs yet, in the winter before I finished I stumbled onto an ad for the job I currently hold, and I knew it was the one for me. As a field biologist, it's great to have 1,200 acres of farms, forest, and river right outside my building to use as a classroom (Figure 7.40).

Evelyn Gaiser (b. 1967; PhD 1997; aquatic ecology; fieldwork primarily in Florida and the Caribbean; Professor, Florida International University; Sustainability Award, FIU, 2014; Champion Partner Award, Deering Foundation, 2017). I was lucky to grow up with parents who care about nature. By an early age, I realized it would be incredible to study lakes for a living. I did my undergraduate education at Kent State University, majoring in biology with an inspiring, demanding, and caring advisor, Dr G. Dennis Cooke, an aquatic ecologist. He encouraged me to apply to Iowa State University to work with his limnologist colleague Dr Roger Bachmann for a master's degree. I took courses at Iowa Lakeside Lab and became enchanted with small

FIGURE 7.41   Evelyn Gaiser. (Photo courtesy of and photographed by Rafael Travieso.)

aquatic life forms, Lakeside Lab, and its people. Dr Bachmann taught me a lot about limnology and being a scientist, and he encouraged me to study with Dr Barbara Taylor for my PhD at the University of Georgia/SREL. I got to tromp around Carolina bays with Dr Taylor, from whom I learned much about how wetlands work, particularly the importance of planning in all aspects of field work (including having functional waders) and taking detailed notes even in the presence of snakes. For my dissertation research I sampled algae and retrieved sediment cores on foot from hundreds of alligator-, feral dog- and hog-, poisonous snake-, and shotgun-bearing landowner-infested but beautiful and mysterious marshes, cypress swamps, and Carolina Bays in South Carolina and Georgia. I am continuing to study lakes and wetlands, particularly trying to pick up the subtle cues of environmental change provided by their beautiful microscopic algae communities – the 'canaries in the coal mine' of aquatic ecosystems. Currently, I work in the vast wetlands of the Florida Everglades and similar habitats in the Caribbean (Jamaica, Bahamas, Belize, Mexico)—where they are called by a better term—"morass." I am a trained musician and have created works to express science through music (Figure 7.41).

Diane Srivastava (b. 1969; PhD 1997; experimental community ecology and ecology of species diversity; fieldwork primarily in Costa Rica and Brazil; Professor, University of British Columbia; Director, Canadian Institute of Ecology and Evolution; E.W.R. Steacie Memorial Fellowship from the National Sciences and Engineering Research Council of Canada, 2010). I grew up in Nova Scotia and spent a lot of my childhood exploring nature: hiking through Acadian forests, canoeing lakes, camping, and cross-country skiing. That, coupled with the fact that my parents

FIGURE 7.42 Diane Srivastava. (Photo courtesy of and photographed by Tania Lynne Zulkoskey.)

were both PhD marine scientists, probably sealed my fate as an ecologist, but there was a diversion into biochemistry and molecular biology first. I studied aquatic plants in Nova Scotian lakes for my honors thesis at Dalhousie University, and then spent three summers doing fieldwork on snow geese grazing in the subarctic salt marshes of Hudson Bay for my master's work at the University of Toronto. The latter was pretty adventurous fieldwork, walking around in a giant parka with a shotgun in case of polar bear attacks (we often saw one bear/day). I did my PhD with John Lawton at Imperial College in London. Lawton encouraged me to study the aquatic insects in "treeholes," natural bark-lined cavities in trees that fill with water and leaf litter. I also spent time in Cameroon, studying the effects of silvicultural practice on butterfly communities and cataloguing the insects that consume Bracken fern. At the end of my PhD work, I learned that the University of British Columbia had just advertised some postdoctoral positions. I had 3 days to create a research proposal for the job. I liked working on tree-holes, as they were a nice contained system for experiments, but in Britain treeholes had just one trophic level of invertebrates. A postdoc in my building had told me that water-filled bromeliads in the Neotropics probably had more trophic levels, so I wrote a proposal on bromeliad food webs—sight unseen. I got the postdoc, and I've been studying this system now for 23 years. I was offered a job at UBC in 2001. In 2011 I formed the Bromeliad Working Group, an international consortium of more than 30 researchers working throughout Central and South America on various aspects related to bromeliads. In 2017 I became the Director of the Canadian Institute of Ecology and Evolution, the national synthesis center in Canada. My research addresses four key questions related to species diversity: How is ecosystem function related to trophic and species diversity? How important is the regional species pool for maintenance of local species richness? How does habitat affect local species diversity? and How do human activities affect species diversity? (Figure 7.42)

**Jennifer Pramuk (b. 1970; PhD 2004; biodiversity surveys and conserva-
tion of amphibians and reptiles; fieldwork in Latin America and Africa; Senior
Program Manager, Amazon Horticulture, Seattle, Washington).** My professional
story began in the 1970s as a little girl who adored nature, especially toads and
turtles. I grew up in semi-rural suburbia in northeastern Ohio and had lots of toads
in my backyard. My family encouraged my interests and were amateur naturalists
themselves. At 6 years old I was holding an American toad, sketching it at my desk,
when I had the revelation that I would study amphibians and reptiles when I grew
up. *The Frog Book* written by Mary Dickerson, a female herpetologist working in
the late 1800s and early 1900s, inspired me to pursue a career in herpetology. I
began volunteering for zoos at the age of 15 and later received my BA in Biology at
the University of California Santa Cruz, where I worked in an insect ecology lab as
a research assistant. I took off a couple of years and worked as a zookeeper in the
reptile house at the Audubon Zoo in New Orleans. Soon after, I began missing the
process of science and wanted to contribute to conservation, so I started graduate
school at the University of Kansas. There my early love of toads came full circle,
for my thesis focused on evolutionary relationships of New World bufonids (toads).
After my PhD, I did a postdoc at Brigham Young University and then in 2006 I
became Curator of Herpetology at the Bronx Zoo—the first woman in that position
in the zoo's 108-year history. In 2010 I left the Bronx Zoo and accepted the position of Animal Curator at the Woodland Park Zoo in Seattle, Washington. In graduate school and as a zoo herpetologist, I performed fieldwork in Latin America and
Africa. My earlier fieldwork focused on biodiversity of amphibians and reptiles (e.g.,
biodiversity surveys) and their taxonomy and evolutionary biology. Later on, fieldwork initiatives in Colombia, Tanzania, and Madagascar were related to amphibian
and reptile conservation projects. In 2019, after working in the zoo field for 16 years,
I made a major career pivot, which I have enjoyed very much. I currently work as
Manager for Amazon Horticulture in Seattle, Washington. I decided to leave herpetology and academia in part because I did not feel like I was treated as an equal to
my male colleagues (Figure 7.43).

**Jeanne Robertson (b. 1970; PhD 2008; evolutionary biology, behavior and
speciation of amphibians and reptiles; fieldwork mostly in Costa Rica, Panama,
California Channel Islands, and Santa Monica Mountains; Associate Professor,
California State University, Northridge).** I grew up in California, the daughter
of parents who loved nature. As a premed student at UC Davis in the Physiology
Department, I never took any organismal courses. After graduating, I worked at the
UC Davis Medical Center in the Cardiology Department. My research mentor often
asked for literature reviews. It was in the library stacks that I discovered journals in
evolution and ecology and read through the titles and abstracts, fascinated. I didn't
really understand the science, but it sparked the idea that I could study organisms
outside and ask questions about their physiology, behavior, and genes. Scientists
actually studied dung beetles and such! My mentor suggested that I take an intensive
field course in ecology at Bodega Bay Marine Lab to learn more about this newfound
interest. After the 6-week field course I quit my job at the Medical Center and moved
to Bodega Bay. I volunteered for anyone who would give me the opportunity to do

FIGURE 7.43   Jennifer Pramuk. (Photo courtesy of Jennifer Pramuk, photographer unknown.)

fieldwork—watching bottlenose dolphins, tracking moths and their parasitoid wasps, measuring the stolon lengths of wild strawberries, and learning how to collect and analyze genetic samples from fish. I worked as a field assistant in Papua New Guinea on a study of the evolution of parental care in direct-developing frogs. While I think of this fieldwork in PNG as the experience that led me to study frogs, according to my mom I've loved frogs since discovering them one summer day in 1977 in a stream in upstate New York. I did my master's work at Southern Illinois University with Dr Karen Lips and earned my PhD at Cornell working with Dr Kelly Zamudio. During that time, I took the Organization for Tropical Studies field course in Costa Rica and developed my PhD thesis on *Agalychnis callidryas* (red-eyed treefrogs). After 18 years, I still work with *Agalychnis* in Costa Rica. I did a postdoc at the University of Idaho and then one at Colorado State University before beginning my position at California State University Northridge. My research interests include natural and sexual selection; spatial patterns of genes, phenotype, and behavior; the evolution and functional significance of color pattern in amphibians and reptiles; population genetics; ecological speciation; and biogeography of Central America (Figure 7.44).

Tiffany Doan (b. 1971; PhD 2002; evolutionary biology and ecology of reptiles and amphibians; fieldwork in Central and South America, United States, and South Africa; Instructor, New College of Florida). I grew up in Jacksonville, Florida, with a backyard full of green anoles. I loved to spend time outdoors, camping with the Girl Scouts, scuba diving, fiddling with my mini microscope to look at small creatures, and reading about astronomy. My mom tells that I was in a "fashion show" in elementary school, and they asked all the little girls what they wanted to be when they grew up. Most answered with a typical female career of the time (secretary, nurse), but I said zoologist. In high school my favorite class was AP Biology. I settled

FIGURE 7.44   Jeanne Robertson. (Photo courtesy of and photographed by Joe Kolowski.)

on herpetology in my senior year as a biology major at the University of Miami. I worked on a master's degree at the University of South Florida for 3 years, but did not finish. Instead, I moved to Peru and worked with the Ministry of Agriculture on Project Tambopata, assessing the effects of tourism on reptile and amphibian populations of southeastern Peru. After I returned to the United States, I did a PhD at the University of Texas Arlington under Jonathan Campbell; my dissertation involved taxonomy and systematics of gymnophthalmids (a family of lizards often called "spectacled lizards" because their transparent lower eyelids allow them to see when their eyes are closed). I worked at Vassar College as visiting assistant professor (2002–2004), and then accepted a tenure track position at Central Connecticut State University. I was unhappy there and left, despite having tenure and not having another position lined up. I coasted for a year doing consulting and teaching martial arts before becoming an adjunct at State College of Florida, Manatee-Sarasota. I later became a visiting assistant professor at the University of Central Florida and then New College of Florida before getting a permanent instructor position in 2019 at NCF. My main research projects are worldwide biogeography of squamate reptiles, which I work on with the Global Assessment of Reptile Distributions working group (I am responsible primarily for Peru and Bolivia); ecology and evolution of gymnophthalmid lizards, mostly in the Andes and Amazon of Peru; and malarial disease ecology of *Anolis* lizards in Central and North America. I love taking students with me to Peru and introducing them to international field biology (Figure 7.45).

FIGURE 7.45   Tiffany Doan. (Photo courtesy of Tiffany Doan, photographer unknown.)

Kate Jackson (b. 1972; PhD 2002; herpetology, particularly venomous snakes; fieldwork primarily in Central Africa; Professor, Whitman College; Woman of Discovery Award for Courage, WINGS WorldQuest, 2011). I was born and raised in Toronto, Canada. My career in herpetology began with a childhood love of frogs, which led to reading about frogs, which led to wearing out the amphibians part of the Peterson Field Guide and venturing into the reptiles half, where I learned that snakes are not villains for eating frogs, but rather unappreciated, misunderstood, and in need of study. I spent 2 years at Dalhousie University in Halifax, Nova Scotia, where I joined Richard Wassersug's lab as "tadpole care girl," because my goal was to be a zoo curator of amphibians and reptiles. I later transferred to the University of Toronto. The summer after my junior year, I interned at the USNM (Smithsonian) working on snakes, an experience that opened my eyes to museum herpetology. I was assigned a project on Brown Tree Snakes, but when I expressed interest in the opis-thoglyph fangs, my advisor let me train on the SEM to examine fangs of other snakes for comparison. Later, as a PhD student at Harvard, I had access to funds for field-work to add to the collections of the Museum of Comparative Zoology. This allowed me the opportunity to leaf through an atlas and choose a site to collect snakes. I wanted to visit a place that would have interesting snakes that had not been much studied, so I chose tropical rain forest. Looking at tropical rain forests of the world I then chose by language. I'm Canadian and the only languages I can function in effec-tively are French and English, so I chose central Africa and worked in the Republic of Congo (Congo-Brazzaville). My Congo team's work has mainly been basic herp field surveys to determine what species of amphibians and reptiles exist in an area

FIGURE 7.46   Kate Jackson. (Photo courtesy of Kate Jackson, photographed by Etienne Bokobela.)

not previously studied. Often these have been funded as environmental impact assessments for oil and mining industries. They also serve as part of geographically broader studies where we can provide the Congo piece, for example, tissue samples from Congo for molecular studies of African taxa across their range for taxonomic revisions, and surveys of amphibian chytrid fungus across Africa (Figure 7.46).

Leslie Ruyle (b. 1972; PhD 2012; conservation and human-wildlife coexistence; fieldwork worldwide, mostly in Africa and Latin America; Associate Research Scientist, Bush School of Government and Public Service, Texas A&M University, and Assistant Director of the Scowcroft Institute of International Affairs; recipient of UN's Equator Prize Initiative). I was born in Telluride, Colorado. My adventurous biological dad was traveling through town on his way to exploring South America; my biological mother was a small-town girl and daughter of miners, too young and poor to keep me. I was adopted by wonderful parents and grew up on a turkey farm in rural Colorado. I loved being outdoors as a child, walking in the woods, watching ants, and exploring my surroundings. My first scientific discovery was realizing that when you pick up a toad, it pees. And if you pick up the toad head-first and point it at your little sister, it makes a great squirt gun. It seems like my whole life I was meant to be a field biologist. Hoping to study wolves, I went to Montana State University-Northern on a sport (volleyball and basketball) scholarship. After I received my undergraduate degree, I worked at the Baltimore Zoo as a zoo keeper, but craved more wild places, so I joined the Peace Corps, serving in

FIGURE 7.47   Leslie Ruyle (and Wechiau Naa, chief of Wechiau). (From the personal archives of Leslie Ruyle, photographed by Gurungu Naa.)

northern Ghana on a community-based hippo conservation project. I then studied otters at the Savannah River Ecology Laboratory and beetles at the Smithsonian Tropical Research Institute in Panama. My PhD dissertation (University of Georgia) was on spiny tailed iguanas in Cayos Cochinos, Honduras. My current work at The Bush School of Government & Public Service at Texas A&M University allows me to search for creative solutions for conservation. I enjoy bringing together minds of various disciplines to create space for new perspectives, approaches, and novel solutions. Much of my research focuses on working to break down barriers that stand in the way of solving some of the world's most challenging conservation problems, for example: How can conservation provide benefits to both humans and wildlife? How can development promote conservation and at the same time better lives for people? and How can we support entrepreneurship and economic development in regions of conflict and conservation concern? My work on these questions has taken me to about 90 countries thus far (Figure 7.47).

Michelle Carlisle (b. 1974; MA 2004; habitat management, wildlife surveys, banding waterfowl; fieldwork in Wisconsin; Wildlife Biologist, Wisconsin Department of Natural Resources). I grew up in Wisconsin, playing outside every moment I could. When I wasn't outside, I was an avid reader and read about the outdoors. I wanted to be a Wildlife Biologist since before I even knew this career existed. I didn't have the funding to go to a university right after high school, so I worked fulltime for 5 years to save money for college while attending community college where I earned two associate degrees. When I finally made it to the University of Wisconsin-Stevens Point to study wildlife, I was so grateful to be there. After earning my BS, my spouse's job took us to Georgia and then to Washington, D.C., where I earned a Master's of Natural Resources from a satellite campus of Virginia Tech. While moving around, I found natural resources jobs that I was happy with, but I really wanted to focus on wildlife. I landed my first position in the Wildlife program with the Wisconsin DNR in Madison as the Wildlife Rules and Regulations Assistant. My next position with the WDNR came in 2006 as a Wildlife Biologist stationed in Balsam Lake, WI. Aside from the regular duties as a county biologist, doing wildlife surveys, managing deer and public lands, etc., I worked on the Western Prairie Habitat Restoration Area, a landscape scale grassland project. In 2018, I relocated to Crex Meadows where I maintained and restored pine-oak barrens habitat that supported species such as sharp-tailed grouse. I am currently the Wildlife Biologist for Polk and Barron counties. During my career, I have worked on the translocation of elk from Kentucky to Wisconsin and the translocation of sharp-tailed grouse from Minnesota to Wisconsin. I also volunteered with the Ministry of Natural Resources and Forestry banding geese and goose nest monitoring, both in Northern Ontario. By far, I find this type of fieldwork the most rewarding. There is nothing better than being in the field as part of a team, where we all have to work hard and work together for a goal that we all care about (and pulling it off in adverse conditions is a bonus) (Figure 7.48).

Priya Nanjappa (b. 1974; MA 2000; ecology and conservation; fieldwork primarily in Iowa, Indiana, and Maryland; Commissioner, Colorado Oil and Gas Conservation Commission; Visionary Leader Award, Partners in Amphibian and Reptile Conservation, 2019). Born and raised in Cedar Rapids, Iowa, I was always interested in the outdoors and animals. My dad loved animals and beautiful places. He would point out birds, bunnies, and frogs and toads in our yard, and eventually I would point them out to him. He took our family on long road trips to visit National Parks across the United States, places he had visited during the 1960s as an immigrant graduate student from India. My best friend in junior high and high school, Emily, invited me to go birdwatching with her dad one day. It was a life-changing experience. After that, she and I often hiked and birdwatched together (and still do), identifying plants and animals we encountered along the way. But when I expressed an interest in pursuing a degree in wildlife biology, my dad (an engineer) instead encouraged me to put my outdoor interests toward environmental engineering. After calculus-based physics, organic and analytical chemistry, and linear algebra at Luther College, I shifted toward ecology and transferred to Iowa State University, which had a better wildlife program. I took two courses at the Iowa Lakeside Laboratory, which provided an immersive field experience. After graduation, I took a field internship surveying birds at the Upper Souris National Wildlife Refuge in North Dakota. I pursued my master's degree at Ball State University with an emphasis in conservation biology. I landed a job with the U.S. Geological Survey in Maryland, primarily doing fieldwork as part of the new Amphibian Research and Monitoring Initiative. As that term-limited position ended I cast the net wide for doctorate programs as well as jobs in ecology and conservation. I accepted a position with Partners in Amphibian and Reptile Conservation, which I stayed in for 13 years.

FIGURE 7.48 Michelle Carlisle. (Photo courtesy of and photographed by Shannon Badzinski.)

I also took on an invasive species policy portfolio (beyond amphibians and reptiles), and various other broad wildlife initiatives. In 2018 I became Director of Operations of Conservation Science Partners, a smallish nonprofit science and research organization. Recently I was recruited into my current position as Commissioner (specializing in wildlife, environment, and reclamation) for the Colorado Oil and Gas Conservation Commission. I am now a decision-maker, applying science directly to policy regarding wildlife issues (Figure 7.49).

Elaine Harvey (b. 1976; MS 2010; fisheries biology; fieldwork in Washington State; Fisheries Biologist, Yakama Nation Fisheries Resource Management Program; PhD student, University of Idaho; Biologist of the Year Award, Native American Fish and Wildlife Society, 2018). I grew up in Klickitat County, south-central Washington State. As a child I loved wading through the creeks, but I would get into trouble because I would walk in the creek with my moccasins made by my grandmother. I felt a connection to the creek and the pebbles in the creek. I loved the small fish and crayfish. I come from a traditional Yakama family and community who continue to practice our longhouse "Washat" religion, speak our language "Sahaptin," have many ceremonies, and gather our traditional foods. Salmon is significant to our people and is one of the most sacred traditional foods to the Yakama Nation. When I was 14 years old, no longhouse in our region was able to catch salmon for our annual spring salmon feast. It was a very sad time, and for our elders and tribal leaders a turning point where we needed to focus more energy and resources on salmon recovery. I decided on that day I would go to college to become a fish biologist. I earned my BS degree from the University of Washington and my MS degree from Central

FIGURE 7.49 Priya Nanjappa. (Photo courtesy of USGS ARMI, Patuxent Wildlife Research Center.)

Washington University. I am currently a third-year PhD student in the University of Idaho's Natural Resource Department, focusing on the impact of climate change on our traditional foods, primarily fisheries. I started working for the Yakama Nation as a forestry/fish biologist in 2006, where I worked to protect fish resources on Yakama Nation forestry lands that were open for harvest or forest health management. Since 2008, I manage the Rock Creek Fish and Habitat Project to study anadromous salmon and steelhead populations in the Rock Creek Subbasin of eastern Klickitat County. The project began as collecting fish and habitat baseline data and monitoring salmonid populations and has transitioned into preserving and restoring fish and their habitat in the midst of climate change. In 2019, I was recruited by the Columbia Land Trust (CLT) to be a board member to help conserve properties in Washington and Oregon for fish and wildlife. As a tribal member, the CLT wanted Indigenous perspective and involvement on lands they plan to acquire that are important to local tribes. Recently, the Yakama Nation Tribal Council made me one of their four Columbia River Intertribal Fisheries Commission (CRITFC) tribal fisheries commissioners for the tribe. I will assist with harvest management, fishing law enforcement, international fisheries treaties, and fishing and treaty rights issues (Figure 7.50).

Emily Taylor (b. 1976; PhD 2005; environmental physiology of rattlesnakes and lizards; fieldwork mainly in California; Professor, California Polytechnic State University; Meritorious Teaching Award in Herpetology, joint award from American Society of Ichthyologists and Herpetologists (ASIH), The Herpetologists' League, and Society for the Study of Amphibians and Reptiles, 2016; Margaret Stewart Award for Excellence in Ichthyology or Herpetology, ASIH, 2020; President, ASIH). I was not particularly into science or nature as a child. I was into soccer, in a major way. I went to UC Berkeley on a soccer scholarship but quit after 2 years, when it was clear that the demands of the sport were

FIGURE 7.50   Elaine Harvey. (Photo courtesy of and photographed by Elaine Harvey.)

impacting my academic performance. I majored in English in college but took science classes, planning to become a physician or veterinarian. Then I took a class called Natural History of the Vertebrates and fell in love with reptiles, especially rattlesnakes, under the tutelage of Dr Harry Greene. I became the curatorial assistant at the Museum of Vertebrate Zoology, and in 1997 I spent a summer doing reptile and amphibian surveys in South Dakota and attended my first herpetology conference. I started an independent research project with Dr Greene when I returned and took his herpetology course. By then I had "found my people," realizing for the first time the possibility of doing research for a career. I did my PhD studies at Arizona State University working with Dr Dale DeNardo and Dr Gordon Schuett, where I studied sexual size dimorphism in rattlesnakes. Arizona is where I truly became a field biologist. I spent more nights doing fieldwork on rattlesnakes in my first few years than I did in my own bed. I absolutely loved the desert and the privilege of peeking into the private lives of rattlesnakes with radiotelemetry. I landed my job at Cal Poly the same year I finished my PhD. I love my job, in part because my university places its focus on student hands-on experience and success. So, mentoring students in research is what I am supposed to do, not necessarily land large research grants. My students are now in charge of all fieldwork and I get to go out with them and give advice. In addition, I founded a field herpetology workshop 10 years ago with my colleague Dr Steve Mullin, which allows me to teach people how to find and observe amphibians and reptiles in Arizona and New Mexico each summer (Figure 7.51).

FIGURE 7.51   Emily Taylor. (Photo courtesy of and photographed by Emily Taylor.)

Lesley de Souza (b. 1977; PhD 2011; ichthyology and conservation biology; fieldwork in the Amazon Basin; Conservation Ecologist, Fishes, Keller Science Action Center, Field Museum of Natural History). I have always loved the outdoors and felt a need to connect with nature. As a child growing up in the United States and in Brazil, I spent time in many different natural environments. I loved jumping in creeks, running in the woods and climbing trees with my siblings and friends, and looking for critters. But I never knew that I would become a biologist. A combination of things led me where I am today. I did well in the sciences in school and pursued field experiences during my undergraduate college years at Auburn University (e.g., aquatic ecology, herpetology, botany). After college I spent 3 years doing field assistant jobs in wood duck research (Alabama-Georgia), fish genetics and aquatic ecology (Alaska), and herpetology (Alabama, Georgia, Panama, and Costa Rica). I began to realize that I could make a career out of this type of work and that it was essentially an extension of my childhood in the outdoors. I returned to Auburn for graduate school, working with Dr Jonathan Armbruster, and the rest is history. I did my dissertation research in Guyana, exploring remote areas to better understand Neotropical fish diversity. After I completed my PhD, I did a postdoc at the Shedd Aquarium in Chicago. During that time, I worked with an Amerindian community in Guyana, studying the movement patterns of endangered *Arapaima* (freshwater, air-breathing fish that can reach up to 8 feet in length). My fieldwork consists of exploring remote areas of the Amazon basin to better understand Neotropical fish diversity and describe species new to science. I use molecular tools to understand

FIGURE 7.52 Lesley de Souza. (Photo courtesy of and photographed by Zachary James Johnsten.)

their relationship to other fishes and biogeographic history in river systems. I have worked as a conservation ecologist at the Field Museum in Chicago since 2016, where my focus is using scientific research to protect wild places, wildlife, and the livelihoods of local people (Figure 7.52).

Madeline (Nikki) Grant-Hoffman (b. 1977; PhD 2009; wildlife habitat studies and restoration; fieldwork primarily in New Zealand and Colorado; Department of the Interior, Bureau of Land Management's National Landscape Conservation System; Excellence in Education Award, Bureau of Land Management, 2015). As a child growing up in Georgia, I was always outside, swimming in the ocean or riding my bike. I thought I would be a veterinarian because I liked animals. I attended Florida Atlantic University on a swimming scholarship. There, I took an ornithology and plant taxonomy class and realized there were other possible careers for me beyond becoming a vet. I did a study abroad with the School for Field Studies in Kenya and got my first experience with fieldwork. I also did a 1-year internship with the School for Field Studies in Costa Rica after college and at that point decided I wanted to pursue a career in ecology. I got my master's degree from Colorado State University working on the effect of Gunnison's prairie dogs on the surrounding vegetation in southern Colorado. I earned my PhD from the University of Alaska, with fieldwork on the effects of invasive rats on the vegetation community on seabird-dominated islands in northern New Zealand. I gravitate to applied science—doing field research and then using those data in concrete management actions on the ground. For this reason, I chose federal work over academia. I worked for the National Park Service and the U.S. Fish and Wildlife Service in Alaska to gain experience. Then I moved to Western Colorado where I am ecologist and science coordinator for two National Conservation Areas with the Bureau of Land Management, involved with management-driven field research. My work with the BLM is diverse, ranging from vegetation restoration to bighorn sheep habitat studies and the effect of introduced bullfrogs on native amphibian fauna. As a Black

FIGURE 7.53   Madeline Grant-Hoffman. (Photo courtesy of and photographed by Aaron L. Hoffman.)

woman, I feel that I bring a different perspective and approach to some of the issues focused on by the BLM (Figure 7.53).

Amanda Subalusky (b. 1977; PhD 2016; large wildlife and ecosystem function; fieldwork mostly in East Africa, the United States, and Latin America; Assistant Professor, University of Florida). My parents were outdoorsy, and I grew up spending a lot of time outside, fishing and camping. When I started college, I planned to become a doctor. I loved international travel, and I thought about working with rural communities around the world. But as an undergraduate at Vanderbilt University, I volunteered in Gary Polis' ecology lab and spent three summers working on desert islands in the Gulf of California and realized how much I loved doing fieldwork. After graduating, I spent 5 years doing seasonal field jobs focused on plants, birds, and mammals. I loved traveling and learning new field skills, and it took me a while to find a system I wanted to study for my graduate research. I studied the role of American alligators in isolated, seasonal wetlands for my master's degree at Texas A&M with Lee Fitzgerald, and I conducted my research in Georgia at the Jones Center at Ichauway, working with Lora Smith. I then moved to Kenya with my partner Chris Dutton and lived for 2.5 years working on a USAID-funded project with Michael McClain on integrated water resources management in the Mara River Basin. After that, we both attended Yale University where we received our PhD degrees working with David Post. After a postdoc at the Cary Institute of Ecosystem Studies with Emma Rosi, I joined the University of Florida as an assistant professor. My research is focused on the influence of large wildlife on aquatic ecosystem function, with a particular focus on the role animals play in transporting resources across ecosystem boundaries. Most of my current fieldwork takes place in the Mara River basin in East Africa, which flows through the Maasai Mara National Reserve in Kenya and Serengeti National Park in Tanzania. I study the role of hippos and wildebeest in transporting terrestrial resources from the savanna grassland into the Mara River, in the form of hippo dung transported during daily feeding migrations

FIGURE 7.54   Amanda Subalusky. (Photo courtesy of Amanda Subalusky; photographed by Paul Geemi.)

and wildebeest carcasses resulting from river crossings and mass drownings that occur during their annual migration (Figure 7.54).

Jessica Ware (b. 1977; PhD 2008; evolution of behavioral and physiological adaptations in insects; fieldwork worldwide, primarily in Guyana and Sweden; Assistant Curator of Invertebrate Zoology, American Museum of Natural History; President, Worldwide Dragonfly Association; PECASE Medal, U.S. Government, 2019). I was raised in Toronto by my grandparents in part and spent a majority of time with them from a young age. My twin and I were introduced to a love of nature through them. My grandmother, though she had little education, valued learning and always told me that I would go to university. But I had no idea what to do! I loved snorkeling in the lake in northern Ontario, so my parents' friend suggested I study oceanography. I enrolled in the University of British Columbia only to find out when I arrived in Vancouver that oceanography involved the study of waves, etc., and not organisms. Embarrassed, I quickly changed my goal to a degree in marine biology. I became discouraged, however, when I realized that all the marine biologists I met seemed to be wealthy and White. No one looked like me. I was the only Black person in my cohort of students. Could I compete on the job market? Then serendipity struck. While taking classes on invertebrates, I learned about the amazing diversity of insects and was offered a work-study job in an ento-mological museum at UBC. I went to Costa Rica and saw some of the diversity firsthand. I was hooked and decided to pursue entomological study. With such a diversity, I knew I could carve out a niche for myself. I wavered a bit on what aspect of entomology I wanted to work on, and during my PhD work at Rutgers University, New Brunswick, working with Dr Mike May, I settled on dragonfly and general insect systematics. I wrote a successful NSF grant to study termite systematics for my postdoc at the American Museum of Natural History, and then was hired at Rutgers University, Newark, in the Biology department. I enjoyed the breadth of knowledge during my 10 years in the department but am glad now to be at the AMNH, where I am surrounded by systematists and evolutionary biologists. I am currently using genomics and morphology to understand the evolutionary history of dragonflies and

FIGURE 7.55   Jessica Ware. (Photo courtesy of and photographed by Patrick Hulick.)

damselflies. Another aspect of my research focuses on the biology of social behavior of termites and cockroaches (Figure 7.55).

Erin Grossman (b. 1980; BA 2002; habitat management and wildlife surveys of game species; fieldwork in Wisconsin; Wildlife Biologist, Wisconsin Department of Natural Resources; Wildlife Federation's Wildlife Conservationist of the Year Award, 2017). When I was young, my family moved around quite a bit between Minnesota and Wisconsin. I love the Upper Midwest! My family spent a lot of time outdoors; we camped regularly and spent frequent day trips hiking and swimming. I have always enjoyed being outside, even if it was just to read a book or have a picnic. I knew I did not want a career that left me indoors. After brief thoughts about becoming a veterinarian, I decided on biology. While at the University of Wisconsin, River Falls, I discovered the career path of wildlife/habitat management. I landed two great summer internships with U.S. Fish and Wildlife Service while in college, and then right out of college I accepted a Limited Term Employment (LTE) technician position with Wisconsin Department of Natural Resources. These LTE positions are fairly standard practice within DNR; they are used to gain experience, contacts, etc. I was in LTE positions for 6 years, working on habitat management to CWD disease surveillance. I had a brief 3 months as a Wisconsin Farm Bill Biologist before accepting a full-time wildlife technician position with the DNR. Currently

FIGURE 7.56   Erin Grossman. (Photo courtesy of Department of Natural Resources.)

I am a wildlife biologist with WDNR. I work with habitat management in Central Wisconsin, including grasslands, native and surrogate, oak and pine barrens, and wetland/sedge meadow. I am responsible for managing several state wildlife areas, which includes planning and implementation of habitat management practices. I also assist customers with wildlife questions and nuisance complaints of game species. Deer are a high-profile species in Wisconsin, so much of my time in spring and fall is spent on deer harvest survey, quota interpretation, and disease surveillance. I have advocated for and participated in prescribed fire for my entire career, and I assist in teaching NWCG fire courses. As a biologist with DNR I am also responsible for several wildlife surveys of game species. I enjoy habitat management work the most (Figure 7.56).

Nicole Smolensky (b. 1980; PhD 2014; ecology and conservation; fieldwork in the United States and Cameroon; environmental consultant scientist with SWCA Environmental Consultants Inc.). Both of my parents are retired physicians, but they are from very different backgrounds. My mother grew up climbing mango trees in Cameroon, and my father grew up climbing stairs in Brooklyn. Perhaps because they grew up so differently, they allowed me to develop my own interests. As a kid, I was always interested in wildlife and biology. As an early undergraduate I planned to become a veterinarian and volunteered in a vet clinic. But then someone suggested that because I was always asking questions, I might consider research. I had opportunities to do research through minority research programs at the University of California Santa Cruz. The door to herpetology opened up to me through my work on lizards in Barry Sinervo's lab. I earned my master's and PhD degrees at Texas

FIGURE 7.57   Nicole Smolensky. (Photo courtesy of and photographed by Nicole Smolensky.)

A&M University. Both my thesis and dissertation were on the conservation biology of reptiles, the former involving an endemic lizard of eastern New Mexico and west Texas and the latter a cryptic dwarf crocodile species in Cameroon. I discovered my passion for teaching and research, but I was not interested in academic positions at tier 1 universities based on my perceptions of expectations, responsibilities, and work-life balance on the path to tenure. I did not see many women and certainly few to no women of color, in tenure-track positions in ecology and evolutionary biology. I taught at a community college in Houston while completing my dissertation and seriously considered a visiting assistant professor position in Oregon. But I was unenthused about the cost and challenges of relocation for a temporary position. My husband was working as a consultant, and his company offered me a position. I took the position to provide time to continue my job search, publish, and pad my CV. But I had a change of heart when I found I could do fieldwork, technical scientific writing, attend conferences, publish research, mentor and teach staff, and have a work-life balance. I was in a community of professional scientists developing scientific and technical solutions for clients and found strong talented female scientists who supported me. As a consultant I primarily guide clients in project planning, permitting, mitigation, and conservation. This involves evaluation of the environment subject to proposed development and scientific evaluation and critique of available research on threatened and endangered species in the United States. I wish more women with PhDs were exposed to career opportunities outside of academia (Figure 7.57).

Christine Figgener (b. 1983; PhD 2019; marine conservation biology, primarily sea turtles; Footprint Foundation Director of Science and Education; Marine Conservation Biologist, Costa Rican Alliance for Sea Turtle Conservation and Science; Next Generation Leader, TIME Magazine, 2018; Footprint Ocean Hero Award, Footprint®, 2019). I grew up in the small town of Marl, in western

FIGURE 7.58 Christine Figgener. (Photo courtesy of Christine Figgener, photographed by Audrey Castillo MacCarthy 2019.)

Germany. I have vivid memories of vacationing at the ocean when I was almost 3 years old. I was terrified of going into the water because of what might be below the surface that I couldn't see. After I threw a temper tantrum, my father walked up to the concession stand and bought me a little pair of diving goggles so I could see underwater. He put them on me, I floated on the surface, watched the fish, and fell in love with the ocean. At a very early age, I decided to become an ocean explorer (marine biologist). While in high school, I volunteered in the dolphinarium at the Allwetterzoo Münster and became intrigued with dolphins and whales. When I went to the University of Tübingen for undergraduate work, I thought I might study the songs of humpback whales. But then I volunteered for a sea turtle conservation project in Costa Rica, which fueled my desire to understand and protect sea turtles. I enrolled at the University of Würzburg for my master's degree and studied the mating system of Atlantic leatherback sea turtles in Costa Rica. After getting my degree in 2009, I moved to Costa Rica and continued working with sea turtles for several years. Over the years I became frustrated, because of my bosses' idea of gender roles and expectations based in cultural differences, insisting I do things their way although I had more than enough experiences to make my own decisions. Adapting to these expectations, I felt myself slipping away as a person. I needed to go beyond what my bosses allowed me to do. I went to Texas A&M University in 2014 to earn a PhD and that way be perceived as more serious and become autonomous as a female scientist. My dissertation focused on the feeding ecology of olive ridley turtles in Costa Rica. Since then, I have been active with my grassroots non-profit Costa Rican Alliance for Sea Turtle Conservation and Science (COASTS). I am also working as Director of Science and Education for Footprint Foundation, which is based in Arizona. My passions include conserving sea turtles, fighting plastic pollution, and empowering women in STEM fields (Figure 7.58).

Tally Hamilton (b. 1983; MA 2012; restoration and enhancement of wild-life habitat; fieldwork in Wisconsin, Nebraska, Colorado, and North Carolina; Farm Bill Biologist, Pheasants Forever; Friend of the Family Farmer Award, Wisconsin Farmers Union, with colleagues Becky Brathal, Julie Peterson, and Gretchen Skudlarczyk, 2018). I grew up in a small town in Indiana. My grandfather hunted, had bird dogs, and did field trials. I frequently went in the field with him. He gave me a choice—camera or shotgun. I chose a camera. I always loved birds and became hooked on birdwatching while in college. I received my bachelor's degree from Northland College and my master's degree from Chadron State College. My early career focused on wildlife research, with positions involving radio telemetry tracking and data collection of bighorn sheep in Nebraska, pronghorn and coyotes in Colorado, whooping cranes in Wisconsin, and red wolves in North Carolina. I worked 3 years as a biological technician for U.S. Fish and Wildlife Service—1 year for Morris Wetland Management District (Minnesota) and 2 years for Upper Mississippi River National Fish and Wildlife Refuge in LaCrosse, Wisconsin. During my time with USFWS, I began to focus more on the habitat needs of wildlife and realized that since much of the land in the United States is privately owned, without engaging private landowners we cannot maintain or increase wildlife populations. I am currently working for Pheasants Forever, a non-profit conservation organization dedicated to the conservation of pheasants, quail, and other wildlife through habitat improvements, public awareness, education, and land management. I cover eight counties in southeast Wisconsin, where I work with farmers and private landowners to restore and enhance wildlife habitat and improve soil and water quality on their property through federal Farm Bill programs and other conservation programs. I also conduct outreach including running Women Caring for the Land workshops and speak to students about conservation and habitat (Figure 7.59).

Sarah Lemer (b. 1983; PhD 2011; marine invertebrate genomics; fieldwork mainly in the Pacific Ocean region; Assistant Professor, University of Guam; Prize for Outstanding Young Researchers, The Bettencourt-Schueller Foundation, L'Oréal, France, 2011). I grew up in France, Indonesia, and the French Caribbean Islands, always by and in the ocean. My parents are not scientists and struggled to answer all the questions I had about marine animals and ecology. I knew very early that I wanted to be a scientist, convinced that it was the only way to find answers to all the questions I had about nature, evolution, and more. I studied marine biology at the University of Aix-Marseille on the Mediterranean Sea in the South of France. I moved across the world to New Caledonia for my master's thesis. During that time, I studied the phylogeny of rabbitfishes. In 2008 I started a PhD at the Faculty of Science and Literature of the Sorbonne University in Paris but was based mostly at the University's CRIOBE marine stations in the south of France and on the island of Moorea in French Polynesia. During my PhD work, I studied the genetic connectivity of fragmented populations of black-lipped pearl oysters in French Polynesia. For 4 years I traveled around the remote islands of the Tuamotu Archipelago collecting pearl oysters. I spent an extra 5 months in French Polynesia on a short postdoctoral contract identifying the genetic basis underlying albinisms and shell color variations in pearl oysters. In 2012, I left the turquoise blue lagoons

FIGURE 7.59 Tally Hamilton. (Photo courtesy of and photographed by B. Petersen.)

of French Polynesia for the white snowstorms of Cambridge, Massachusetts, where for 4 years I was a postdoctoral researcher at Harvard University OEB and MCZ. I developed expertise in Next Generation Sequencing approaches, molecularly dissecting the evolutionary history of various marine invertebrate taxa and traveling to the Pacific Ocean and the Caribbean for fieldwork. In 2016, I joined the University of Guam Marine Lab faculty. My research interests focus on coral-reef associated species in the context of fragmented and changing environments. I am particularly interested in using genetic tools to estimate population and individual responses to natural or manmade disturbances such as increased sea temperature level, habitat destruction, or aquaculture. I actively work toward advancing women, people of color, and Pacific Islanders in my lab and in the classroom with opportunities such as science education, laboratory experience, conference attendance, and outreach (Figure 7.60).

Mercedes Burns (b. 1984; PhD 2014; evolutionary biology, invertebrates; fieldwork mainly in the U.S. Appalachians and Japan; Assistant Professor, University of Maryland, Baltimore County). I grew up in Minneapolis, Minnesota. I wasn't allowed to wander far from home, because my neighborhood had gang issues. Nature was my backyard, where I was fascinated by insects and other arthropods. Birds and mammals came and went, but no matter what the weather, I could always find cool insects. As I grew older, I became interested in the biodiversity of animals, especially invertebrates, but I didn't think I could get a job studying animals. I planned to be pre-med while an undergrad at Macalester College, but

FIGURE 7.60 Sarah Lemer. (Photo courtesy of and photographed by Dr Serge Planes, CRIOBE EPHE-CNRS.)

that went out the window when I took a comparative physiology class and loved it so much I decided I wanted to study animal traits and behaviors and how they had evolved. My senior year I participated in a 4-month field biology study abroad program in Monteverde, Costa Rica, focused on fieldwork. During that time, I studied the mating behavior in a local millipede species for my senior thesis. After I received my BA degree, I took 2 years off from academia to make money (I worked for the U.S. Bureau of Land Management and the University of Minnesota, Chemistry Department) and figure out how to apply to graduate school. My PhD dissertation at the University of Maryland, College Park, focused on mating system evolution in eastern North American harvestmen (Opiliones—non-spider arachnids commonly called "daddy-longlegs"). I got an NSF postdoctoral fellowship and worked in Japan on the role of male/female ratio in the evolution of opilionid mating systems. I began my faculty position at UMBC in 2017, where I am continuing my research on the evolution of sex and sexual conflict, especially in Opiliones. People often express surprise at what I am doing out in nature, as if they didn't expect someone like me (a Black woman) to do field biology. This can be frustrating, but now I try to use these moments as opportunities to show others that scientists are in fact a diverse group of individuals. As the first Black female arachnologist in the United States, I was recently honored as the namesake of a newly described trapdoor spider species, *Ummidia mercedesburnsae* (Figure 7.61).

Becky Brathal (b. 1985; BA 2007; land and conservation best management practices; fieldwork in the United States and Canada; Watershed Technician,

FIGURE 7.61 Mercedes Burns. (Photo courtesy of and photographed by Sarah D. Stellwagen.)

Resource Management Division, St. Croix County, Wisconsin; Friend of the Family Farmer Award, Wisconsin Farmers Union, with colleagues Tally Hamilton, Julie Peterson, and Gretchen Skudlarczyk, 2018). I grew up spending a lot of time outdoors, playing with lizards in Arizona, and then exploring the local creeks and fields after I moved to Wisconsin/Minnesota. We moved around a lot, and we were fairly poor, so spending time outdoors was cheap entertainment. Up until high school, I wanted to be a veterinarian. But then I had a biology class that sparked my love for biology. I started college (University of Wisconsin, River Falls) intending to become a biology teacher. My college advisor, however, told me that the field was highly competitive and suggested I change my major, which in hindsight was a good thing for me. In my first wildlife class, when I learned there was an overpopulation of deer, I made up my mind to do my part and become a hunter. I began my career as a seasonal worker for the U.S. Fish and Wildlife Service, as a fire tech. My job was primarily land management, prescribed fire and wildfire suppression. This job sparked my love for prescribed fire that has yet to be extinguished. I worked spring to fall for 3 years before they lost funding for my position. The next season I worked for the University of Windsor, Canada, gathering baseline data of the Canadian wetlands of the Great Lakes. For the following 4 years, I worked as a limited term employee for the Wisconsin Department of Natural Resources, with Science Services, Bureau of Natural Heritage, and Wildlife Management, where my main job was habitat restoration, mostly for prairies and oak savannas. My focus changed a little when I began work with Pheasants Forever Inc. as a Farm Bill Biologist in central Wisconsin helping landowners manage their land for wildlife and navigate Federal Farm Bill programs to receive cost sharing to accomplish their habitat goals. I recently moved back to St. Croix County, Wisconsin, to try and make a positive impact in the places where I grew up. In my current work, as a Watershed Technician, I work with the

FIGURE 7.62   Becky Brathal. (Photo courtesy of and photographed by Lesa Kardash.)

agricultural community giving technical assistance and cost sharing assistance to implement conservation best management practices. I also work with a group of local stakeholders, St. Croix County Water Network (Figure 7.62).

Claire Ike (b. 1985; MA 2010; natural resource ecology; fieldwork in Georgia; senior environmental specialist, Southern Company). I grew up on the coast of Georgia, in Savannah. As a child, I loved horses and had my own horse for a while. My interest in science started early and continued throughout high school. I had an influential high school science teacher who encouraged me to pursue environmental science as a career. I earned my undergraduate degree at Georgia College & State University as an environmental science major. As part of my undergraduate program, I completed an internship at a consulting firm and became interested in consulting as a career path. However, job availability was low in the environmental field when I graduated (2008 recession), so I continued my education and obtained a master's degree in Forestry and Natural Resources at the University of Georgia. My research focused on soil carbon sequestration in the longleaf pine ecosystem, at the Jones Ecological Research Center at Ichauway, Georgia. My experience in graduate school was eye-opening: I realized that although I enjoyed it, research in academia was not a life path that excited me. Given my previous consulting experience, I knew that was a career path I wanted to pursue. My first professional role after graduate school was at Oak Ridge Institute for Science and Education, a 1-year

FIGURE 7.63 Claire Ike. (Photo courtesy of and photographed by Travis Hale.)

position with the Conservation Branch of the Army Environmental Command at Fort Benning, Georgia. This role included a partnership with The Nature Conservancy and included fieldwork engaging with landowners and working to enroll tracts of forested land into conservation agreements. After my year at Fort Benning, I relocated to Atlanta and began working as an environmental consultant with Arcadis, Inc., where fieldwork included wetland and stream delineations, aquatic species surveys, threatened and endangered species surveys, and mitigation monitoring for transportation development projects. After consulting with Arcadis for 6 years, I took a new opportunity with the energy sector at Southern Company and now focus on natural resource research, conservation, and policy. In this role I work to ensure compliance with federal regulations regarding natural resources such as water and migratory birds. My job is dynamic, exciting, and energizing—no 2 days are alike! I had no idea that such roles exist for biologists with natural resource expertise (Figure 7.63).

Sinlan (Sheila) Poo (b. 1985; PhD 2014; behavioral ecology, amphibians; fieldwork in United States, Central and South America, Singapore, Sri Lanka, Thailand, and Taiwan; Research Scientist, Department of Conservation and Research, Memphis Zoo). I was born and raised in Taipei, Taiwan, where I received a public school education. In Taiwan, we picked our major areas of study during the first semester of high school, which then determined what we are allowed to major in as college students. Although I had an interest in both history and biology, I chose to be a humanities major, since it was the norm for girls. Luckily, I had decided to

FIGURE 7.64   Sinlan Poo. (Photo courtesy of and photographed by Cindy Poo.)

pursue a college degree in the United States and later realized that I could study both subjects in the States. What was unexpected, and what really peaked my interest, was the first time a biology professor answered a student question with "I don't know, no one knows yet." Coming from a traditional East Asian education system of mostly memorizing facts and figures, of memorizing discoveries made and the names of historical figures who made those discoveries, I did not realize there were things that were still unknown and areas to be explored. I decided to become a biologist. I did fieldwork in Panama and Ecuador as an undergraduate at Boston University. After graduation, I worked as a Field Biologist for a non-profit in Southern California and surveyed threatened and endangered species within the region. After 2 years as a Field Biologist, I went to the National University of Singapore to pursue a PhD because I wanted to do fieldwork in a new region, specifically in Southeast Asia. My dissertation work, carried out in the tropical dipterocarp forests of Thailand, focused on a novel parental care behavior and reproductive ecology of a treefrog. After receiving my PhD, I worked as a lecturer for a summer field course in Sri Lanka for 2 years, before returning to the United States. I was a postdoc in the Department of Conservation and Research at the Memphis Zoo for 3 years, and now have a position as Research Scientist in the same department. My current work involves local fieldwork with captive-release programs around the midwest and southern United States. My research interests broadly focus on parental care, behavioral ecology, reproductive physiology (sperm cryopreservation), and conservation of amphibians (Figure 7.64).

Briana Abrahms (b. 1986; PhD 2016; large vertebrate ecology; fieldwork primarily in Botswana, Africa; Assistant Professor, University of Washington). I loved to read as a child, but I also loved being outside in nature. If my parents didn't know where I was, they knew I was most likely in a tree. In middle school, I became intrigued with astronomy and a little later astrophysics and knew I wanted to become

FIGURE 7.65   Briana Abrahms. (Photo courtesy of and photographed by Briana Abrahms.)

a scientist someday. I got my bachelor's degree in physics at Brandeis University. During my senior year, however, my interests took a turn away from physics when I took an ecology course on a hunch and fell in love with the field. I had never realized that studying the ecology and behavior of animals could be a career. I decided I wanted to pursue a graduate degree in ecology, but I had gotten no field experience as an undergraduate. I felt I needed that to get into a competitive PhD program. So, after I graduated I spent 4 years exploring different subfields and gaining field experience. I worked with Americorps/The Nature Conservancy for 11 months, on prairie restoration. I spent 6 months in Argentina as a member of the field crew working on Dr Dee Boersma's study of Magellanic penguins. I did an internship with Conservation Northwest, learning about GIS conservation planning. My final field experience before graduate school was as a research assistant on a carnivore monitoring program in Botswana, which is where I ended up doing my PhD research at the University of California, Berkeley. After getting my PhD, I did a postdoc with NOAA studying marine top predator ecology with northern elephant seals and blue whales. I started as an Assistant Professor at the University of Washington in 2020, where my current research focuses on the impacts of global change on large vertebrate ecology and conservation (Figure 7.65).

Erica Hoaglund (b. 1986; BA 2008; wildlife biology; fieldwork in Minnesota; Minnesota Department of Natural Resources, Division of Ecological and Water Resources, Nongame Wildlife Program). I see myself as an odd duckling in my area of field biology because I grew up in the heart of urban Twin Cities and did not come from an "outdoorsy" family. I was interested in animals from a young age, but because my high school geared vocational I was unaware of career paths like wildlife biology until college. I was a Girl Scout camp counselor the summer before my freshman year at the University of Minnesota and had to register late for classes. As I was filling my schedule I noticed a class called "Introduction to Fisheries and Wildlife Management" that sounded interesting, so I took it. On the first day of class, as the professor introduced the subject and laid out the class, I thought to myself, "This is it. This is what I want to do with my life." It was a high learning curve, since many of my fellow students were farm or Department of Natural Resources kids who could already drive boats and handle animals, but I loved it, caught up, and still love it! I work in the center and southeast quarter of Minnesota and am responsible for about 26 counties. I have spent most of my career so far working with sand prairie and oak savanna ecosystems and their associated rare wildlife species. I work mostly with reptiles (especially Plains hog-nosed snakes and Bullsnakes) and amphibians and insects, but I have also done a fair bit of work with songbirds, raptors, bats, and other small mammals. The primary goal of my research and survey fieldwork is to understand the distributions and habitat needs of the rare and listed species in Minnesota so that I can best advise others on how to manage habitat, plan developments, and mitigate impacts on these species.

Kate Langwig (b. 1986; PhD 2015; infectious disease ecology, white-nose syndrome, and bats; fieldwork in Midwestern United States; Assistant Professor, Virginia Tech University). My parents were hippies living in an urban setting in Portland, Oregon. After they inherited a 100-acre farm in upstate New York, they moved to the east knowing nothing about farming. I was born a year later. Growing up in a rural town, I had free range to explore the outdoors, including flipping rocks in the creek in search of crayfish and other wonders. I loved animal behavior, so I majored in neurobiology as an undergraduate and figured I would go to medical school. While at Union College in Schenectady, NY, however, I was mentored by a fantastic disease ecologist, Prof. Kathleen LoGiudice. We took a fieldtrip into a cave where the bat population had decreased by 90%. This cave was a mile from where I grew up, and it really brought this serious wildlife crisis to my own backyard. I started working with the bat group at New York State Department of Environmental Conservation right after college. I think they were looking for someone small who could squeeze into crevices in caves. At the time, we thought the bat declines might be an anomalous event happening in a handful of caves, centered around my hometown. That winter, however, we saw an explosion of mass die-offs consistent with the spread of a pathogen and realized this issue was going to be a huge problem for bats. I got my PhD from the University of California Santa Cruz, and then spent 2 years as a postdoc at the Harvard T. H. Chan School of Public Health. I joined the faculty at Virginia Tech University in 2017. My research focuses on the role of host-pathogen interactions on population dynamics and community structure, and I work on the virulent fungal pathogen of bats called white-nose syndrome. This pathogen has recently invaded North America. The fungus infects hibernating bats only during the winter, the time period when bats are hibernating deep in caves and

FIGURE 7.66   Kate Langwig. (Photo courtesy of and photographed by Ryan von Linden.)

mines. I currently have a long-term project in the upper Midwestern United States (Wisconsin, Michigan, and Illinois). The combination of extremely deep snowdrifts in the winter and the caving itself make for challenging fieldwork! (Figure 7.66)

Caitlin Nagorka (b. 1986; MA 2018; improvement of wildlife habitat; fieldwork in Wisconsin and elsewhere in the Upper Midwest United States; Private Lands Biologist, U.S. Fish and Wildlife Service, St. Croix Wetland Management District). I grew up in rural southeast South Dakota on a 20-acre hobby farm. This experience peaked my interest in animals and the outdoors. During my undergraduate work at the University of Minnesota, I joined the Fish, Wildlife, and Conservation Biology Club. This organization, in addition to my classes and professors, provided opportunities that solidified my career choice as a biologist. During this time, I started volunteering with the U.S. Fish and Wildlife Service, in the Minnesota Valley National Wildlife Refuge & Wetland Management District. I was hired for that position because I had experience with heavy equipment and was a "farm kid." Field projects mostly consisted of maintenance work—mowing, installing parking lots and signs, and fixing fence. I was rehired the following year and nominated for the Service's Student Career Experience Program (SCEP). Through this program, I enrolled in graduate school. I moved around the Upper Midwest for work at different refuges for my graduate fieldwork. Once I completed my SCEP obligations, including receiving my graduate degree in Natural Resources Management from North

FIGURE 7.67   Caitlin Nagorka. (Photo courtesy of Caitlin Nagorka, photographer unknown.)

Dakota State University, I was placed at St. Croix Wetland Management District in a full-time position. My fieldwork focuses on improvement of wildlife habitat on private lands. This includes the enhancement and restoration/reconstruction of prairie, wetland, and oak savanna habitat throughout eight counties in northwest Wisconsin. I conduct site assessments, work with a wide range of partners to leverage funds and resources, compose environmental clearances and necessary paperwork, coordinate with contractors, and provide project oversight. I firmly believe that a "Partners Biologist" is the absolute best job in the Fish and Wildlife Service! (Figure 7.67)

Jessica Murray (b. 1991; BS 2014; canopy soil ecology; fieldwork in Costa Rica; PhD student, Utah State University; 2014 Newman Civic Fellow, Campus Compact; 2021 Fulbright Scholar (Fulbright Student Program). I grew up in Snellville, Georgia, where I spent a fair amount of time outdoors exploring our backyard and neighboring woods. I am not from a particularly "outdoorsy" family, but I often visited my grandparents on their farm in rural Missouri. Those experiences were very formative for me, I think, because I spent a lot of time outside exploring. I became interested in biology during high school when I took the Environmental Science class as a senior. That class was a total life changer for me, as I learned about human–environment interactions and field ecology. As an undergraduate at the University of North Georgia, I studied biology and Appalachian studies. I started my research path conducting "arts-based" research on the movement of heirloom seeds across space and generations in the rural community around my university. I presented our work along with my professors and other students at conferences in Washington, D.C., West Virginia, and North Carolina. My primary advisor for the project and I were invited to present on the project as a form of place-based pedagogy in Ukraine in 2013. After that experience, I worked with one other student and a faculty mentor studying the effects of hiking trails on salamander communities in northern Georgia. After I graduated, I was accepted into a Resident Naturalist internship

FIGURE 7.68 Jessica Murray. (Photo courtesy of and photographed by Marissa Devey.)

program at the University of Georgia field station in San Luis de Monteverde, Costa Rica, where I spent 7 months learning about the tropical forest, guiding hikes, learning Spanish, and volunteering on research projects at the station. Through an ecology mailing list, I connected with Dr Sybil Gotsch, an ecophysiologist who studies epiphyte communities in Monteverde. I worked on her project studying the vulnerability of cloud forest epiphytes to climate change for about 18 months. This was my introduction to tree climbing and canopy research, and I was hooked! I realized that canopy soils were poorly understood and decided to study them for my PhD dissertation research. My fieldwork takes place in a number of different sites in Costa Rica, including a lowland rainforest, a cloud forest, and a high elevation oak forest. I study carbon cycling and decomposition in canopy soils. I climb trees to conduct my fieldwork, which largely consists of soil collections, in situ monitoring of soil respiration, and nutrient addition experiments (Figure 7.68).

Anjana Parandhaman (b. 1991; MS 2015; herpetology, conservation; fieldwork in Texas, California, Nevada, Utah, and Arizona; PhD student, University of Nevada, Reno). I am a first-generation immigrant to the United States. My first experiences with the outdoors happened through my middle and high school in Chennai, India. The school, located on a 100-acre wooded campus in the middle of the city, focused on experiential learning and exposing children to nature at a young age. We took many fieldtrips to understand social and environmental justice and my

FIGURE 7.69   Anjana Parandhaman. (Photo courtesy of and photographed by S. Hromada.)

school shaped the person I am today. I volunteered for a few nature conservation organizations and zoos in India in high school and during my undergraduate degree in zoology, I learned how to track sea turtle nests, clean crocodile/snake pits at a zoo, and track macaques and king cobras in tropical rainforests. Obtaining a graduate degree in ecology and pursuing a career in this field is not easy in India, so I moved to the United States to pursue a master's degree in wildlife ecology; here I worked on the conservation of Texas Tortoises at Texas State University with Dr Michael Forstner. I spent some time during and after my master's working on various herpetology projects and then started a PhD program in 2018 with Dr Ken Nussear at the University of Nevada, Reno, continuing to work on tortoises. I am currently studying the impacts of land use and climate change on Mojave Desert Tortoise habitat and landscape connectivity across the Mojave Desert. I'm also deeply involved and passionate about justice and equity work in my program and community and hope to one day inspire historically underrepresented students to be involved with conservation work (Figure 7.69).

Kristina Chyn (b. 1992; PhD 2019; road ecology, conservation; fieldwork in Taiwan and Australia; Postdoctoral Research Associate, Texas A&M University). I am a somewhat unlikely ecologist in that I was born and raised in a city as a child of Taiwanese immigrants who avoided the outdoors. I have first-hand understanding of why and how there are few Asian Americans, especially Asian American women, in environmental and field-based sciences. Growing up, my outdoors time was limited to the school playground; I was rarely allowed into our own backyard as my parents feared I would encounter snakes and insects. I explored my fascination with the natural world on my own, mainly through books and nature documentaries on PBS. My first

FIGURE 7.70 Kristina Chyn. (Photo courtesy of Kristina Chyn, photographed by Hung-Nien Chen.)

hiking and camping experience was an eighth grade fieldtrip to Big Bend National Park, Texas, where I felt great joy in learning about and identifying plants and wildlife and witnessed the Milky Way for the first time. I have carried that joy in learning about the natural world into my career. Throughout my undergraduate studies at Cornell University, I spent free time on curation of the insect collection, conducting field-based research on salamanders, performing genetic research on bees, and taking a study abroad course on tropical field ecology in Kenya. Field ecology was most immediately and deeply joyful for me, so I pursued field-based ecology in graduate school. At Texas A&M University, I designed a dissertation that wove my passions for biodiversity conservation, collaborative and engaging field science, and connection with my family and cultural heritage. I conducted most of my doctoral fieldwork on wildlife roadkill in Taiwan, where I surveyed live herpetofauna within habitats on a landscape level to detect if there were differences in community and population structure between areas with differing road densities. I also collected road mortality data of wildlife found killed by vehicle strike and collaborated closely with a community science program, the Taiwan Roadkill Observation Network, to collect roadkill data opportunistically and systematically. These data allowed me to compare prevalence and abundance between species found as roadkill and those found within habitats and to model roadkill risk for wildlife groups across the island's entire road network. After earning my PhD, I decided to postdoc on projects in road ecology with more immediate and tangible outcomes. I have researched on Brown pelican road mortality and am currently working on a Houston toad road ecology project (Figure 7.70).

Emily Marie Purvis (b. 1992; BS 2015; ecosystem ecology; fieldwork primarily in California; PhD student, University of California, Davis). Growing up in the oak-studded woodlands of central California nurtured my innate and intense love of the natural world, I was a child obsessed with questions and have always had an intense and boundless curiosity. In college at the University of California Berkeley, I sought to answer as many questions as possible by double majoring in biology and philosophy. The moment I fully grasped the enormity of what we don't know about the natural world, my career trajectory as a scientist was fixed. I believe that field science is at the crux of our human identity. At our core, we ourselves are just animals: infinitesimal acts of biology bumbling about here on Earth. Every aspect of the human experience is fundamentally tied to the world around us. What could be more human than investigating the world we interact with every moment of every day? I graduated from Berkeley without an idea of career options, so I spent 5 years working as a field biologist for local and international conservation nonprofits, academic institutions, and government agencies to see what kind of job I might want. These jobs took me across four U.S. states and to the Smithsonian Tropical Research Institute in Panama, in ecosystems spanning coastal wetlands, arid deserts, tropical jungles, freshwater streams, conifer forest, and alpine meadows. I discovered that my favorite ecosystem is the one I grew up in: forests of the Sierra Nevada mountains in California. As a PhD student at the University of California, Davis, my general research interests focus on the impacts of climate change and land management on forest disturbance. I am particularly curious about how disturbance regimes are being altered as a result of management strategies and changes in climate patterns; how these disturbance regimes are interacting with each other within forest ecosystems; and how these factors are transforming forest community composition and ecosystem function. I am currently working on a project evaluating the interacting effects of wildfire and drought inside and outside Giant Sequoia groves in the Sierras (Figure 7.71).

Mary Jade Farruggia (b. 1993; BS 2015; ecology, amphibians; fieldwork in California; PhD student, University of California, Davis; Biological Science Technician, U.S. Geological Survey, Aquatic Ecology Group). I grew up with an almost unexplainable obsession with frogs. I rarely saw any frogs where I grew up in the suburbs of Sacramento, but I was an avid reader and fascinated by books about the natural history of amphibians. Math and science classes were always the most challenging subjects for me in school. I honestly didn't believe I had what it took to be a scientist. I entered the University of California San Diego as an environmental policy major, thinking that maybe I would go into environmental law. In 2013 I signed up for a study abroad program in tropical ecology in Costa Rica, thinking it would be a great opportunity to better understand some textbook examples of environmental issues and I was excited to see some of the frogs I had always read about. It turned out to be much more field-based than I realized when I signed up, and it turned me into a field biologist in 4 short months. I discovered I loved the outdoors, natural history, and asking questions. I changed my major to ecology after returning from Costa Rica, started getting into hiking and nature (in

FIGURE 7.71 Emily Marie Purvis. (Photo courtesy of Emily Marie Purvis, photographer unknown.)

real life this time, not just in books), and haven't looked back. I still struggled with science and math classes, but with the help of classmates, coworkers, and mentors, it hasn't been a barrier to my success. For my PhD research, I am studying mountain lakes and ponds in the Sierra Nevada Mountains of California. I'm interested in linking physical habitat characteristics to biological outcomes—in this case, the communities of invertebrates and populations of frogs in lakes and ponds in remote, backcountry sites. With the USGS, I primarily do fisheries research in the Sacramento-San Joaquin Delta of California. I work on a boat crew of 3–6 people running various fishing gear, primarily trawls, gill nets, boat, and backpack electrofishing (Figure 7.72).

Dakota Keller (b. 1993; BA 2016; environmental science; fieldwork in the United States; U.S. Fish and Wildlife Service, Alaska). As a kid I lived in the woods behind my house, building elaborate bark shelters where I would cozy up with zero regard for frostbite. I dug through trashcans for soda and beer cans (Iowa has nickel deposit still), and I saved all my money for years to get a dog. I had guinea pigs that went with me everywhere and a pet hedgehog. My mom tells about the time she watched me pet a wild goose as a kid, and the time she watched me walk into the middle of a herd of deer with a buck snorting in my face. Animals make sense to me, and as an adult they are one of the few things that still makes sense. It seems like a

FIGURE 7.72   Mary Jade Farruggia. (Photo courtesy of and photographed by Veronica Larwood.)

cliché way to describe a start to a professional story, but the most keen observations and experiences I have had in nature have come from moments where I forgot I was working for a federal agency, private consulting firm, or non-profit. They are the moments where I was just being myself, trying to understand the living things around me. I like to believe that the first few entries in my professional story are defined by what I observe, not who I worked for. To be able to step back from the politics of BLM ranchland, Pebble Mine's development of salmon habitat in Alaska, and Lyme disease ecology in one of the most developed and yet rural states is harder than I ever imagined, but a favorite personal challenge. I received a BA degree in Sculpture and a BA in Environmental Science from the University of Iowa. Most recently I worked at The Cary Institute of Ecosystem Studies in Millbrook, NY, studying the relationship between Lyme (and other tick-borne illnesses) and ecology. I just started a job with the U.S. Fish and Wildlife Service in Alaska, assessing the productivity of salmon-bearing streams north of Anchorage. As a child, I was wildly introverted and independent. I've grown out of the shyness, but I still seek peace in the woods. I have a deep respect for living things, especially the living things we can't control (Figure 7.73).

Sidney Woodruff (b. 1994; BS 2017; herpetology; fieldwork in the United States; PhD student, University of California, Davis). Although I grew up in rural Alabama exploring and tromping around in the woods around home, I

FIGURE 7.73   Dakota Keller. (Photo courtesy of and photographed by Stephanie Dea.)

never considered myself to be "outdoorsy" in the way that most people think—hiking, camping, and other forms of outdoor recreation. My mother's field was psychology, and when I started at the University of Georgia, I figured I would do that too. During my freshman year, I took a small course on sustainability and was exposed to environmental science. That course got me hooked (plus I liked the professor's British accent). For a while, I majored in Natural Resources and Recreational Tourism, then Environmental Economics and Management, and then I found Wildlife Sciences. I graduated double majoring in Forestry and Wildlife Sciences. During my senior year, I completed the Mosaics in Science internship with the Greening Youth Foundation and the National Park Service, which aims to provide underrepresented college students in STEM with natural resource science-based work experience within the National Park System. With that internship, I worked at Yosemite National Park with the Aquatic Restoration Program. That experience provided the steppingstone I needed to be exposed to this career path and recognize my passions as a wildlife biologist and outdoor enthusiast. I am studying Western pond turtle conservation for my PhD work, focusing on how the turtles respond to changing communities and ecosystems. I work as a Park Ranger at Yosemite National Park during the summers, where I am involved in a variety

FIGURE 7.74 Sidney Woodruff. (Photo courtesy of Sidney Woodruff, photographed by Laura van Vranken.)

of management and conservation projects concerning native reptile and amphibian species, including Western pond turtles, California red-legged frogs, Sierra Nevada yellow-legged frogs, and Yosemite toads (Figure 7.74).

In the following chapter, these 75 women share their experiences and thoughts regarding being a woman in field biology today.

8 Experiences and Perspectives

The voices of the 75 women interviewed (Chapter 7) offer a lens through which to understand what it is like to be a woman field biologist today. They reflect passion in tales from the field, frustration from the challenges they have encountered and still encounter, and optimistic insight to improve the situation for women in the field. Each woman's perspective is unique, reflecting the times, specific field of research, type of employment, personality, influence and support from family, friends, and mentors, and an element of chance.

Each interviewee responded to a set of written questions that included, among others: (1) Were there any particular hardships or challenges you had to overcome, as a woman, to work in your chosen area of field biology? (2) Were there any particular advantages you feel you benefitted from, as a woman, to work in your chosen area of biology? (3) Do you feel like you need to prove your worth as a woman in field biology? (4) Do you feel like you're treated as an intellectual equal to your male scientist colleagues? (5) Have you had any particularly positive or negative experiences, as a woman, doing fieldwork? and (6) Have you seen changes in attitudes toward women in field biology during your professional career? Interviewees were encouraged to respond only to those questions they felt comfortable answering. In follow-up meetings, either in person or through Zoom, discussions ranged from expanding written responses to exploring new topics. It was a genuine privilege to meet and talk with these women. We offer our sincere thanks and appreciation for their willingness to share their perspectives and experiences—the positive and the negative.

ROLE MODELS AND MENTORS

When asked to identify what women (living or past) have served as role models or mentors and in what ways, Jane Goodall was most often mentioned—for her determination, perseverance, passion, and excellence. Sarah Lemer sums up what Goodall meant for many of us: "As a child, Jane Goodall seemed to me like the most badass woman scientist ever! I often imagined myself in remote field stations working with wild animals because of her." Other inspirational women scientists commonly mentioned included Rachel Carson, Marie Curie, Sylvia Earle, Mildred Mattias, and Rosalind Franklin.

Themes emerged in the qualities of outstanding women mentors and role models. Qualities included compassion, inner strength, resiliency, competence, independence, and passion. Outstanding mentors provided opportunities, freedom to grow, and space to explore; accomplished a positive balance between work and personal life; and were involved in scientific outreach and activism. Motivation, work ethic,

DOI: 10.1201/9781003311508-10

leadership, and courage were other traits often mentioned, for example, as expressed by Jeanne Robertson:

> Dr Karen Lips is perhaps the most hard-core field biologist I know. We endured some pretty rough field conditions in the early years of El Copé, before a field station was built. We lived under a tin roof with a semi-finished cement floor and playground fencing as walls. This site experienced so much rain that we were often cold and soaked while sitting on the floor working on our data and eating dinner. One night she was bit by a kissing bug—the insect vector that carries *Trypanosoma* and leads to Chagas Disease. Luckily she was not infected, but the few days that we waited to see if symptoms would develop were nerve-wracking. Her work ethic, motivation, and mentorship were as strong as ever. I saw a courageous leader. She never cut a corner, took a night off, or left a single frog un-caught.

Several interviewees highlighted role models who helped them learn to be themselves. For example, Claire Ike wrote:

> Raina Singleton and I were office mates in graduate school. Raina is tough as nails and exudes confidence. She doesn't take sass from anyone and is always 100 percent her authentic self. It wasn't until I worked with her that I began to feel comfortable being my authentic self as a woman in biology. I am a girly person. I wear makeup; fix my hair every day; and wear skirts and high heels regularly to work. When I was in school, I felt like I stood out compared to my female peers. This made me feel unsure about the path I chose and I questioned if I really belonged. Raina helped to change that.

Another theme was the value of being accepted and acknowledged, as expressed, for example, by Lora Smith:

> Kristin Berry [field researcher who has focused on desert tortoises] gave me much-needed encouragement and advice after my first presentation at an international meeting. Even though I had just finished my master's degree, she put me front and center on a discussion panel as part of her effort to be inclusive to early career women.

A similar sentiment was expressed by Emily Taylor:

> Dr Maureen Donnelly noticed me, paid attention to me, remembered my name, and encouraged me at some of my first herpetology conferences when I was a young student. She acknowledged me when acknowledgement of my existence from established herpetologists was rare. This made me feel like I belonged at the meetings, like I belonged as a herpetologist.

Rather than naming an individual, many tropical field biologists emphasized the impact that course instructors and participation in an Organization for Tropical Studies (OTS) course had on their lives. For example, Julie Denslow wrote:

> OTS admitted women to participate with men in its field courses where living conditions were often rough and at a time when many lobbied that it was inappropriate. Women among course faculty were role models, and we were given no reason to think that women couldn't contribute to the field. I remember being questioned after a talk I gave in Japan many years later about why there were so many female tropical biologists working in the Neotropics. I think that OTS played a very strong role.

Many interviewees highlighted their mothers and grandmothers as primary sources of female strength in their lives, providing unconditional encouragement. The direct or indirect message was "You can be and do anything you want." For example:

Martha Crump. My mother showed me by example that women are fully as capable as men in the field. She was a geologist back when that career choice was unusual for women. She received her master's degree from the University of Wisconsin in 1939. The following year she participated in a 6-week geology field course to study the Black Hills, Devils Tower National Monument, Crater Lake, and Yellowstone, Glacier, and Grand Canyon National Parks. They were a ratty-looking group of two women and nine men, hiking long distances, sleeping little, and rarely showering. She loved it, kept up with the guys, and felt fully accepted as part of the team. They slept together on the ground, and they dressed and undressed in their sleeping bags. She told me, with great pride, that she learned how to take off and put on a girdle inside her sleeping bag. In those days "women simply had to wear girdles, even with slacks." Thanks to my mother, I never doubted I could be a field biologist.

Julie Denslow. I have never felt that I am less capable because I am a woman. I grew up in a family of strong, engaged, and capable women, including an attorney, an artist and office manager, a medical doctor, a newspaper reporter, a Second World War pilot, and a Second World War army nurse who was with the troops when they landed in Africa. It never would have occurred to me that being a woman made me less than capable.

Evelyn Gaiser. The most important female role models in my life have been my mom and my sister. My mom voices progressive opinions, anguish, and delight about the worldly and other-worldly things as a professional artist. My older sister is a pianist and musical theater professor. I am astounded that I have had the fortune of having my life molded and accompanied by such exceptionally brilliant, loving, beautiful, hilarious, strong, and self-actualized people. They taught me how to stand up for myself, not take things too seriously, forgive and love, have fun, and the importance of knowing the names of plants (my mom) and birds (my sister).

Madeline Grant-Hoffman. Both of my parents earned PhDs (mother in counselor education, father in physics) at a time when higher education was not attainable for many African Americans, so they have always challenged me to do whatever I want. My mother strongly influenced the person I am today. When I was young, she told me that in some situations I had two strikes against me: being female and being Black. She taught me to *respectfully* question authority, because an authority figure might not always have my best interests in mind (such as in grade school when all the little Black girls in my class were placed in the "snail" reading group regardless of whether they were good or poor readers). This advice has served me well in my career. I frequently question the status quo and am less likely to assume that just because something has been done one way historically, that we should continue along those lines. I am more likely to think outside the box.

Elaine Harvey. My grandmother played an important role in my life. She was 1 of 11 children (all girls except for 1 boy). Her parents did not allow her to go to school, because they chose to groom her to learn about Yakama tradition and pass that knowledge on to the next generation. She told me she never regretted not going to school, never learning to read or write. She became leader of the longhouse and took on the role of chief. She taught me much about our traditions and how we should live with and be a part of nature. We must protect all those who cannot speak for themselves and protect those resources for our future generations. Her teachings molded me to become the person I am today. She supported me 100% in my career goals.

She told me that it was important for me to go to school and get a good education to protect the treaties and rights of the Yakama. Just before she died, while I was finishing my master's degree, she said I had to come home and share what I had learned with future generations of Yakama. I promised that I would, and I did.

Erica Hoaglund. My grandmother was an independent lady who worked building bombs during the Second World War. She always told me to be my own woman, not get "tied down," and not let the haters change my mind.

Dakota Keller. My mom is a force to be reckoned with. She is polite, unbelievably kind, and gentle, and I have never once heard her say she can't do something. She's never made a single excuse for not knowing how to do something. She always finds a way. Her work ethic has instilled in me the value of not asking permission if you want to learn. You just read, research, ask questions, find the people who know how and ask them, and just keep on figuring it out until you've mastered it. No fear of failure or shame. You want something, you do it—period.

Gwen Kolb. My mother has been by far the biggest influence in my life. She told me to be what I want to be. That I was a Black woman and that I should have faith in myself and keep going forward. I stood on her shoulders; I hope other people step off mine.

Kate Langwig. My mother grew up in the suburbs, but before I was born, my family inherited a farm. We moved to the farm where my suburban mother raised chickens and horses, painted the house, carried firewood, trailered horses, and stacked hay. I don't remember ever hearing her say there was a farm task she couldn't do, which was very influential when I was learning fieldwork.

Nicole Smolensky. At the age of 18, my mother came from Cameroon to the United States by herself to get a better education. She broke through the ceilings of racism and gender during the 1960s and 1970s to become an oncologist. She not only succeeded but also became the chief of medical staff presiding over the quality and care administered by all doctors at her hospital. Despite all the naysayers she rose to her aspirations. From her, I learned to never settle for less.

Diane Srivastava. My mother was the first woman to earn a PhD in the Zoology Department at the University of British Columbia—the department where I am a faculty member. She struggled to be recognized as an equal researcher, and in so many ways paved the way for me. She had gotten her master's degree in fisheries biology in England but was unable to get a job. Men had returned from the war and the belief was that men should get the jobs. She emigrated to Canada in the 1950s, where she found work with the Fisheries Research Board in New Brunswick. She quickly realized that to do the sort of work she wanted to do, she would need a PhD, so she enrolled at UBC. There she met her future husband, a fellow graduate student, from India. One faculty member told her she might as well give up because the department had never awarded a PhD to a woman. The social constraints of loving a man from India as well as being a woman PhD student were difficult to navigate. But she did it.

Amanda Subalusky. My mother studied acoustical engineering and was one of the first women allowed on a naval research vessel. Her groundbreaking experience in science has always inspired me. She and my father raised my sister and me to know that women could be anything they wanted to be. My mother has the mindset of a lifelong student. She became interested in water issues many years ago and was a volunteer for Adopt-A-Stream for over a decade. She inspired me to study freshwater ecology.

TALES FROM THE FIELD

The following vignettes reflect the passion, excitement, adventure, inspiration, and wonder of doing fieldwork. They range from the pure joy of being surrounded in nature to the thrill of scientific discovery.

Briana Abrahms. My fieldwork is in the stunning Okavango Delta in Botswana. One night, I was sleeping in my tent on a raised wooden platform about 5 feet off the ground, when I was awakened by loud snuffling, snarling noises coming from under my tent. I peered through the wooden floorboards with my flashlight and saw two leopards vigorously mating. I couldn't believe their audacity. They seemed undisturbed by my flashlight, the sound of me moving around above them, or the fact that they were surrounded by human structures. With all the excitement, sleeping more that night was a lost cause (Figure 8.1).

Joan Berish. Gopher tortoises do not reveal their secrets easily, so I felt extremely fortunate to observe one of my radio-instrumented females laying eggs. I quietly knelt on the ground and watched her swing her hind legs back and forth to excavate a nest cavity. She slowly dropped seven Ping-Pong ball-sized eggs into the shallow hole. Instead of heading directly back to her burrow in the adjacent woods, she turned and walked clumsily toward me. She looked into my face as if to say, "That was a real pain!" Having never given birth or laid eggs, I couldn't empathize, but I could sympathize about her exertion in the Florida heat. Perhaps she was curious and was checking me out, but part of me would like to think that this was a more mystical and spiritual interaction.

FIGURE 8.1 Briana Abrahms' tent in Botswana, under which a pair of leopards mated. (Used with permission, Briana Abrahms; photographed by Briana Abrahms.)

Dee Boersma. When I first saw the amazing abundance of Magellanic penguins on Punta Tombo, Argentina, I thought "Wow! I've died and gone to Heaven!" The penguin colony is noisy, bustling with activity and excitement. Penguins give me energy. They are cute, comical, and fascinating. One has become a friend. Turbo (so-named because his original nest was beneath our Ford F-150 Turbo truck), often marches out to greet me. During the 13 years I've known him, Turbo has never had a mate. I don't know why. Maybe his beak is too short, or maybe he isn't very nice. He has a well-shaded nest more than 0.5 km inland, so the long walk may be the problem. He often follows me when I check nests. Sometimes he enters nests, and that is a problem! When we walk past other penguins that would normally try to peck, when Turbo is walking with me the other penguins move aside. It's like he conveys the message, "It's OK. She's with me." He sometimes knocks his beak against the door of the field house to be let inside. He wants to be with us. I don't think Turbo sees himself as a person. I think he sees us as penguins and potential mates. Turbo has taught me that penguins are sentient beings.

Amy Boyd. For 25 years, I've been working as part of a four-woman team studying an endangered cactus in the Sonoran Desert. The project began my first year of graduate school at the University of Arizona, in 1995, when my advisor Lucinda McDade announced, "We should set up a long-term population field study." We chose Nichol's Turk's Head Cactus, in the Waterman Mountains west of Tucson. Several of us set up study plots, tagged plants, and since then we have followed individuals, photographing every cactus each year, measuring height and diameter, recording flowering and fruiting, and tracking condition and mortality. We get together in February for 2 days of fieldwork and a day of discussing the project. After all these years, it's a part of my work that feels uniquely empowering and egalitarian (Figure 8.2).

Kristina Chyn. It's hard not to have an incredible first field biology experience on a 16-day undergraduate course in Kenya, surrounded by megafauna, falling asleep in safari tents to the calls of hyena, and led by knowledgeable professors and enthusiastic graduate students. Prior to this course, I had only a vague understanding of field biology research, and as a child of modest-income immigrants, I had few expectations of travel for education. I applied to the course on a whim and was ecstatic to be accepted. I learned how to observe wildlife and went "herping" for the first time. Because we relied heavily on the moon for light beyond our campfire, I found that it rose later every night, something I never noticed having grown up in the city, and I gained a deeper understanding of how it could drive wildlife behavior. The course allowed me to conduct my first mini-field research project, where we proposed research questions, designed experiments, gathered data, conducted analysis, and wrote research papers over a week. These experiences launched my life's trajectory into field biology and ecological research.

Lesley de Souza. Some of my most memorable field experiences have been working with Macushi Amerindians in Guyana. During my postdoctoral work, I lived in the village of Rewa. The villagers consider the fish I was studying, *Arapaima*, to be sacred and the "Mother of all Fishes." *Arapaima* grow to 10 feet and are obligate air-breathers. Most of my field assistants were local villagers, who took great pride in the data they helped to collect. We bonded over cooking, camping, and conversations. They taught me much about how they interacted with the environment.

FIGURE 8.2 Nichol's Turk's Head Cactus Team. From left to right: Margrit McIntosh, Amy Boyd, Lucinda McDade, and Betsy Arnold. (Used with permission, James Kitchens; photographed by James Kitchens.)

My relationship with the forest and the river changed. I became more sensitive to my surroundings and began to see the environment in the ways I learned from the Macushi.

There was a bird that gave two types of calls. One call brought good luck, the other bad luck. If we heard the bad luck call, we stopped work for the day. I listened to stories of mermaids that live in the river. To see a mermaid brings good fortune. A mermaid may appear during difficult times, to encourage and reassure that all is well. A part of me thinks that what the villagers are calling mermaids could be *Arapaima*, rolling out of the water to gulp air, slapping their tails against the water surface in acrobatic moves, revealing their gorgeous red-specked scales. But another part of me wonders: do mermaids live in the river? I felt powerful, spiritual connections with the forest and the river. People who are evil in life are believed to become jumbees after death, creatures that haunt the minds of the living. When we set up camp along the river, we did it in such a way that would ensure that we were on the jumbees' good side. My journey with the Macushi has transformed my life.

Janis Dickinson. One day we were doing a mate-guarding experiment with Western Bluebirds, which involved removing the female's mate for an hour and watching to see if other males would intrude on the territory and try to gain extrapair mattings with the unguarded female. First a young male came in and tried unsuccessfully to copulate with the female twice, then departed. Shortly, another male—the dad of the male we removed—came in relatively quickly. The female accepted all four of his copulation attempts and then he took off toward home.

This started the wheels turning. Sons winter on their parents' territories and often mate with unrelated females that join their winter group. Usually these sons will nest next door to the parents, but if they don't get a mate or they lose their mate mid-season, they help at their parents' nest. Sometimes they even help while they have nests of their own. So I wondered, what if dads benefit, not just by having sons as helpers, but by being able to get extrapair paternity with their sons' mates next door? This would be better for sons than losing paternity to a nonrelative. And given that helpers are always related to the young in their parents' nest through their mothers, if older males are more likely to be accepted as extrapair mates than are young males, does this mean that sons are sometimes more closely related to young in their parents' nest than in their own nest? This added a whole complex of questions about age, paternity, and female choice of extrapair mates to the project, culminating in a model of how extrapair paternity influences the benefits of helping.

Mary Jade Farruggia. Like most recreational fishermen, my days working on a fisheries crew with U.S. Geological Survey often begins early morning at a boat ramp. We're often in the queue of trucks with fishing boats attached, waiting to back our trailers up and get the boat in the water. Most of my co-workers are White men, as are the fishermen at the ramps. When fishermen see me, a relatively petite and young-looking Asianish woman, hop into the driver's seat to back the boat into the water, it's not uncommon to hear fishermen poking fun or laughing at the prospect of me successfully maneuvering a huge truck with a huge boat down "their" boat ramp. It can be discouraging to hear these comments, but there's nothing quite as satisfying as doing your job perfectly under the disbelieving gaze of those who don't expect you to succeed!

Rosemary Gillespie. I began working with nocturnal spiders in North Carolina and was usually alone while doing this fieldwork. It was a wonderful feeling, completely engaged in watching spider activity in my flashlight beam. Later, I started fieldwork in Hawaii, working with nocturnal spiders. One night I went out to see what the spiders do in the dark. Imagine my surprise when I found the same group of spiders I had worked on in North Carolina. The spiders were everywhere, and there were many different species. It was exhilarating to find something new—something that no one else knew about. I ended up studying adaptive radiation in this group of spiders.

Madeline Grant-Hoffman. I did my PhD fieldwork on small islands off the coast of New Zealand. These were generally small, steep, rocky, uninhabited islands. It was often a scramble to reach vegetated parts where I was collecting data. On one island after climbing up a steep area, I was surprised to see a little blue penguin strolling by, as if it was no big feat that a flippered, web-footed bird had scaled this island!

Elaine Harvey. As a child I loved wading through the creeks. Now this feeling of walking through the creek is special and is an escape from worries or responsibilities. It is also a memory of time spent with my cousins and siblings as we grew up. Today I get to wade through the creeks to count adult salmon and steelhead or catch juvenile salmonids. I get to relive my childhood memories through my fieldwork.

I lost my younger brother in 2008 and cousins over the years. I feel a connection to them as I work at the same creek we played in as children.

Sarah Lemer. My first remote fieldtrip was in the Tuamotu's in French Polynesia. I was the only woman. Other participants were my PhD advisor, another student, a field technician, and four crew members (fishermen). On the first day, our small 30-foot boat was out at sea, and we couldn't see land anywhere. I was sitting on a barrel of gasoline (all the space was taken by food and field supplies), clutching the dry bag that contained the satellite phone and GPS. The waves were huge. I was excited and terrified. But then we neared the atolls and it was paradise for 2 weeks straight. We slept in tents on the beach of a new atoll every night and swam all day to sample. I collected bivalves from untouched populations in isolated and uninhabited remote lagoons, golden material for my dissertation. I knew then and there that I would be doing it again and again and again!

Sue Moore. While spending many hours in a small plane surveying for bowhead whales, which are considered a pagophilic (ice-loving) species, I couldn't help but notice a pattern: The whales did not "follow the ice" but rather moved onshore into ice-free habitat when prey were available there. This was entirely counter to accepted thinking. It gave me great pleasure, as a graduate student, to demonstrate this discovery by calculating habitat selection indices for bowheads (also beluga and gray whales) while completing my dissertation.

Debra Moskovits. Three colleagues and I were working with monkeys in Peru. I was about 19 years old at the time. The monkeys reacted differently to each of us. A pregnant female capuchin we were following left the group and had her baby in the tree above my tent. The mother often came to the ground near me and allowed her baby to play with my pen. Sometimes she left the baby by me while she foraged in the trees. One day, a male colleague came charging through the forest, hacking the vegetation with his machete. The mother grabbed her baby and fled to the top canopy, returning only when my colleague had left. And there was the large alpha male leader of another group we were following. When I lagged behind the monkeys in densely tangled vegetation, this big guy waited for me to catch up. I had many personal interactions with the monkeys that my male companions did not experience.

Erin Muths. My PhD research focused on lactation in red kangaroos, in part determining the composition of milk at different stages of lactation. Red kangaroo females can have a very small young in the pouch, pretty much attached to the teat, and another "young at foot" that is still nursing but not living in the pouch. The milk from each teat is quite different, perfectly tailored for the almost opposing nutritional needs of the two offspring. The most efficient way to get lots of milk samples was to go into the field with kangaroo shooters. These were men licensed by the state to shoot kangaroos in outback Australia. In the early 1990s, kangaroo meat for human consumption was not legal. Meat could be sent overseas for pet food, but because that market was inconsistent, carcasses were rarely kept. The return was in the kangaroo hide.

I would drive to the Queensland outback and meet Neville the "roo shooter." Neville was swarthy and built like a truck. He was a crack shot and a fast talker. He had an off-sider, and sometimes his kids would come along to fold the skins. Neville and his mate were cordial, and either their initial sexist jokes weren't very good or maybe they didn't make them. I can't remember. They were politely curious about what I was doing, but didn't pay me much attention at first. My job was to wait for a kangaroo to be shot, then collect blood from a tail vein and milk samples from a teat. It didn't take long for me to realize that I'd completely won them over without even trying, because I could collect my samples, then hop in the HiLux, and be ready to help "skin up" when we'd filled the truck with dead kangaroos.

The camaraderie was perhaps odd but palpable and comfortable. After skinning the carcasses and folding and tagging the skins, the scene was out of an outback vacation advertisement. We'd boil the billy, make tea, and a purple bar of Cadbury milk chocolate would appear, to be shared around the fire. The stars, wheeling above in those quiet hours before dawn, reflected on the huge pile of entrails just to the side of the fire where the butcher ants were already starting dinner.

Nalini Nadkarni. One of my most exciting discoveries happened in 1979, when I was a graduate student. I took an Organization for Tropical Studies course in Costa Rica and became intrigued with tropical forests and the tree canopy. When I returned to the United States, I told my advisor I wanted to work in the canopy. He suggested I start small and local, measuring the biomass of epiphytes in the Pacific Northwest. I climbed big-leaf maple trees and noticed vascular root systems in the canopy soils—the dead organic matter comprised of epiphytes that die and decompose in place. I followed the roots within the soil mat and found they emerged from the maple tree itself. The moment of tracing that first root system, seeing that the roots were connected to the branch was an "Oh my gosh!" moment. I rented a chain saw, climbed the tree, cut the branch, and showed my advisor. I said, "This is what I'm talking about!" I had discovered a previously unknown pathway of nutrient transfer in the canopy: The adventitious roots of host trees provide access to canopy nutrient resources that are normally not available to the host tree until they fall to the forest floor and enter the soil system. I later discovered the same transfer pathway in tree canopies in the cloud forest of Monteverde, Costa Rica. I was thrilled to publish my results with the cover picture in the journal *Science*. Most importantly my discovery made people realize that the canopy is not just a Tarzan and Jane thing, but a place where real science can be done!

Emily Marie Purvis. Every time I'm out in the field, I'm captivated by a sense of childlike wonder at our extraordinary world. Earlier in my career I was worried that getting caught up in scientific details would destroy my sense of awe and amazement. I thought that learning biology might fundamentally alter the way I relate to my surroundings, that I would begin to see diagrams and equations instead of boundless beauty and mystery. On the contrary, becoming a field biologist has enhanced my enchantment with the natural world. Trying to untangle the infinitely vast and complex web of the natural world is my way of celebrating the sublime.

Jeanne Robertson. Discoveries are treasures that bring wonder and awe, enrich our lives, and challenge assumptions. Sometimes these discoveries change the direction of our research. One of the most exciting nights of my life as a biologist came with a question that had been haunting me for almost two decades: Why do some nocturnal frogs have bright and colorful arms and legs? They are asleep and hide their colors during the day, instead of flaunting them to deter predators. Can they see colors at night? Are those colors relevant to *them?* All of our mate-choice experiments with red-eyed treefrogs up to this point presented a female with a choice between two simulated males, sculpted and painted to resemble different color morphs. Each stood in front of a speaker that played its song. Herpetologists have long recognized that male song is an important means of attracting a mate, but we didn't know if our colored frog sculptures also played a role in mate choice for red-eyed treefrogs. If we turned off the song, would these sculpted models be enough to attract a female? I held my breath as we released the first female into the behavioral arena, presented only with a choice of two sculpted male frogs. In silence and in the dark. The answer came within minutes. We repeated the experiment 40 times, with the same result. Yes! Frogs can distinguish color at night (albeit as shades of grey), and they make choices based on these visual cues alone.

Leslie Ruyle. As a Peace Corps volunteer, I had been living in a small community in northern Ghana for about a month and was completely enamored with all the new wildlife. I spent my days learning the local language and walking around the big mango trees on the outskirts of town using my binoculars to identify the resident lizards, birds, and bats. One day, while strolling along with my neck cricked up at the trees and my eyes focused through my long lenses, a trio of women walked by. After giving me a thorough once-over, they demanded, *"poga-la be dow-la,"* which in the local language translated to "Are you a man or a woman?" While I am a 6-foot tall, athletic, blonde woman, I had never before had my gender questioned. Surprised, I looked at them and pulled on my pony tail, emphasizing, *"poga-la!"* They laughed and pulled up their shirts exposing their breasts while shouting *Poga-La!* A bit surprised, yet wanting to fit in to the village, I reciprocated the gesture, flashing them my bareness, proclaiming, *Poga-la!!!* They started clapping and smiling while crowding around me to engage in conversation. After trying to fit in as a member of the community with nights spent drinking *akpeteshi*, the local moonshine, with the older men in the village, it was this simple interaction with the women where I finally felt like I had made a connection. From then on, the village women would greet me and help me find new creatures to identify, ask questions about why I found them interesting, and include me in their joyful discussions while cracking groundnuts (peanuts) under the mangoes. My success in the community depended on these relationships built by sharing what we had in common; from there, anything was possible.

Donna Shaver. In the mid-1970s, the National Park Service recognized the dire plight of critically endangered Kemp's ridley sea turtles. It organized a panel of experts and agency representatives from the United States and Mexico to develop measures to help save the species and began a bi-national recovery program in 1978. One goal was to form a secondary nesting colony at Padre Island National Seashore (PAIS), off the coast of Texas. No one knew if head-started turtles from this experimental program would survive to maturity and return to PAIS to nest. After a decade of patrols conducted, little Kemp's ridley nesting recorded, and no confirmed returnees from the experimental project found nesting yet, many nay-sayers recommended that I move on and work with other sea turtle species that had a more promising outlook for recovery.

May 29, 1996, was the most important day of my scientific career. It began when I received a radio report from one of my "turtle patrollers" that she was with a Kemp's ridley that was crawling ashore to nest about 13 miles south of where I was. National Park Service Law Enforcement Officer Randy Reader also heard the call and offered to drive me to the site. Randy had a booming voice and exuberant demeanor, and I can still remember how animated he was as we drove to the site and received updates about the turtle. When we arrived, the female had completed egg-laying and was covering her nest. I leapt out of the vehicle and brushed sand off her top shell, revealing a living tag, a large white spot on the darker background of the shell. I could not believe my eyes. She was the first confirmed returnee from the experimental imprinting and head-starting project, and I had hatched her at PAIS in 1986.

I jumped up and down, and then thought about my mother who had passed away 6 months earlier. I wished I could tell her this tremendously exciting news. I thought about how happy she was that I had found my calling before she passed on. I did not know the heart-wrenching struggles that would lie ahead for me to help save these turtles and our program in the coming years. I am glad I took the time to revel in this phenomenal day and to reflect with pride and gratitude on my National Park Service colleagues and the other pioneers who had the foresight and courage to begin this ambitious program to save a magnificent species that was almost lost during the course of one human generation.

Diane Srivastava. I have worked with some amazing women in the field and developed lifelong friendships with them. So much happens during fieldwork that you have all these intense shared experiences with the other researchers on your team, whether it's a close scare of almost stepping on a venomous snake or the amazing sight of the Northern lights. And that creates a real sense of camaraderie. Part of that is the shared experience of defying gender norms to be tough and sweaty and covered in mud. And then there are the discoveries.

I had spent several years working in Costa Rica in the rainy season, roughly October to December, when the water-filled bromeliad tanks contain the charismatic top predator in the system—a damselfly larva. One summer I taught a field course at my site. This timing overlapped with the few months where the damselflies were adults, so my PhD student and I netted and marked a few to see where we recaptured them. Netting them is harder than it sounds, because these are some of the world's most maneuverable damselflies. They are called helicopter damselflies because their asynchronous wingbeats allow them to pluck a spider off a web without getting caught. We quickly discovered that many adult males were establishing territories around the largest bromeliads. If another male approached, they would both hover in midair, like two cowboys with guns drawn, before one either fled or attacked the other. If a female approached, she checked out the bromeliad with an anxious male hovering above, trying to shepherd her to the water-filled tank. The systematic observations we serendipitously made that summer helped us discover why there were so many damselfly larvae in the largest bromeliads—because their parents actively chose these. Later we discovered that large bromeliads were so desirable because they didn't dry out before the larvae had fully developed. I guess that makes these adult damselflies the ultimate helicopter parents!

Melanie Stiassny. I love the excitement of pulling a seine full of fish from the Congo River. It's like you receive 100 birthday presents all at once. Many of the fish will be surprises, and some will undoubtedly be species new to science. When I go to bed in my campsite along the river, soon after the sun goes down, I daydream about the unknowns that will surface tomorrow. The thrill of the unknown is exhilarating.

Emily Taylor. As a graduate student, I radio-tracked rattlesnakes for years in the Arizona desert. One night I kept smelling a musky, feral, funky scent. I couldn't place it. I tracked a snake deep into a rocky area I had never been in before and came across a den where a large mammal had bedded down. The smell was extra strong and I saw large prints in the dirt. I, the fearless herpetologist studying rattlesnakes, was completely freaked out at stumbling into a large creature's empty bedroom. I left the area and tracked elsewhere. I returned to my truck about 2:00 am and threw my sleeping bag on the ground. Just as I was drifting off, I heard an unearthly, loud piercing wail from a wash on the other side of a group of mesquite trees nearby. A curdling, guttural cry. I scrambled out of my sleeping bag, and the predatory odor punched me in the face. I couldn't see past the mesquites, but I didn't need to. The creature was about 20 feet away. I threw my gear into my truck and hit the gas pedal. Several miles away I stopped to breathe. Was it the Wendigo, an evil man-eating creature from creepy tales Dad told me when I was little? The Chupacabra that sucks goats' blood at night?

The next day a mammologist friend confirmed that I had heard and smelled a mountain lion—a large adult, possibly in search of a mate judging by its mournful wails. Or maybe a tasty meal. That's when it dawned on me that the smell had been with me throughout the night. The cat was tracking me. I will never forget that piercing wail, feral scent, or the feeling of complete and utter exposure and helplessness. The feeling of being prey. I never heard those wails again, and that rattlesnake never took me near the den again. But occasionally I saw a large cat paw print, and, very rarely, caught a waft of that smell. Although I never did see the cat, I'm pretty sure it was watching me.

Sandra Vehrencamp. In 1974, when I had finished my fieldwork but hadn't yet written the dissertation, I was invited to attend the International Ornithological Congress in Australia and participate in the first symposium on cooperative breeding in birds. There was one other woman there, Veronica Parry, who had studied kookaburras. She and I were the only ones who had successfully captured, color-banded, and sexed our subjects and then followed the behavior and success of known individuals. The male researchers either didn't know individual birds or shot all of their birds at the end of their study to determine their sex. Veronica stood up and asked: "Is it only the women who can capture and color-mark their birds to conduct a proper behavioral study?" At that meeting, and the discussion session afterward, the dominant males asserted that helper-at-the-nest systems were the "normal" form of cooperative breeding and that it was up to me to figure out why only a few species, such as my anis, had a communal system. I vowed to take on that challenge, and I *did* come up with a theory to explain the conditions under which the two systems were likely to evolve.

JOYS AND CHALLENGES OF MOTHERHOOD AND FIELD BIOLOGY

Many field biologists interviewed have chosen not to have children. Some who have combined motherhood and field biology emphasized that compromises sometimes must be made, for example, concerning choice of field site and amount of time spent per day engaged in their research. Most of the mothers who have taken their children into the field found that doing so enriched their research experiences and provided positive opportunities for their children, from expanding their willingness to try new foods to learning about diverse cultures and to appreciate nature. Many mothers emphasized the importance of partner support.

Jeanne Altmann. In 1963, my toddler Michael and I joined my husband in Kenya for Stuart's planned 15-month study of communication in baboons. Our study site was in Amboseli, at that time a remote Maasai Reserve. The isolation of the reserve and the fact that it was difficult to find anyone willing to live in a tent with lions, leopards, and elephants nearby meant that for much of the time we were on our own for cooking, water-hauling, and diaper washing. We multitasked childcare, daily maintenance, and searching for and watching baboons. My time in the field was abruptly cut short when a virus paralyzed our son from the neck down. He was kept near an iron lung in Nairobi for the next 2 months before he had recovered enough for me to return with him to the states. What a thrill it was to watch Michael take his first steps after this life-threatening paralysis!

Amy Boyd. I thought combining motherhood and fieldwork would be easier. I had a friend in graduate school who would pop out babies and immediately take them to Borneo with her, living in the wilderness with her husband and young children. I assumed I would be the same, but my child didn't make that possible. I did take him in the field with me during his first year, stowed in my Ergo carrier as I set out insect traps or collected plants. I was on sabbatical that year and envisioned taking him with me everywhere. He was fussy, and if I didn't keep moving constantly, he would cry and scream. So I shifted from ID-ing the plants as I collected them, to sticking them in a bag on the fly so that I could keep moving, and ID-ing them later when Leo was sleeping. He would sometimes sleep in the infant bouncy chair sitting on the lab bench while I worked but never for very long. Occasionally I managed to get him out in the field with me as an older child, helping to reset traps or check phenology of trees, but those moments are few and far between. My son has a temperament that makes that mostly difficult. For me, motherhood and fieldwork has really been about separating and compartmentalizing the two.

Judith Bronstein. I became a mother after I gained tenure. At the time, my fieldwork took place on the other side of the country. It was not possible to continue work there after I gave birth; my husband was a graduate student and couldn't come with me to provide child care, and I wasn't prepared (or financially able) to hire help. This led to a big shift in my career. I stopped focusing on field research and instead started some big conceptual papers and initiated collaborations with theoreticians. The success I've had in my professional life is directly related to that shift. The new direction of my work was ultimately seen as much more interesting and is now viewed as the most impactful contribution I've made. I resumed fieldwork when my son was about seven and my husband and I started collaborating on local plant/animal interactions.

Patricia Burrowes. Perhaps the biggest challenge I had to deal with as a field biologist occurred while I was working on my PhD as a mother of a 4-year-old daughter who had a very serious genetic disorder, osteogenesis imperfecta. It was the partnership with my then-husband and colleague that made it possible to deal with our daughter's multiple hospitalizations and surgeries while I tried to conduct demanding fieldwork in caves, finish a dissertation, and deal with the productivity challenges of a tenure-track position.

Robin Chazdon. I got my academic position at the University of Connecticut when our daughter was 6 months old. I never stopped doing fieldwork, even during two pregnancies and child-raising, as my husband and I were able to go to the La Selva field station in Costa Rica together to work on our separate projects. We had a wonderful Costa Rican nanny who took care of the kids during the day until they were in their teens. She became part of the family. Being at a field station made it easy, as meals were provided and we didn't need to do shopping or housework. I don't know how I would have been able to do fieldwork otherwise! I tried one project in Connecticut, but it was too difficult to get time away when I was "home." Having to travel to do my research actually made it easier, as I didn't have to juggle teaching or other duties at the same time.

Martha Crump. Six weeks after my daughter was born, my husband and I flew with Karen to Costa Rica where we spent a sabbatical year in Monteverde. The eldest daughter of two of the original Quaker settlers took care of Karen parts of 5 days/week. I spent many hours that year expressing milk (with a plastic manual pump) while perched on logs or boulders—with army ants swarming nearby, during lightning storms, and at night with my headlamp propped on the ground. The frozen milk became my lifeline to the freedom to do fieldwork. We began our second year-long sabbatical in Tucumán, Argentina, when Karen was 7 years and Rob was 4. We enrolled both kids in a bilingual school. I frequently took the kids with me to my field site, a 6 hours' drive north of Tucumán, where they helped dip-net tadpoles, collect predaceous aquatic insect larvae, and locate mating frogs. Both kids immersed themselves in Argentine culture, learned Spanish, and made friends. Their positive experiences associated with Latin culture and meeting diverse people will last their lifetimes, and I feel tremendous satisfaction at having combined long-term intensive fieldwork and motherhood.

Rosemary Gillespie. I have two boys, now 23 and 25 years old. I took them into the field often. Although neither went into biology, both are very willing to engage in discussions about the environment and both appreciate the environment. At times during my fieldwork, the boys complained relentlessly, but now they say they were the best experiences of their lives. Sometimes I took the boys with me when I led fieldtrips. It was a good opportunity for the students to see that one can manage kids in the field and carry out fieldwork.

Rosemary Grant. Career advice in the United Kingdom in the 1950s and early 1960s, from both outside and inside the university, was that a scientific career was incompatible with marriage and children; it would be detrimental to your husband's career, which would in turn negatively impact your marriage. Surprisingly, this strong negative advice came as frequently from women as from men. I thought if a man could combine a scientific career with marriage and a family, why could not a woman do so as well, if family tasks were shared? When I decided to do it, I had enormous support and encouragement from my husband, Peter. The key: Marry the right person!

With two small children in Montreal, and in those days no good daycare facilities, very little money, and no relatives nearby, I stayed at home until they went to school. On Mondays, I hired a babysitter and instead of spending that day catching up on shopping and appointments I spent it in the McGill University library catching up on research articles. This prepared me when I finally was able to return to research. The hiatus was not entirely wasted; it increased my ability to be more organized when juggling family and research.

We began our fieldwork in the Galápagos Islands when our daughters were ages 6 and 8 years. Every year we took them with us. Their schoolwork took a couple of hours each day, after lunch, and somehow they were always ahead when they returned to school. Being a family on an uninhabited island was magical. Now adults with children of their own, they both say that going to the Galápagos was an immensely rewarding and enjoyable part of their childhood, one that they would not have changed for anything.

Nalini Nadkarni. The years when the kids were young were my most challenging times as a field biologist. My science was something I wanted to do because it gave me such a sense of contribution and fulfillment. But having the kids with me in the field was also rewarding. So I had to combine the two. Having kids with me in the field opened up possibilities in Costa Rica that might not have been possible otherwise because I was able to interact with the locals, especially women, and talk about kids. The negative was the conflict I felt—the screams in the back of my head: "Mom!" (when I was doing fieldwork) and "Work!" (when I was with the kids). Although my husband is very supportive, he said that he did not feel those conflicting tugs. My children grew up knowing that mom is a field biologist.

Jeanne Robertson. I decided to become a single mom during my postdoc. I was terrified that I couldn't have a child and a career as a single parent. I thought that being a new mom would make fieldwork difficult if not impossible and that being a single mom would be even more challenging. Fieldwork was a part of my identity. I didn't know if I could give it up. But, I told myself that I could re-focus my research on lab work and data analyses and that I would eventually return to the field. That plan didn't play out as I imagined. I never stopped going into the field. I just had to reimagine fieldwork. Field seasons became shorter. Although logistically complicated and more costly, I have found ways to be a single parent and conduct fieldwork. When my baby was 4 months old, I strapped her to my chest and waded through a pond in Central Florida to collect fish. When my colleague pointed out the alligator nearby, it gave me pause for thought. Should I have a baby with me in the field with large predators close by? We kept going into the field, and it has been a part of her life. Without a spouse to stay home and look after her, it meant I needed to recruit friends/family to come to the field with me. The specific challenges of having a child in the field changed as she grew. When she was an infant, the near-constant breast-feeding and issues with naps at inconvenient times made fieldwork feel inefficient at best. When she was a toddler, it was difficult to have meaningful conversations with the field team while entertaining her and keeping her from getting into trouble. Looking back, though, I wouldn't change a thing (Figure 8.3).

Michelle Pellissier Scott. While preparing to apply to graduate school, my husband and I adopted a third child so I had to wait until she was in nursery school to take courses I needed. I could usually manage my teaching obligations so that I needed to hire afternoon babysitters only once or twice a week. I was home usually from the time the kids got home from school until they went to bed. After that I returned to my office. While I was responsible for almost all the housework, cooking,

FIGURE 8.3 Jeanne Robertson and daughter in Costa Rica. (Used with permission, Jeanne Robertson; photographed by Jeanne Robertson.)

and child care, my husband was very supportive. Still, if a kid was sick, I was the one to make adjustments. When I was choosing a dissertation project, I had to consider the whole family. It seemed preferable to move everyone to my field site rather than be gone for part of the year, as most of my classmates were doing. So we all moved to Australia for 3 years. The kids took up surfing and "Aussie rules" soccer and sometimes helped with the fieldwork. My husband worked as an architect but took over all the household chores at the height of my fieldwork.

Diane Srivastava. To combine field biology and motherhood, one needs to adapt. I adapted by setting up an international research network—the Bromeliad Working Group—so that I could combine forces with other researchers to pool data and ideas. This meant that I could get much of my research done in intense bursts of 2-week fieldtrips and spend the rest of my time organizing multi-site experiments and global data analyses. In the end, this was probably a much more rewarding direction for my career than continuing on the solo researcher track.

Amanda Subalusky. One of my biggest joys, and challenges, has been balancing being a field biologist and a mom. My fieldwork requires long-distance travel, working in remote places under often challenging conditions, and long, consecutive days of work, none of which is very conducive to taking care of children. I have been incredibly fortunate that my husband and I conduct much of our research together, so we have been able to integrate our family into our research. We took our oldest daughter to the field for the first time when she was 1 year old, and she has now spent three field seasons living in a field camp in a remote part of Kenya. We now have these incredible shared memories of fieldwork and travel together, and I feel so fortunate to be able to share these experiences with her and hopefully my younger daughter one day as well. Balancing the challenges of remote fieldwork with caring for a young child, even with an amazing nanny, was a challenging transition and a whole new level of exhaustion. I could not imagine doing this without a wonderful partner and co-parent and without supportive mentors who value work-life balance (Figure 8.4).

Marvalee Wake. My husband and I often took our son Tom with us into the field. Once when we were collecting aquatic salamanders, I saw that Dave had perched

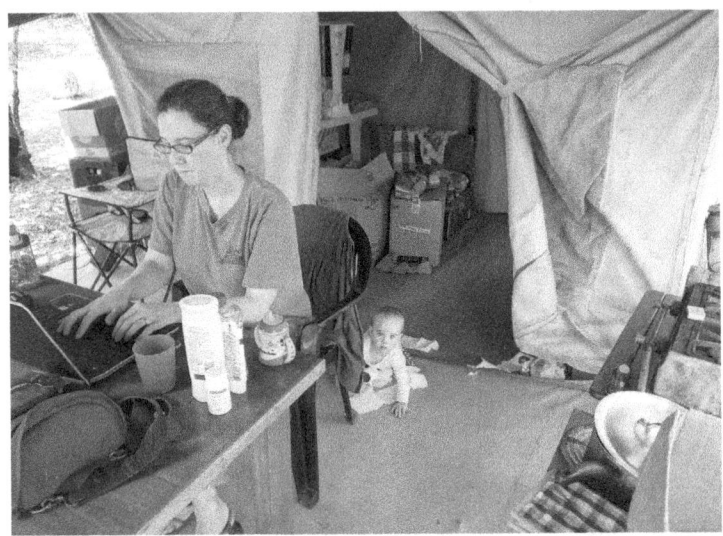

FIGURE 8.4 Amanda Subalusky and daughter in Kenya. (Used with permission of Christopher Dutton; photographed by Christopher Dutton.)

18-month-old Tom on a boulder in a stream. I asked, "What are you doing!" Dave responded, "He's great! His hands are so fast, and he doesn't let go!"

Jessica Ware. I have always enjoyed bringing my children with me to the field. They have collected with me since they were in baby carriers. I had a thrilling month-long trip to Guyana one year that really inspired them in terms of all sorts of things from willingness to try new foods to excitement for science. My son Zach was 7 years old and a very picky eater. I lugged a suitcase filled with various boxed pasta mixes—and took most of the mixes home because he branched out from Mac & Cheese and peanut butter & jelly! Zach went fishing and watched his caught fish being prepared by Amerindians. They were his fish, so he loved them! He ate fresh fruit picked from the local trees. Much of the food was flavored with curry. At home he would have refused to eat it; in Guyana he ate it just fine.

Kathy Winnett-Murray. I loved including my two sons in my fieldwork. I wouldn't trade that for anything because it gave the fieldwork special meaning for me, and it undoubtedly had a deep and lasting impact on my kids. It was very gratifying for me as a mother to feel like I was doing the things in biology that I wanted to do, and I was also providing my children very special educational and cultural experiences that were authentic and could not be replicated.

ADVOCACY FOR WOMEN; SCIENCE OUTREACH; SERVICE TO THE PROFESSION

Many interviewees expressed the importance of service in their lives. Several have been instrumental in establishing foundations, centers, and institutes. For example, Dee Boersma established the Center for Ecosystem Sentinels at the University of

FIGURE 8.5   NamanaBe Hall, Centre ValBio, Madagascar. (Used with permission NRowe/
alltheworldsprimates.org; Noel Rowe, photographer.)

Washington, serves as its Director, and endowed the Boersma Endowed Chair of
Natural History and Conservation at the Center. Margaret Lowman co-founded the
TREE Foundation, a non-profit organization dedicated to tree research, exploration,
and education, and serves as its Executive Director. Debra Moskovits helped launch
the Keller Science Action Center at the Field Museum, Chicago. Anne Pusey established the Jane Goodall Institute Research Center at Duke University. Patricia Wright
spearheaded the formation of Ranomafana National Park, Madagascar, founded the
Institute for the Conservation of Tropical Environment at Stony Brook, and established the Centre ValBio research station at the edge of Ranomafana National Park
(Figure 8.5). Early career and late career women alike provide service through
mentoring women in science, participating in science outreach, and being actively
involved in their professional organizations.

Julie Denslow. My active service to my profession has been not only highly
rewarding, but it also gave me a very wide professional family with whom I interacted over the years. Service takes time away from other activities, such as teaching,
mentoring students, and writing, but it was good for my career. The Organization
for Tropical Studies (OTS) was established in 1963. OTS has made magic for many
thousands of students and faculty, offering logistical support and facilitating collaborations, and being especially welcoming to women students and seasoned scientists. I served as President of OTS (1994–97) and spent many years on the Executive
Committee, Board of Directors, and committees. After I retired, I served as Chair
of the Board of Visitors and of the Fiftieth Anniversary Committee. OTS gave me a
lot also. It gave me mobility; I could change jobs without interrupting my research.

OTS provided me with a secure place to work (La Selva, Costa Rica), safe from logging and other anthropogenic disturbance. I have also been heavily involved with the Association of Tropical Biology (now Association for Tropical Biology and Conservation) and the Ecological Society of America. Contacts made through these service activities have expanded my professional network and, indeed, my view of life.

Mary Jade Farruggia. I serve as a mentor with an educational youth program through the Center for Land Based Learning, in agricultural Woodland, California. I work with a group of students to complete a habitat restoration project, often in partnership with local landowners and the county resource conservation district. Throughout the 3 months, the students learn about the benefits and challenges related to balancing ecological restoration and agricultural necessity and the importance and far-reaching impacts of watershed health.

I believe that improving access to, appreciation for, and diversity in science starts with our youth. Showing young people that studying science and nature is something tangible, exciting, and attainable builds the foundation for them to see themselves as scientists. This is especially true for underrepresented groups who have never seen scientists that look like them. I experienced this first hand in my work with the Center for Land Based Learning when I was matched with a group of students who were quick to point out that all of us were Asian-American, including their mentor (me!). As an early career scientist, I am acutely aware of how great an impact mentors can have on one's confidence to pursue science. Outreach is one of the most impactful things we can undertake as women in science to improve the field. It will always be a core part of what it means to me to be a scientist.

Christine Figgener. In August 2015, while conducting research off Costa Rica's coast, my research team and I captured an olive ridley sea turtle with a 10-cm section of plastic straw lodged in its nostril. I recorded a video of the excruciating process of extracting the straw with pliers and uploaded the 8-minute video onto social media (it is still available on YouTube). The video went viral and thrust me into a world of high-profile advocacy work, including a documentary project, working with community activists, and giving presentations. In 2018, *Time Magazine* named me a Next Generation Leader for my outreach work. The suffering turtle's experience led to a global anti-plastic straw movement. Numerous businesses, including Disney, Starbucks, and American and Alaska Airlines, have phased out plastic straws. Many cities, including Seattle, Washington, and San Francisco, California, moved to ban or limit plastic straws. The European Union moved to prohibit various kinds of single-use plastics, such as cutlery. But eliminating single-use plastic items is just one step in our greater awareness of the need to fight plastic debris in our ocean. The heart-wrenching video gave me a platform to speak up for sea turtles and connect people through the turtle's suffering to our oceans. It also shows how changing our daily habits can make a huge difference in this world.

Elaine Harvey. I visit schools and tell students about the importance of water quality and need to protect fish populations. At the Yakama Nation Wildlife Camp for middle and high school kids, we collect, measure, and ID fish. I emphasize that it's important to know what fish occur in a given area so we can protect them. We also catch macroinvertebrates, and I explain that these are food for fish and that they indicate water quality. I give presentations and lead nature hikes at a Trout Unlimited camp for kids. We talk about, locate, and identify plants for traditional uses. I give presentations at group homes for troubled kids and other places, where I talk about the importance of the Columbia River, salmon, and other resources, local history of the tribes, and ecological changes that have occurred in the region historically. I am passing on my knowledge of fisheries biology to the next generation of people who live in the Yakama Nation.

Frances C. James. At the 1972 Cooper Ornithological Society meeting in Las Cruces, New Mexico, I facetiously proposed the establishment of a new ornithological society, the Phalarope Society. The women would present their research and the men would make and serve the coffee. (For non-birders: Phalaropes are shorebirds with a polyandrous mating system. Females are larger and more brightly colored than males; they fight over males and defend them from other females.)

In a more serious tone, in a Letter to the Editor published in 1973 in the *Auk* (90:488–489) I wrote:

> Dear Sir:
> To a questionnaire asking for information on their employment situation sent last fall to some 70 women ornithologists, 7 of them outside the U.S., 27 responded. Two of these women are not employed because they have children at home, four are over 65 years of age, and the rest are employed. When asked whether they felt they had been discriminated against because of being women, their replies varied. Several said they had always been treated fairly and a few added that women should not make themselves "unhappy" by "wanting money" or "dominance." The majority said that hiring practices were discriminatory, salaries lower, and promotions slower for women than for men. On a more subtle level, they discussed general psychological putdowns and restrictions to job opportunities in towns where their husbands were employed.
> My conclusion is that women ornithologists are being affected by discrimination in employment at least to the same extent as women scientists in other disciplines. Women with advanced degrees in ornithology want to work and are working. But, there is a feeling among some women as well as some men that it is unbecoming for women to ask for equal pay and rank for equivalent work. Reviews of nationwide statistics on employment of women scientists are available from the Scientific Manpower Commission, Washington, D.C.

I also pointed out that in the previous 2 years I had received reprint request cards from 38 institutions that began with the printed salutation "Dear Sir." I suggested, "How about Dear Colleague?"

Margaret Lowman. I recently initiated project "MISSION GREEN" through TREE Foundation, aimed to conserve canopies in highest diversity forests. We plan to build and maintain ten canopy walkways including Amazonian Peru, Bhutan, California redwoods, Florida, Great Smokey Mountains, Madagascar, Malaysia, Mozambique, and Western Ghats India. Four walkways have been completed but need outreach programs. We will establish science partnerships with the host countries and train local people, especially women, as environmental stewards—work that will provide sustainable income from canopy ecotourism instead of from logging the forests. We will also provide fellowships for international students to conduct biodiversity research at these sites.

Nalini Nadkarni. In the 1990s, I began carrying out public engagement activities to raise awareness of the importance of trees to humans. I first did this in traditional venues (museums, zoos, National Geographic), but found that I was mostly "preaching to the choir." That led me to create programs to engage people who do not already seek forests or do not have access to them. These activities include giving sermons about "trees and spirituality" in churches, synagogues, and seminaries, and partnering with artists to create gallery exhibits, dance performances, poetry readings about trees, and rap poetry slams about urban parks. I recruited science faculty to provide research lectures and carry out conservation projects to adult inmates and to youth in juvenile detention centers as well as bringing nature imagery to inmates in solitary confinement to connect them as best as I could to nature—where nature is not. I have also developed a training program for scientists to carry out engagement with "science-unengaged" public groups: the NSF-funded STEM Ambassador Program. All of these activities are knit together in what I call "tapestry thinking"—the interweaving of many ways of knowing to enhance knowledge and appreciation of trees, and nature in general, in a world increasingly disassociated from the natural world.

I never liked playing with dolls as a child. My father was from India, and my family tried to live simply, following the principles of Gandhi. For me that meant tree-climbing for play. I recognized that many little girls love Barbie dolls. For years, I tried to persuade Mattel to develop a "Treetop Barbie." Finally, in 2019, the National Geographic Society partnered with Mattel to develop a new line of "Explorer Barbies," including an astrophysicist, a wildlife conservationist, an entomologist, a marine biologist, and a nature photojournalist. They invited me as an advisor. Explorer Barbies are now sold at Target, Walmart, and other retailers. As a thank-you for my involvement, Mattel sent me a custom-designed Barbie in my likeness, complete with a climbing rope, binoculars, a "Rite-in-the-Rain" notebook, and streaks of gray in her hair (Figure 8.6). The best part is that I hear from girls across the country, asking how they can help trees in their communities. Getting kids excited about protecting our forests is one of the most important things I do.

Michele Nishiguchi. I am passionate about training students from underrepresented groups to increase diversity in science to add different perspectives. I am a third-generation Japanese American; growing up with my parents' perspectives was humbling. My mother and her sisters were sent to Japan before the Second World War to "get cultured" in Japanese ways. The war broke out and my mother couldn't return to the United States until 2 years after the war ended. She was about 250 miles from Hiroshima when the bomb was dropped. She lost many family and friends. By the time she returned to the United States, her parents had lost their farm and had to start again from scratch. My father, too young to be sent off to fight, was interned during the war at Tule Lake Relocation Center in northern California. After the war, he felt ostracized for being Japanese. Embarrassed. He and many Japanese Americans felt they had to work twice as hard to prove they were worthy Americans. They lost a lot of their culture because of trying so hard to fit in. My parents made me work hard in school and always taught me to stand up for my beliefs. Many of the students I mentor from underrepresented groups experience imposter syndrome. They fear they cannot compete. I understand where are they coming from. I use myself as a role model, and the students make a connection and gain confidence when "the squid doctor" tells them about her roots.

FIGURE 8.6   Custom-designed tree-climbing Barbie by Mattel, in the likeness of Nalini Nadkarni. (Used with permission, Nalini Nadkarni. Betsy Eckhardt, photographer.)

Susan Riechert. More than 25 years ago, when my son was in elementary school, his teachers were always asking me to come to their classrooms to share biology. One day I filled a wooden trunk with sand, dirt, and buried fossils, rocks, bones, and pottery shards. I wrote a lesson plan entitled "Fossils: True or False." I delivered the trunk, lesson plan, and an answer key to the school library. Within a month, all 1,100 elementary students at the school had completed the exercise. The librarian asked, "What's the next box?" So I made another, called "Of Skulls and Teeth." These two projects began the Biology in a Box program, still going strong. Objects in the boxes range from bones, seeds, chunks of wood, and seashells, to animal furs and turtle poop. Since its inception, the boxes have provided hands-on science experiences for children in nearly 1,200 schools across Tennessee. There are now hundreds of inquiry-based exercises, developed to meet school curriculum goals. The program has expanded to include a YouTube channel, a card game, and two Flashgames, and thanks to a partnership with the National Institute for Mathematical and Biological Synthesis it has reached out to surrounding states. To date more than 650,000 elementary through high school students have participated in the Biology in a Box program. The materials have facilitated discussion, experimentation, and deeper questioning, encouraging students to explore and better understand the natural world. The most recent box is on biomechanics.

Sandra Vehrencamp. Relatively early in my faculty career at UC San Diego (late 1970s) I was placed on a search committee. While reading through the application materials and articles about the plight of women in academia, I saw some obvious patterns in the male and female applicants. Men were bolder about selling themselves and extrapolating their findings to big questions; women were more careful, less willing to take risks, and more focused on details and descriptive science. Often it was the male letter-writers for the women who pointed out the greater implications of the woman's work. I held some sessions with the women graduate students to talk about these gender issues and help them see what they needed to do to succeed, such as seeing the big picture of their work, selling oneself honestly, and speaking well. Many years later when my husband and I arrived at Cornell (1999), there were very few female graduate students in the department. We studied the ways in which women were discriminated against during the admissions process by our mostly male colleagues. We discussed this openly with the faculty and made some changes, successfully getting more women into the program. Later, we held similar help sessions with the female graduate students. The few recent women faculty didn't seem to have experienced the same difficulties that my generation had to deal with and didn't seem to think there were further issues to discuss. In several cases, they had women mentors and advisors, so they benefited from the efforts of our earlier generation.

Sidney Woodruff. It's difficult to conceptualize what is possible when you haven't had the seed planted in your head. For many people growing up, seeing Steve Irwin and Jane Goodall on TV was the seed that planted in their mind that they could grow up to be a wildlife biologist, environmental manager, or advocate for natural resources. Many young people of color, particularly Black young people, do not see that reflection in the media when it comes to public and visible Black environmentalists. In recent years, that lack of representation has been acknowledged and actively countered through social media campaigns like #BlackBirdersWeek, #BlackInNationalParksWeek, #LatinoInSTEM, and many others.

I am now a mentor and Vice-President of MUSE (Mentorship for Underrepresented STEM Enthusiasts). This program connects historically underrepresented students to STEM professionals, especially those who share underrepresented identities, who can provide support and advice on navigating academia. By having that personal connection to a mentor with similar life experiences and identities, mentees feel empowered to follow their academic pursuits. First-generation college students often have to learn the maze of higher education on their own, and it is easy to feel discouraged. In contrast, if your parents attended graduate school, they pass on institutional knowledge throughout your academic career by telling you "Next summer, you should look up exam times for the GRE," and "When I was in school, I worked in a research lab." Those minor inputs add up quickly and generate knowledge about higher education that is passed down to students. MUSE removes that institutional knowledge gap. Our mentees benefit by connecting with someone who advocates for them and who understands their struggles and grievances as often being the only "other" student of a certain identity. My mentees hear about my academic journey and see the mistakes I made, then circumnavigate that worrisome process by not having to reinvent the wheel on their own.

ARE THERE ADVANTAGES TO BEING A WOMAN IN THE FIELD?

Some interviewees adamantly indicated NO (e.g., "Are you kidding? How could you ask this!"). Others felt there are ways they benefit from being a woman in field biology. One common theme was the perception of a woman as being less threatening, which facilitates collaborations and other interactions. Another frequent comment was that women bring a different perspective and approach to issues related to fieldwork.

Joan Berish. I found that in some circumstances, as a woman I was able to enter scenarios in which a man would engender suspicion, hostility, or downright fear. My gender was likely an asset in entering the inner sanctum of rattlesnake roundup rings in Georgia and in gathering information during interviews and field forays with snake hunters, gopher tortoise harvesters, and softshell turtle fishermen.

Dee Boersma. In Argentina women are not equal to men. As a woman I'm not seen as a competitor (or equal) to men, but as an American I have a higher status than being an Argentine. If I were a White male it would be harder for me to do my work in Argentina because men would compete, but since they feel they are of higher status than a woman it isn't a problem. Sometimes they really are not quite sure what to do with a female.

Michelle Carlisle. My gender benefits the team, just as everyone's background and experiences bring a new lens through which to see issues. Diverse backgrounds bring different ideas and add more tools to our toolbox. Working in wildlife management takes a lot of problem solving, so the more angles we have from which to see problems, the better.

Tally Hamilton. I had an interview for a position out west, and the comment was made that they liked hiring women because ranchers were less likely to pull a gun on women than on men.

Erica Hoaglund. Current hiring practices help me. Minnesota Department of Natural Resources considers women a protected class. I think this has given me a slight leg-up when competing for positions. Being someone different (whether because of my gender or background or both) has helped me connect with other people who are also in that category of "other." I have gotten the opportunity to work with more diverse groups of people, because my gender and background make me the easy person to which to assign work involving diverse parties.

Kate Jackson. The disadvantages of being a woman doing fieldwork in central Africa (being underestimated, not being taken seriously, not being respected, being seen as a person who lacks power or influence) have helped strengthen my bond with my African collaborators, and my relationship with these collaborators has definitely given me an advantage in fieldwork. Our herpetology team (I and my Congolese students/colleagues) have had to struggle through many situations dealing with White expat male project managers and NGO and extractive industry administrators who treat us badly because they don't respect women or Africans and who thus have a big problem with us.

Dakota Keller. Being young and female comes across as unintimidating, which is both frustrating and a blessing. Some ranchers and land owners either don't have the time of day for you or are especially polite and willing to cooperate. You never know what you're going to get, but it is expected that you play the role they think you fit.

Karen Lips. Sometimes it is good to be underestimated and to appear non-threatening. You can slip in under the radar, wander about, and do whatever you need to do without setting people off. Drawing attention to yourself can get you into trouble when you are alone in a remote place.

Margaret Lowman. I prioritized working in low-income countries (e.g., Ethiopia, Cameroon, Western Samoa) for my conservation work. In those countries the village chiefs trusted me more as a mom and female than my male counterparts. And it sure helped to have some photos of my kids to share.

Debra Moskovits. I think that as a woman, I was trained to be a problem-solver, listener, and consensus-builder. Those are essential ingredients for successful conservation results.

Caitlin Nagorka. Sometimes landowners hesitate to partner with the federal government (surprise!). Landowners tend to view a woman, especially a young woman, as less intimidating than a man. I am able to use my gender and unintimidating appearance to initiate conversations and create potential opportunities under circumstances where my male colleagues have encountered resistance or hostility.

Jennifer Pramuk. In my opinion, women can be better at setting aside their egos, which can assist with relationship building, communication, and project planning. These traits came in handy when I was helping to develop conservation projects in Madagascar and Tanzania.

Emily Marie Purvis. In our culture, women are socialized to be more attuned to the needs and desires of others. This quality is reflected in my leadership style, especially in the field. I will never be a commander barking orders. Demonstrating care and concern for my co-workers results in more effective and productive teamwork. In our culture, women are also socialized to be more collaborative than men. Because my research sits at the nexus of conservation efforts, resource management, and scientific investigation, I benefit tremendously from the "feminine" quality of seeking to collaborate closely with others.

Lora Smith. There is an advantage of being female when working with private landowners and in education outreach: people are more trusting of a woman in the field than a man. If I'm not afraid to handle snakes, others are more willing to do so. When I encounter people in the field, they are less concerned that I'm up to no good (i.e., I appear less threatening) than if I were a man.

Mary Jane West-Eberhard. I think that while doing fieldwork in Colombia I got extra help because I am a woman. People were protective and would help me find places to work, help me get there, and do things like help carry heavy ladders I needed to use.

Kathy Winnett-Murray. I think there's an advantage just by being kind of unique. Decades ago, when it was more of a rarity to be a female field biologist, you attracted more attention (whether you wanted it or not) because you were a woman. Even today, people want to interview *you* and invite *you* to be a guest speaker, not your husband, who did all of the same things that you did in the same way at the same time, and just as well. He's a man, and other people think male field biologists are "a dime a dozen." I felt another advantage of being a woman while working in the field. In Costa Rica I often wandered alone all over people's farms and through their woodlots studying birds, and I was happy to have a friendly relationship with so many community members. I know that some of them were "looking out for me" from a safety standpoint and felt more compelled to do so because I am a woman.

Patricia Wright. Madagascar was ruled by queens in the 1800s, and this tradition perhaps has led to the fact that there are many strong professional women in Madagascar today. I have never felt "disregarded" or "not respected" among the Malagasy because I am a woman. When it comes to conservation, being a woman has an advantage. White men tend to have "the answers" and talk down to local people. From the first village meetings in Madagascar, I realized that I had no "answers" for the rural farmers of Madagascar, but we could develop solutions together, once they explained their problems. We became partners to solve the problems of natural resource depletion in Madagascar. Thirty years later, I would say that the system has pretty much worked. I listen a lot.

HARDSHIPS AND CHALLENGES

Some women interviewed feel they have experienced considerable gender discrimination in field biology. Others feel they are accepted as equal to men. Still others place themselves somewhere in between on this spectrum. Why so much difference? Area of study might account for some differences. For example, wildlife biology was long dominated by men, whereas animal behavior has traditionally been welcoming to women and in fact in some areas dominated by women. Type of employment might play a role. Women in academia expressed less overt discrimination than women employed by state and federal agencies. The timing of when the woman was/is active in fieldwork could explain some differences. What was accepted as "boys will be boys" behavior in the past is no longer acceptable. Several women who worked with their husbands/significant others in the field felt sheltered from discrimination. Another non-mutually exclusive factor might be personality and confidence level, in part influenced by parents, siblings, teachers, and colleagues.

Sentiments voiced repeatedly, from early career to late career women and from academics and non-academics alike, include feeling dismissed because of gender and that male colleagues garner more respect. Women often feel invisible, ignored, and overlooked. Men are treated like they have better insights or ideas. Men "talk over" women and "mansplain." Often a male will sum up what a woman has just said before the conversation moves on, or worse, a man is credited for an idea proposed by a woman. (See Chapter 9 for further discussion of these challenges.) Following are some themes of hardships and challenges expressed by interviewees.

SUBTLE OR OVERT MESSAGE THAT FEMALES ARE INTELLECTUALLY INFERIOR TO MALES

Rosemary Grant. My father, a medical doctor, truly believed that men were more intelligent than women. I was brainwashed to believe this, and I believed it up until I became an adult. When I would say something that impressed my father, he would say, "Oh, you have a mind like a man." As an early teenager, I began to think maybe I was changing into a man. At one point, worried after missing my period, I went to the library to learn about testosterone. I told the headmistress at my boarding school in

Edinburgh that I thought I might be turning into a man. She assured me, "Rosemary, there's no way that is happening, not with breasts like yours."

The headmistress was firmly against a girl with two brothers (me) going to university. This was the prevailing view at the time in the United Kingdom (1950s), held by women as well as men. Money should be allotted to the training of sons; daughters married and raised children. After persuading my parents and headmistress to allow me to at least take the university entrance exam, I caught mumps with severe complications at exam time. The headmistress, standing at the end of my bed, declared it was God's will. I circumvented that problem by leaving school, getting a job, and taking correspondence courses that were at a much higher level than those at my boarding school. (My school basically trained women to become ideal wives for professional men.) I took the examination the following year and passed.

Once enrolled at Edinburgh University, I soon learned a trick passed around by the few women undergraduates in science: Never write your full first name on an exam; use initials so as not to reveal your gender. The exams, hour-long essay questions, had very subjective grading. Males generally received higher scores than females.

Sinlan Poo. It is common for people to defer to the judgment of my male scientist colleagues, regardless of our relative knowledge or expertise. And more common for me to wait until a male colleague has finished talking before my opinion is offered or requested. Certain male colleagues with less experience and no publications have gotten better positions and more respect and accolades from our peers. I have been told to seek advice from male colleagues who have little or no concrete proof of their merits and have watched female colleagues experience the same. While I think I am valued as a scientist, I do not think I am always treated as an intellectual equal to my male colleagues.

Diane Srivastava. I was working with a colleague in Kenya. After a full day of fieldwork, we were chatting, and he said quite bluntly that he didn't think women were as intelligent as men. My jaw fell to the ground. Here was a man in his early 30s, talking to a more senior woman, who had in fact chaired the committee that recently hired him, saying baldly that he thought her sex was less intelligent.

PREVENTED FROM DOING SOMETHING BECAUSE OF GENDER

Martha Crump. In 1993, I took two Ecuadorian students (one male, one female) to Yasuní National Park in eastern Ecuador. Maxus Energy had an oil concession; a road was to be built through the rain forest to reach the oil fields. The students and I inventoried amphibians and reptiles in an area being mapped for the road. Ultimately, we hoped to receive Maxus funding for an impact study. After our survey, I spoke with the Maxus supervisor in Quito about our proposal. He expressed little interest in our survey but discussed "The Huaorani Problem" at length. Maxus intended to carve its road to the oil fields through the territory of two hostile splinter groups of Huaorani Indians who had never been contacted by the outside world. The superintendent told me, in a patronizing tone, that until Maxus had solved "The Huaorani Problem," he would not allow me or any other woman to work in the area. He warned, "Huaoranis raid other villages for women, rape, and enslave them. Do you want that to happen to you?" His overbearing and aggressive demeanor and words were, to me, the epitome of male chauvinism. Our proposal was rejected in favor of one submitted by an all-male team.

Mercedes Foster. Getting an opportunity to go into the field when I was a graduate student in the Museum of Vertebrate Zoology at Berkeley (mid-1960s) was a challenge. Every summer the museum sent graduate students to various places to collect specimens or help with faculty research. At that time, I was the only woman graduate student at the MVZ. I wasn't allowed to go on any of these trips, or rather no one would take me. "What would we do with a woman?" "Our wives would object." I think they thought I might faint if I had to pee behind a bush! So after my OTS course, I went by myself to Nicaragua, Costa Rica, and Panama for a month.

After I received my PhD, I applied for a field biology professor/museum curator position at a major university for which I was supremely qualified but did not get an interview. As a personal friend of the person hired, I know that I had twice as many publications as he did, more grants, and both teaching and museum experience, which he did not, as well as equal field experience. I also had at least as good letters according to two people who wrote recommendations for both of us. I was told that two individuals in the department absolutely refused to consider a woman for a field position. Period.

Sue Moore. During the summer of 1978, I was part of a project involving dolphins being caught in tuna nets. I was not allowed to join the crew on the fishing vessel. Women were not permitted on the vessel because they were believed to bring bad luck (due to menstruation) to fishermen. My consolation prize: I was offered the opportunity to work on an acoustics project—with captive dolphins—at a Navy lab in San Diego.

Lynne Parenti. As a graduate student, I was not allowed to accompany AMNH ichthyologists on their field expedition to Guatemala because women in the field were considered a distraction.

Anne Pusey. In 1969, I talked with Dr Niko Tinbergen at Oxford University about doing graduate work with him. He told me, "I don't believe women can work on their own; they need to work alongside a man. But if you do well on your exams, I'll consider it." As an undergraduate at Oxford, which had links to the Serengeti Research Institute through Tinbergen and others, I envied my male colleagues who went to the institute for gap years or summer research projects. Single women were not allowed there. But then I was fortunate to make a connection with Jane Goodall, who needed an assistant and had originally hired only female assistants. So I got to Africa anyway, in 1970!

Marvalee Wake. I was married at the end of my first year in graduate school and pregnant during the second year. My major professor said I could not go with him and my graduate student cohort to do fieldwork in Mexico because I was pregnant. He eventually relented and said that if my obstetrician said "OK," I could come. My obstetrician said "no way."

Patricia Wright. When I decided to study the world's only nocturnal monkey, *Aotus*, the owl monkey, everyone said it couldn't be done. Furthermore, they said a woman should not go into the rain forest at all, but to go out all night alone was ridiculous. I also had a young child. When I requested to study owl monkeys at STRI and Barro Colorado Island, I received a very negative letter of rejection because children were not allowed at the station. To attain my goals, I had to do my first field research in Puerto Bermudez, Peru, where no field biologists had ever worked. I took Amanda to the rainforest, and my husband or local people watched her while I followed owl monkeys all night. Later, Dr John Terborgh offered to include both me and Amanda to study owl monkeys at Cocha Cashu Field Station in Manu, Peru.

NEED TO PROVE SELF

Robin Chazdon. There have been times when people questioned my dedication and serious intentions for no reason other than my gender. Fortunately, I worked with outstanding mentors and gained their respect and trust. But I always knew that they doubted me at first. There was a strong male bias in my field (plant ecophysiology) in graduate school, and I surprised everyone by making and wiring my own light sensors and using the first ever field-ready datalogger to get the first detailed measurements of light availability and sunflecks in the understory of wet tropical forests in Costa Rica. I felt I had to prove myself much more than if I had been a male student.

Maureen Donnelly. Dudes do better than women. It's a man's world. Field biology is still a man's world. The burden is on women, because they are not considered equal to men. The supposition is always that a woman is going to quit.

Dakota Keller. I am aware that my mistakes, attitudes, and work abilities are perceived through the lens of gender. If I'm having an impatient day, I know that it's often perceived that it is because I'm a woman and not just because sometimes people have impatient days. I hold myself to a high standard to make sure that if I'm being judged it is for my own individual actions and not attributed to a stereotype.

Sinlan Poo. I feel like I have had to prove my worth, but being a young, Asian female, it is usually difficult to know if people are dismissing me because of my age, race, or gender. Out of these three things, there's only one that, in time, I might know what it feels like to be on the other side of. The other two, I may never have the chance to disentangle. I've done most of my fieldwork in Southeast Asia, where people almost always defer to older, White, male biologists as higher authorities regardless of their intellectual merit. While my worth can be "proven," the worth of my White male counterparts is "given" freely and willingly.

Emily Marie Purvis. Even though the number of women in field biology has been steadily increasing, we are still viewed as less capable and less intelligent. Much of my work occurs in close collaboration with state and federal resource management agencies like the U.S. Forest Service and Calfire. These agencies, especially the wildland fire crews, are pervaded by an extreme and urgent sense of masculinity. In these spaces, women are generally looked down upon as weaker and less capable. I've been the only woman on field crews where I'm not sure I was ever taken seriously by my male peers. I'm embarrassed to admit that I often combatted this misogyny by playing along with it, by trying to prove my worth by "keeping up with the boys." I've always received the implicit message that I need to be smarter, stronger, and more hardworking than my male peers to be viewed as an equal. I feel like I've had to prove my worth at every stage of my career in a way that my male peers have not.

Donna Shaver. I am often questioned and challenged by others. Opinions of men are still taken over mine. Men with PhDs are referred to as Dr, but I am referred to as Donna. Credibility seems short-lived and I am required to prove myself time and again. When I was young I felt that I had to prove that I had the physical strength and endurance necessary to keep up with any man in the field. This harmed me physically. While working with the U.S. Geological Survey, I had no funding or staff. Sea turtles can be quite heavy. Trying to lift these by myself was difficult, and I injured my back. I have had five back surgeries to date and am in pain daily. If I could advise my young self it would be to ask someone for help, and if I could not find someone then tell the collaborators that I could not perform this aspect of the work.

SAFETY ISSUES/VULNERABILITY AS A WOMAN

Anonymous. While working for a graduate student from the United States in Latin America, I shared his small house and worked with him long days alone in the forest. I slept with my door locked because I didn't trust him not to hurt me, given his expectations for me outside of work and his severe anger issues. I am confident he would not have behaved that way to a male field assistant nor had the same expectations. It was a very stressful and, in some moments, scary situation. He was extremely possessive and expressed irritation when I chose to visit friends or my partner in the community instead of staying home to eat with him each evening. He came from an abusive family and put an emotional burden on me to gain emotional support for himself. I quit the job after 1 month because the dynamic made me so uncomfortable. At that point, he articulated what I had already known—that he had inappropriate expectations for me as a field assistant with regard to our relationship and the amount of time spent outside of work with him. I explained this to his PhD advisor, who supported me in leaving. I was disappointed to see 2 years later when I returned for my own PhD fieldwork that this student was, once again, working and living with a female research technician. While I attribute that to negligence on the part of his male advisor, who apparently did not internalize that this student's behavior was targeted at women, I still worry that other women may have had similarly negative experiences and whether there was anything more I could have done to prevent it.

Patricia Burrowes. I worked alone in the Colombian southern Andes for my master's fieldwork during the late 1980s, where I often encountered *guerilla* fighters dressed as *campesinos*. Although the encounters were frightening, after I convinced them I was just a university student studying frogs, some of the guys helped me find frogs!

Julie Denslow. I carried out my PhD fieldwork in Colombia just as the country was holding its first open elections 20 years after the end of La Violencia. Civil violence had depopulated the countryside, leaving many fallow fields that were to be my study sites. Political feelings still ran high, however, and the Colombian army was present in the countryside to maintain peace and to discourage the activities of *guerrilleros* who were beginning to coalesce into extramilitary operations. On more than one occasion I returned home to the company town to discover that the army had sent helicopters searching for a *guerillera norteamericana* seen on the road. Fortunately, I never had to prove that I was not a CIA spy.

Tiffany Doan. It is more dangerous for me to travel by myself in South America than it would be if I were a man. I have been mugged twice in Peru and robbed nonviolently more times than I can count. These experiences might have happened to any gender, but I think I looked like an easier target because I am a White woman.

Mercedes Foster. I had three scary experiences that all resolved themselves without serious results—scary because as a lone woman I wasn't sure how they might turn out. First, I was doing fieldwork in a remote place in Bolivia, a dicey area with coca growers and cocaine processing plants, where locals told me not to follow certain trails. I was a bit uneasy whenever I met strangers in the forest, but everyone was friendly. While I was there, a coup occurred in La Paz. Soldiers had been sent to protect a hydroelectric generating plant on a river near my field site. I was collecting frogs in the middle of the night, coming down the side of a mountain. When I stepped onto the road, I was met by a group of armed soldiers who had followed my light moving down through the forest toward the plant. After questioning me, they realized that I was the "weird biologist" who was in the area studying animals and not there to sabotage the plant. Second, While I was owl hunting with a shotgun in the middle of the night in northeast Paraguay near the Bolivian border, I walked out of a ravine and was met by armed soldiers. Unbeknownst to me, Anastasio Somosa (in exile in Paraguay after a coup in Nicaragua) had been assassinated in Asuncion. All the borders of the country were sealed off in a search for the killers. They took my gun and me back to their camp for questioning, but also recognized me as that strange bird lady they had heard about. Third, I flew in from Paraguay en route to my field site in Bolivia and changed planes in Santa Cruz. While crossing the tarmac to board the plane, I was forcibly pulled from the line by two policemen. They took me and my baggage downtown to an underground police facility where they went through all my stuff and suggested I was smuggling cocaine in bird specimens. After a while, they took me to a hotel and said I could leave the next day. Whew, I thought, until the colonel (or whatever he was) opened the door to my room (with his own key) and walked in to "spend the evening." Anyway, all's well that ends well.

Dakota Keller. I keep rolling around nauseating memories of places I've worked where I tolerated conditions I should not have. My biggest fears and stressors while working in the field are not bears, swift water, or rattlesnakes—but being trapped in close living quarters, without cell service, with unpredictable people. It is difficult to know when it is important to persist and when it is important to leave an unhealthy environment, but I've never quit. Pound for pound, I'm getting pretty good at taking up space and making space for the ones that come after me.

Karen Lips. There is always the need to have outstanding "situational awareness," especially in regard to humans (rather than physical or environmental threats). There is the need to be on constant alert for developing situations, unsafe conditions, and to try to identify threatening or dangerous individuals. This is most important in areas of human habitation rather than in remote field sites. I always say the most dangerous animal in the field has 2 feet, not 4 or zero.

Debra Moskovits. While doing fieldwork in China, I had set up my tent far from others because I like to experience solitude in nature. One night a Chinese worker tried to get into my tent (he seemed drunk). I grabbed my machete and yelled in English for him to leave until he did. After that I felt I had to keep safety on my radar all the time. Men don't have to deal with this sort of issue.

Diane Srivastava. Many researchers experience barriers through multiple axes of their identities. For example, I identify as a cis-woman, queer and mixed race, and my experience as a researcher is a tangle of how all these aspects of my identity intersect. There are places in the world and decades within my life when it has not been safe to be out as queer in the field, and this inability to share my personal life has only reinforced my outsider status in science (and western society) as a woman of color. Moving forward, we need to think of field safety not just in terms of wrapping sprains, but also in terms of creating a safe space for all types of diverse researchers—even when some aspects of that diversity may be hidden to others.

Judy Stone. Traveling alone as a woman in Latin America can be very difficult. Strangers are conspicuous, and a female stranger is especially so. I regulate my behavior more than I would if I were a man. I always need to know where I will stay in advance, and I try very hard to make sure that I am not traveling after dark.

Mary Jane West-Eberhard. I was working on a farm near Cali, Colombia, and I knew that people saw me alone when they passed along on a path between two poor barrios. I was robbed there and thrown to the ground, completely helpless. The two guys could have done anything, but didn't. However, I fainted for a whole hour after they left, as I could tell from my notes. The university had offered to send a driver to accompany me, but I did not want to have somebody hanging around for hours while I worked. This was poor judgement, and I abandoned that site and became more cautious as a result.

Sexual Harassment/Assault

Several women reported they had experienced instances when men touched them inappropriately. One commented that sexual harassment was rife in her workplace. Following are four vignettes.

Kristina Chyn. In the United States, there have been multiple instances at conferences where I have felt uncomfortable by content in presentations or in one case sexual assault by a more senior researcher. At the time I did not feel empowered to speak out, but as general support for victims has grown and after earning my PhD, I now feel empowered to speak more openly about the assault so that other women in the same position feel supported. The #metoo movement definitely reached field biology and has empowered many of my peers and myself to feel safer speaking out against sexual harassment and abuse.

Kate Jackson. I can think of at least three fairly serious situations of sexual harassment/assault in which I was either the victim or the main support person to the victim (e.g., as the only other woman there) during remote foreign fieldwork. In all cases the perpetrator was an expat in a position of small-scale power (e.g., over a survey or expedition) in a developing country. As an expat this person was not being governed by the laws of their home country because they were abroad but was also not being regulated or restrained by local customs/laws/standards of decency because they were a foreigner in a remote location.

Caitlin Nagorka. It is not uncommon to get "hit on" when meeting with private landowners. Some advances are more serious than others. I have felt unsafe a couple of times. One time I got called out to a place, and when I got there, the guy seemed nice enough. He invited me into his house, but then he offered me a doobie [a joint], which I refused. As we walked his property he made more and more inappropriate sexual references and innuendos. Then he got obscene and hit on me. I put an end to it on the spot and left.

Donna Shaver. I had a supervisor in the National Park Service who told me "You do not know how to have fun Donna; you should stay late after work sometime." At one point he exposed himself. He was wearing shorts and sitting in a chair facing me when he lifted his leg, put one of his feet on the desk, and it all hung out. He told me "I am going to hire someone who knows how to be nice to me." He hired someone else instead of me. Back then there were no structured programs to report harassment. It would have been his word against mine, and my career would have been sunk before it even began. So, I ignored him. Sometimes you just have to put on blinders and keep going. Fortunately, the National Park Service now has a strong program to deal with sexual harassment and someone acting this way would be severely disciplined.

BULLYING/HARASSMENT/JEALOUSY/STEALING

Lesley de Souza. When I began my doctoral program there were few women doing fieldwork in remote areas of the Amazon Basin. I was the only female on most of my expeditions. My supervisor was a tremendous supporter of me, but many other men in the field (graduate students and male faculty members in other departments) bullied me throughout my doctoral work and even into my postdoctoral work. I made efforts to incorporate women on my expeditions and support other women in their fieldwork.

Maureen Donnelly. Biology intrigued me as a career because, as a hippie, I love the communal nature of science—working together, sharing data, and communicating ideas. But before I began as an assistant professor, I had been scooped three times, each time by other women. The first time, it happened after I had my dissertation project mapped out. While participating in a scientific meeting, one of my committee members talked with a woman colleague with similar interests to see if she was planning to work on a similar question. She indicated that she did not plan to focus on that topic. Some months later, my advisor showed me a grant proposal he was sent to review. It was my dissertation idea, written by the senior woman herpetologist. I was so disillusioned I almost quit my PhD program, but I eventually re-focused my research. A few years later, the thunder of my postdoctoral research was diminished when two tenured women herpetologists published papers based on my ideas using their own minimal data after I gave a symposium presentation at a professional meeting. One woman even used one of my figures!

Margaret Lowman. A challenge I've experienced is "Tall Poppy Syndrome," a cultural term used especially in Australia, describing a behavior that undoubtedly occurs worldwide and in diverse professions. Once a woman reaches mid-career or senior status and is successful and visible, insecure peers (mostly men, but sometimes women) are threatened by her achievements and try to chop her back. The term comes from the expectation that poppies in a field together will grow to the same height. If one grows too tall, it is cut down to size.

Sue Moore. I was the only woman on our marine mammal aerial surveys in the early days, working alongside male colleagues and two male pilots in remote settings in Alaska. In 1988, when funding ran low, my male supervisor at the Navy lab accused me of "trying to take over the project" and had me evicted from the base without any means to appeal the unwarranted charges. A short time later he showed up intoxicated at the office where I worked as a contractor loudly proclaiming the same allegations. I faced him and quietly asked him to leave while walking him toward the door. Once gone, several co-workers came out of their offices to "congratulate" me on my courage, but none assisted during the event and there was no thought at the time of pressing harassment charges.

Debra Moskovits. I started in conservation biology at a time when the field was less competitive and very open to women. Starting with a low profile was advantageous, but once the program at the museum became successful, I got much pushback. I wonder how gender may have played a role. I encountered jealousy from some who felt I was drawing too much attention and resources to the program. One person advised me to slow down. I asked him, "Would you ever tell one of your graduate students to slow down, publish less?" He thought about it and said no but still maintained that I was moving too fast.

Donna Shaver. I was accused by a male coworker of achieving my goals because of bedroom politics. It was my pure hard work and determination, but he minimized how others viewed my contributions. He was jealous of my accomplishments and lied to his supervisor and others, saying that he had obtained the grants I brought in. I have had to deal with the syndrome "There can be only one woman on the mountain." During the last decade, two women I had known for many years worked to bring me down. It was pettiness, jealousy of my credentials and success, and retaliation because I reported some things they did.

APPEARANCE

Mercedes Burns. I had one research position where I realized I was always the one taking notes on procedures in the lab, bringing snacks to lab meetings for everyone to enjoy, and organizing instrumentation schedules for others. I didn't like the optics of that and started being more selfish with my time and expertise. Despite my job, I really enjoy fashion, wear makeup, and do my nails when I'm not in the field. I think there are times my image doesn't square with peoples' expectations of a scientist.

Patricia Burrowes. I am a Latin American woman who loves fashion, and *that* did not bode well in the 1980s in the Herpetology Division at the University of Kansas. I was taken as a "bimbo" until my fieldwork proved me differently. I still like fashion, and it has never kept me from being the muddiest field biologist when I must!

Jennifer Pramuk. The greatest challenge I've had is that men and women often saw me first for my physical state: a slightly built, blonde female. This was confirmed when fellow scientists would tell me as much to my face. For example, a gentleman at a professional meeting asked me during the Q&A following my presentation, "Are you single?" The negative, long-lasting effects of even a seemingly simple comment such as this in a professional venue cannot be overstated. Even more damaging were the multiple instances where male graduate students propositioned me in a professional setting. At the time, these sorts of experiences felt like a crushing blow and I felt I had to work at least twice as hard as my male colleagues to receive recognition for my intellect. As with any suffering, I think that adversities make us stronger, and in my case, they definitely made me work harder.

OTHER CHALLENGES

Anonymous. There have been times during meetings when I've offered my opinion and felt I was being ignored or overlooked. Because of this, and not wanting to be blacklisted, I constantly monitor the tone of my voice. I don't want to come off as shrill, or emotional. I'm always careful about what I say and how I say it. There seems to be a fine line between being perceived as an assertive, rational, thinking human versus being perceived as a bitch. I wonder if men feel the same way when offering their opinions.

Anonymous. When I applied for my banding and scientific collecting permits in 1986 after starting my assistant professorship, my husband was also submitting applications for the same permits. After mailing in our applications, I got a phone call from a state permit specialist who explained that he would prefer not to issue us separate permits, but that he would issue my husband the permit and I would be listed as an authorized sub-permittee on his permit. I explained that I was the primary user and that sometimes we worked in different locations, and that is why we had requested two permits. I also told him I understood that if this might save their office some extra paperwork, it would be fine if he issued the permit in my name, and my husband could be listed as the authorized sub-permittee. He replied that it would need to be the other way around. And that's the way it has been for 34 years now.

Jane Brockmann. When I started my research at our island field station, there were quite a few things I didn't know how to do, such as driving a boat. One can learn these things if trained, but the station manager did not like women working on the island alone, so he made it difficult for them. He provided no training, and if something went wrong, he found it hilarious (until I became his supervisor as Zoology Department Chair at the University of Florida).

Judith Bronstein. Within my department and university I've consistently been labeled as a "good teacher." I certainly like being thought of as a good teacher, but this phrase is particularly likely to be applied to women who take education seriously and to lead to some dismissiveness (and lower salaries) compared to professors who are highly productive but substantially more selfish with their time. I have gained University of Arizona's highest recognition for my teaching, a University Distinguished Professorship. It comes with no perks, unlike the highest university recognition for research (a Regents Professorship), which comes with a yearly slush fund. Basically, I've been rewarded for good teaching with more teaching.

Michelle Carlisle. A male colleague once told me that he didn't think we needed to pay any special attention to making sure we hired females any longer. He said, "Every time I walk into a meeting now, there is a woman there." I replied, "Every time I walk into a meeting I think, Huh, looks like I'm the only woman here again." It's all about perspective.

Maureen Donnelly. I was sexually abused, beginning as a toddler, by grandpa (who fortunately for me died before I was 6 years old). As they say about hardship and challenge, "If it doesn't kill you, it makes you f***ing stronger." I believe my imposter syndrome tendencies stem from these early traumas, which are a defining part of who I am today. I often imagine someone knocking at my door, announcing that I am done. "We're coming for your ass tomorrow." Having poser syndrome makes me exceedingly careful and cautious when gathering and interpreting data. It pushes me to be excellent in all my endeavors, including being fiercely protective and fighting for other people's rights. Don't shit on someone I care about! I am fearless. No one could be worse than grandpa.

Evelyn Gaiser. In decision-making roles, I usually have to try harder than my male colleagues to be convincing about a process or plan. One time, a thoughtfully prepared plan I had proposed was quickly dismissed in a formal meeting, and then immediately became reality when the men converged in the restroom afterward, with credit given to the lead man for the idea. (This was communicated to me by a friend who happened to be in a stall!)

Tally Hamilton. I feel that when working with landowners my knowledge and expertise gets questioned more than a man's. I've had times when I'll give a landowner technical guidance and they will disagree and question me, but when a male co-worker gives the same guidance, the landowner accepts it. I've also been in the field with male interns that I'm training, and landowners address comments or questions to them instead of me.

Claire Ike. While pursuing my master's degree in Forestry and Natural Resources at the University of Georgia, I noticed the friends and family of my male classmates easily understanding this career choice. In contrast, when I outlined my degree path to those close to me, I was met with many questions and few understood. It was disheartening to notice this polarizing difference in acceptance for the same degree, male versus female.

Sarah Lemer. The diving community (recreational or professional) used to hold a reputation of being quite sexist, and although it has changed quite a bit since my student years, it is still mostly dominated by men. On fieldtrips it sometimes translates into two types of sexisms. One is paternalistic attitudes where people think I need extra special attention and help with everything. This attitude usually fades away after a couple of days when people realize I can fend for myself. The other is the "you don't belong here" attitude. This one is more complicated, because it sometimes translates into people actively trying to make my life more difficult to prove their point. This type of attitude was difficult to deal with as a student. Fieldwork is stressful and this attitude was adding to the stress. Since my postdoc years I can't be bothered with this anymore and I speak up. The main difference is that most fieldwork I conduct now is with and for my students (all women or underrepresented minority) and I need to stand up for them.

Karen Lips. I feel there is a particular perception of a field biologist (or let's call it an explorer for those of us who were really off the grid and had to be completely self-sufficient), and it rarely includes women. I have done some pretty extreme fieldwork, living by myself in the middle of nowhere for an extended period of time, and I did some good work. I don't think most people have any idea of what was involved.

I see reference to men who are celebrated as amazing field biologists. I don't often see the same celebration of women who do the same. Possible exceptions are Drs Jane Goodall, Birute Galdikas, Dian Fossey, and Sylvia Earle.

Ellie Prepas. In high school I had no encouragement to pursue advanced education, no career counselling as an undergraduate or master's student, and much resistance to being a doctoral student by numerous professors who felt women could not handle the challenge. Fortunately, my supervisor was supportive.

Leslie Ruyle. I was told by a chief at my Peace Corps conservation project in northern Ghana how disappointed he was that a woman was sent to the site: "I know it is not how it is in the United States, but here in Africa, we know men are better than women."

Michelle Pellissier Scott. Once when I was at a board meeting of my professional society, I tried for more than 5 minutes to get a word in, but was prevented by a guy who dominated the discussion. I finally had to just butt in. Afterward, he told me that I had interrupted him and that he thought it was very inconsiderate of me. His reaction spoke to how unaware he was of his domination and that the women of the group got less air time.

Melissa Sparrow-Lien. I have been with the Department of Natural Resources for 26 years and still deal with people who don't seem to value my knowledge and experience. Because most of our customers are hunters, and most hunters are male, they don't think I know the answers. It really frustrates me when people direct questions to young male co-workers with very little experience versus asking me, when I have many years of experience on many topics.

Diane Srivastava. It is hard to fight against the tide of assumptions. Fieldwork requires a lot of very practical skills, like constructing field equipment with power tools or securing boxes on vehicles with rope—things that I had grown up doing. However, sometimes people assumed that only the men in the field crew could or should do these things. I think having a woman of color trying to jump in to do these things was even more baffling to them.

Marvalee Wake. I was my major professor's first woman graduate student, so the dynamic was a new one for the group. I had "isolation" from certain kinds of interactions, for example, some drinking and carousing, and lawn football, which was fine. But the isolation from scientific discussions and lab work was not helpful. I tried politely but persistently to take part.

CHALLENGES ASSOCIATED WITH BEING A WOMAN OF COLOR IN FIELD BIOLOGY

Kristina Chyn. In addition to the barriers I have faced as a woman, there are intersectional challenges I have faced as an Asian American woman in field biology and it's hard to tease apart these identities. There is little representation of Asian women in environmental fields, and I can count the number of women on one hand that I've met who look like me or have similar backgrounds to me in the wider ecological field. I know only one Asian (immigrant) woman who holds a faculty position in field biology. Asian Americans are underrepresented in ecology, but funding and programs targeted at minorities in ecology do not usually include Asian Americans partly because Asian Americans are not minorities in the wider biological fields.

I have found that Asian Americans are often stereotyped as non-outdoorsy and meek, traits that don't fare well in field biology. I experience imposter syndrome as a young woman of color in field biology. Often, I feel overlooked or have my ideas co-opted, or I second guess my ideas and thoughts and feel less motivated to pursue them.

Madeline Grant-Hoffman. As an African American woman, I often feel like I'm not taken seriously and have to prove I have interesting and relevant things to say. I was once asked if I was the entertainment when I walked into a professional meeting. I had to ask for an embarrassing photo of me to be removed from a national manual. It could have been just a lapse in judgement on the part of the person who submitted the photograph. I don't know. I was in an awkward position trying to identify a very small plant, and I didn't know someone was taking photos. I wonder if a photo of a male counterpart in the same position would have been used.

Anjana Parandhaman. There are very few Indian women in herpetology, let alone in field biology. I had no family members or other people who could serve as inspiration growing up. We didn't camp or hike, so I had to learn how to do these things on my own. These combined challenges made it a struggle to overcome and succeed as an Indian woman, and it can still be a struggle in academia without role models who look like me.

Nicole Smolensky. I deal with imposter syndrome as a Black woman with a PhD in ecology and evolutionary biology. The sense of being out of place and unconfident makes it challenging to communicate and participate in scientific debate, presentations, and other professional settings. Through time and experience, I've learned that silence stymies confidence and that growth and confidence develop with practice.

Diane Srivastava. I think the need to prove one's worth as a woman in field biology is even more accentuated for a woman of color. As a postdoc or assistant professor, I would go to conferences and introduce myself to some established researcher who would say, "Oh, and whose lab are you doing a master's in?" It got so bad, I had to start putting Dr before my name on my conference name tag because it was clear that everyone was assuming that a woman of color couldn't be a professor. Sometime after tenure, my hair flecked with gray, people started recognizing my name, and this stopped being such an issue.

Jessica Ware. My first day as a graduate student at Rutgers University my then-advisor (I switched labs shortly after this) told me I had to choose between having a family and having a career since I was a woman. That same year, the tech in my new lab told me I should drop out as he felt that women and Black people were not meant for science! Despite these few negative experiences, I persevered and even went on to have my children during graduate school. I felt like I had the support of my peers and that gave me strength to carry on despite the often-pervasive systemic racism and sexism I encountered.

Sidney Woodruff. For my senior thesis, I worked with another female biologist, estimating occupancy for eastern spotted skunks. It was very fun work, hiking around the Blue Ridge mountains together, setting up baited camera traps, but I thought about my safety every single moment. Our truck got stuck in the mud once, deep in the national forest. Of course, we didn't have cell phone service, and everything we tried to do to get the truck unstuck made it worse. One of the most viable options to get out safely was to hike out to a nearby rural home, knock on the door, and ask to use the phone to contact help. I was immediately terrified of the possible consequences of being in rural north Georgia and knocking on the door as a young Black woman. That would likely be the first and easiest option for a man, even solo, but I knew the female biologist and I could not choose that option.

A few years later, I was working as a Park Ranger at Yosemite National Park. I was driving home from a hike on my day off, so I was not in uniform, but I had my park radio with me. I encountered a "Bear Jam." People were walking toward the bear near the road, taking photos. I called into dispatch to ask for someone to come and direct traffic. I asked people to move on, saying that this was not an official parking area. A woman and her teenage son started cursing at me, and they blocked my truck with their vehicle. She yelled, "I've got your license plate and I will find you." She called me a bitch and the N-word. At the time I was not thinking of myself as a woman or a Black woman, instead just as a park ranger trying to defuse a potentially dangerous situation and keep the people and bear safe. I was in one of the most beautiful places I knew, with the privilege of not having to think about my identities. Then overt prejudice hit me. I still think about it. When this sort of thing happens, I feel like an outsider, like this field wasn't meant for me.

MICROAGGRESSIONS (AND SOME NOT SO MICRO)

Microaggressions matter. Inappropriate behaviors or statements, even those that don't necessarily reflect malicious intent, can be deeply offensive. The impacts of microaggressions range from hurt and insult to severe psychological damage. Microaggressions can make a person feel insecure, excluded, alienated, and discriminated against. They can induce and foster imposter syndrome. Following is a sample of microaggressions experienced by interviewees. As you read each one, imagine how you would feel if this statement had been directed to you.

- A senior professor at a university in Michigan listened to my interests and aspirations and responded by saying "Wouldn't you just be happier back at your stove?"—Joan Berish
- A former supervisor told me, "My first choice for your position was a guy, but I was told I couldn't hire another guy. You were the best female."—Becky Brathal
- In 1981, when I told a senior male colleague that I was expecting my first child, he responded, "Well, I guess we'll see *your* productivity drop."—Martha Crump
- I was yelled at publicly by a boat captain who accused me of "wasting time putting on lipstick instead of helping clean up the top deck." In reality, I was sorting samples on the lower deck while other people were available to help on the top deck.—Sarah Lemer
- In the late 1970s, a professor in the Botany Department at Duke University told a couple of female graduate students that they should stand back and let the guys get the jobs. His take was that, given that men could not bear children, it was their duty to be the bread winners and women needed to let them do it.—Lucinda McDade

- In 2003, after I was elected President of the American Society of Ichthyologists and Herpetologists (the first female ichthyologist elected President), a male ichthyologist asked me, "How did you pull that one off?"—Lynne Parenti
- A male field assistant of mine would never do any of the cooking. He didn't even know how to wash dishes. After I showed him how, he refused to do it. He argued, "No, that's what my mom does. All you need in life is a cold beer and a pretty woman. Females belong in the kitchen."—Sinlan Poo
- Sexist jokes while I was a graduate student in the early 1970s were common. Nudie ice cubes were offered by a professor in front of a class of fellow male graduate students after I had suffered a head injury in a car accident on a field trip. The men had a spring phenology event that entailed guessing when the first female sunbathers would show up on the roof of a dorm that could be viewed from the Zoology Building. I learned that male faculty members were guessing my bust size.—Susan Riechert
- I have had culturally insensitive expressions used in a room where I (a Black woman) was present. Though not necessarily directed at me, I feel an internal sting every time I hear them: "Gosh, you're being a real slave driver!" and "I don't mean to crack the whip, but you really need to get this done ASAP."—Nicole Smolensky
- At the Cornell Lab of Ornithology, the director and several staff members were strongly male chauvinistic. At a staff recognition party, the director was honoring my 5- or 10-year service, and said, "I don't know how Sandy managed to get that NIH grant!" I was insulted. I wrote a brilliant grant proposal for a very interesting project, and I understood the NIH review process.—Sandra Vehrencamp
- I overheard the "She'll never finish now" conversation when members of my graduate student cohort learned I was pregnant during my second year.—Marvalee Wake

POSITIVE CHANGE

The times, they are a-changing. In 1964, one of us (MLC) was told that women did not become wildlife biologists. Now, in 2021, many of the women field biologists interviewed here have become successful wildlife biologists—and they love their careers. When Jane Brockmann was in grade school in the 1950s, her father told her that she could not be a veterinarian because there were no vet schools that accepted women. Now the majority of vet students are women. Not only have women been given expanded "permission" to enter male-dominated professions, but also men's attitudes toward women in science have improved. Many interviewees described positive changes they have seen during their careers in field biology.

Julie Denslow. These days women seem to dominate the applicant pools to graduate programs, field courses, and faculty positions, so clearly our programs are producing well-trained, enthusiastic, creative scientists, and teachers. I do wonder, however, what is happening to the men. Perhaps the tropics and field research have lost the allure of adventure they had years ago and men are seeking other challenges. It would be good to know. Science profits by a diversity of viewpoints.

Tiffany Doan. The accepted attitudes of men are slowly changing. In my experience men have said crude things, touched improperly, etc. so much my entire life that it was "normal" and expected. The only thing to do was to laugh it off as best as possible. Now there is at least some awareness that those behaviors are not acceptable. We should give a lot of credit to the younger generations who finally said no, this is not OK. The culture is changing.

Mary Jade Farruggia. My professional career has been relatively short, but I have seen deliberate efforts to change peoples' perceptions of what it means to be a scientist and a woman in field biology. The "me too" movement, along with other high-profile events such as the women's march in 2017, helped bring more conversations about women in science to the table, and it became more acceptable to talk about our collective experiences. As someone who has faced harassment in science, I was initially afraid to speak about my experience. However, when I did, almost every person I opened up to responded with a story of their own or that of a friend. I think movements like these have helped women feel empowered to share their stories about discrimination and harassment and have brought to light how common it is, particularly in field situations. The unfortunate abundance of stories from women in science about discrimination and harassment seems to have helped people think about how they might be contributing. That said, I think much of the empowerment and advancement that came from this particular wave has primarily benefitted White women. Field biology (and science in general) still has some work to do to improve the attitudes toward people of color, and particularly women of color.

Rosemary Grant. When we first arrived in Princeton in 1986, the faculty in the EEB department was 100% White male. Today about 50% are women, mostly hired in the last 10 years. These young women are without exception confident, well-liked, and excellent teachers and field researchers. It is wonderful to see such a change in 30 years.

Lucinda McDade. It is so much better now that there are women in science—successful women—who have lives that are as full as they want them to be, including the personal side of things. When I was a graduate student, the basic assumption was that developing a personal life stood a very good chance of, if not ending, then definitely slowing down or reducing a woman's prospects as a career scientist. That's not the case now.

Michele Nishiguchi. I think men of this generation have been taught how important women are in science and how much they contribute to the understanding of how organisms live and evolve. There is less sexism in this younger group of scientists.

Diane Srivastava. I have seen positive changes in my acceptance in field biology over the years, although I cannot separate the experience of being a woman in field biology from that of being a queer woman of color. It's not like there is a straight White version of me that can address the change, but on all those levels it has become easier. Some of that is just me becoming more established in my career, becoming tenured, and receiving recognitions. And some is a genuine change in society, although it feels like change in gender equality has been glacial at times.

Having worked in a number of countries, I am also deeply aware that the pace of change and the obstacles faced are very different around the world.

Marvalee Wake. I have seen extensive changes in attitudes toward women in field biology during my 60 years of involvement, from deliberate exclusion in the past to complete acceptance now. As the numbers of women in the field increase and women demonstrate their capacities for hard work and excellent ideas, collegiality and acceptance have dramatically increased.

Jessica Ware. When I applied to graduate school in 2003 I fretted about what to wear. Would a skirt imply I couldn't do field biology? I was afraid of appearing too feminine. I remember talking with my fellow female graduate students about whether we would be taken seriously based on how we behaved or what we wore. I was told that I had to choose between family and a career. I hope that women today are not given the same poor advice, and I am pleased to see science embracing diversity in a way that differs from my graduate school experience.

Patricia Wright. When I started, few women participated in field biology. Jane Goodall paved the way to the possibility of watching wild primates, and National Geographic Society was instrumental in changing the public image of women in the field. I spent decades in rain forests with men and very few women. Now more than half of my graduate students are women and there is no longer any doubt that women are capable and effective field biologists.

INCREASING DIVERSITY AND INCLUSIVITY

Several women of color offered insight concerning ways we can support and increase diversity in science and in field biology.

Madeline Grant-Hoffman. Internships and other types of positions that give people experience are critical in order to enter the job market with an advantage. We need to make these opportunities more accessible to anyone who is interested, including women of color. Everyone needs to feed themselves, so young women whose parents cannot support them during volunteer positions often are left out of these valuable opportunities. We need to think about financial barriers and figure out ways that everyone, including women of color, can afford to participate in volunteer/low-paying internships to gain experience.

Priya Nanjappa. Although I didn't notice it as much in the early years of my academic pursuit, the more I got into the wildlife and conservation emphases and jobs, the more I was told about, and the more I noticed, the ways in which I differed from the typical field biology or wildlife conservation archetype. By that, I mean: Not from an outdoorsy family. Not having grown up with hunting and fishing. Not having been camping until college. Not having the right clothing and gear. Not hardy enough for outdoor work (file this under "frequent misperceptions and underestimation of ability or potential"). Not White.

I didn't fully realize how my own identity differed from how others perceived me. I didn't think of myself as not White. My upbringing was in a mostly White, suburban, and Midwestern region of the country, and my desire to fit in resulted in denial of my cultural heritage for many years. But others saw me as different and let me know in both subtle and overt ways. My experience is unique and yet relatable to other people of color who grew up in similar surroundings. The other aspects of being different from the typical field biologist are many of the reasons that wildlife conservation, as a field, has tended to be exclusive.

Much of our conservation history, including philosophy espoused by our field biology heroes—Muir, Thoreau, even Leopold—is problematic. Core to our heroes' views is prizing a land ethic of "wide open spaces" where one can't see a person for miles. But this country was not uninhabited when White men began to explore it. In nearly all of our open spaces, the violent removal of Indigenous peoples who inhabited and stewarded the land was a preceding event. In some parts of the country, some of those same spaces cleared of Indigenous peoples made way for large agricultural operations that exploited slave labor. Furthermore, the occupation of the Western United States was facilitated by the Homestead Act of 1862, which granted 160 acres to citizens who were heads of households and who could afford the filing fee. Most people of color (particularly southern Blacks and Native Americans) were not universally considered citizens at the time of that act. Blacks, once freed and recognized as citizens, had few resources to succeed with land grants; therefore, very few benefitted from these privileges.

Many of those wide-open spaces where we often conduct our field biology studies are battlegrounds, burial grounds: places of loss. The violence is not limited to people of color. People who identify as LGBTQ+ have also been harmed in forests and open fields. Yet, few professors currently teach this in their wildlife conservation courses, despite this history being immensely useful in understanding the current state of the landscape, literally and figuratively. Too many of us do not acknowledge the ways in which our efforts to gain ecological knowledge contributed to the erasure of Indigenous and Black traditional ecological knowledge.

The very reasons that people of color are less comfortable in outdoor spaces or that some families don't take their children into the woods is related to these histories and the fears that have been perpetuated (and that are still very real). It is related to the ways those histories contributed to fewer advantages, lower income and fewer opportunities, and thereby exclusion of future generations. An exception exists with Asian Americans. We, myself included, benefitted from intentional policies that allowed for "high-achieving" immigrants to come to America. As such, STEM fields are not lacking for Asian American representation. Field biology and conservation, however, are.

That said, field biology has become far more diverse in the ~28 years since I started my training. Yet, conservation agencies and organizations remain predominantly White. Some reasons include: (1) A lack of leadership directly engaging in evaluating those barriers that they can influence and change; (2) the use of metrics that tend to prioritize candidates who are "great on paper" and have the "right pedigree" instead of seeking ways to create opportunities that catapult someone's excellence simply by giving them a chance; (3) consideration of those who are "a good fit" without examining and understanding how these more likely reflect people who are like us and our unconscious biases (e.g., valuing particular experiences of hunting, camping, outdoor exploration), over our mutual admiration and passion for the natural world; and (4) unwillingness to embrace the uncomfortable histories as part of acknowledging past wrongs and fostering inclusive conversations and spaces that welcome unique experiences and approaches, enriching our collective knowledges.

Imperative in making conservation more representative of our diversity of identities as a nation is acknowledging that these efforts cannot be handled simply by a Diversity Committee. These efforts will require sustained attention, learning, and working to understand and appreciate how past decisions have systematically affected our field's demographics today. I have been fortunate to have been put in leadership roles where I can more directly make change, and I am driven to do so, as a member of an underrepresented group. Although people today are not at fault for the impacts of past decisions, the predominantly White leadership within the conservation community has the ability, influence, and responsibility to make changes in internal policies and practices to right these past wrongs.

Sinlan Poo. Reducing discrimination should not be the responsibility or burden of the minority being discriminated against. Despite the privilege and advantages I've been given, I have never felt secure enough to confront these issues directly, without fearing negative consequences. This says something about the system in general.

Diane Srivastava. To increase diversity and inclusivity in science, we cannot be passive in recruiting undergraduates in research. Bolder students are more likely to become involved because they take initiative to seek out opportunities. It shouldn't be this way. Positions should be openly advertised so that the less bold students know about opportunities and say, "Oh, this is something I can apply for." We also need to have a funding system that is based on more than grades. The NSERC student fellowships in Canada are based on grades and it's general knowledge that one must have an A-average. This gives an advantage to privileged students who don't have to work two jobs to make ends meet. Many students struggle with financial issues that affect their ability to maintain a straight-A grade average.

Jessica Ware. One need not look far to find excellent researchers in field biology who are also women of color. At this point, to not have diversity in a seminar series or a field research team is a choice. There is ample evidence that there are women of color who are interested in nature, who are presenting ground-breaking research in their fields, and yet we sometimes hear the specious claim that a lack of diversity in academia is due to a small talent pool. This is simply untrue, and thanks to social media it is even easier to discover that there is truly a wealth of scientists of color out there.

"WORDS OF WISDOM" FOR THE NEXT GENERATION

One of the most frequently suggested bits of advice for the next generation of women field biologists offered by interviewees was a version of "Surround yourself with supportive women, and then support each other all the way." A second theme expressed by many was "Follow your passion."

- Take risks. Follow your enthusiasm, because you will be both more productive and creative when you are working on something you are passionate about.—Briana Abrahms

- The times are past when you can do just fieldwork. It's important to combine fieldwork with other components of your research toolkit, for example, lab work and modeling. Form close collaborations with other scientists.—Jeanne Altmann
- You have to really want to do fieldwork because you will face discrimination. Women are not equal to men, Black people are not equal to White, etc. What is important is how much you want to pursue your passion. You can't have it all, and there will always be trade-offs. You have only one life, so think carefully about how to use it.—Dee Boersma
- Don't be afraid to be you. Wear what you want, earrings and all. I was told too many times that I needed to be less of myself in order to go anywhere. "You need to turn the 'Becky' down." Being authentic will get you farther than hiding who you are. Your gut is right! Be aware of who you are with and where you are. It's okay to speak up. You are capable of doing more than you think and you should go for it! We are here to support you!—Becky Brathal
- Do what you love and do what you are good at.—Jane Brockmann
- I recently advised a promising limited term employee that she should take advantage of an opportunity to get certified on a heavy dozer; she was going to pass it up since it was not a skill she would use in the future. I told her with that on her resume, she would dispel preconceived notions about her abilities. It is best to leave nothing to chance when competing in what is a historically male-dominated career field.—Michelle Carlisle
- The best advice I've heard that I remind myself of when feeling imposter syndrome is something along the lines of "do not change yourself for science, change science to fit you." The sentiment has stuck with me and has been empowering when I feel out of place.—Kristina Chyn
- Go for it. Don't let anyone tell you that you can't do it. But be prepared. Try to imagine every possible scenario of what could go wrong and plan for how you would deal with it. When you go to a new field site, get to know the local people and explain what you are doing. People gossip, they have heard about the strange woman in the area, so confront it head-on and make yourself known. If possible, speak the local language. In a non-U.S. country, visit the embassy; sometimes they will give you a letter of introduction.—Mercedes Foster
- Field biology is an area where women can excel. Learn your system of choice. As others have said before, if you know one taxon and one region well, you begin to see patterns in the rest of the world. I think that women can often use the ability to develop a deep understanding of a system to their advantage.—Rosemary Gillespie
- Do it! It will be fun, challenging, exciting, scary, and amazing. Be yourself. Don't cave to the pressure to fit in or be like everyone else. You need to develop your own voice and stick up for yourself. Having sample language for responding to discrimination on the spot is really helpful. What will you say when men talk over you or cut you off? What will you do when you are the butt of jokes in the field? Be ready so you don't regret what you did or

didn't do. Find other strong women and build a network, support each other. Lean in. Climb the ladder.—Erica Hoaglund

- Don't wait for your circumstances to be perfect to get started. Do what you can now with what you have where you are. Play a long game: rest when you need to. The timeline that makes sense for you may look very different from the timeline we are used to expecting in this (traditionally male) field.—Kate Jackson
- Don't ask permission. You want to be the first to check traps, hold animals, and drive the truck? Then go get the keys. Enthusiasm and initiative are sexy. Being too shy and polite to say what you want is a waste of everyone's time, but most importantly your own.—Dakota Keller
- Embrace who you are. You know what's right, do it!—Gwen Kolb
- More than ever we need smart, dedicated, adventurous types to get out and collect data on organisms in the field. Be strategic in what you do. Try to think in advance of where you'd like to end up, what work you'd like to do, and how you'd like to live. Choose the projects, collaborators, and topics that will have the greatest chance of providing you with the results and experiences to get there. Think big and push hard early on so that you keep as many options open as possible. It takes a village: we need to band together and help each other out. Take advantage and ask for help and guidance from the Old Girls' Club.—Karen Lips
- Follow your passion. Find a species or ecosystem that really resonates with your love of nature, and enjoy the ride of exploration and discovery.—Sue Moore
- Get experience doing different types of research. My most valuable experiences were finding out what I didn't want to study. Look for other strong female biologists and connect with them, ask questions about their experience, collaborate with them.—Jessica Murray
- Our field is still full of "good old boys" and some young ones that act the same way. Attitudes are shifting, and I have plenty of male colleagues who are first in line to correct those that are out of line. Our job is to do good science, and call out what is wrong, but also to roll with the punches if it is ignorance and not malice driving the behavior.—Erin Muths
- Be confident in your abilities and potential to lead others. Seek opportunities for experience (internships, volunteer days, job shadows, and reverse interviews). These experiences will give you confidence. Be willing to learn from others (especially Mother Nature) and keep an open mind. Be aggressive when seeking jobs, and do your homework when applying for positions. Connections, connections, connections.—Caitlin Nagorka
- Learn to detach your emotions from negative experiences, whether they occur in your personal or professional life. While mastering this, however, speak up when someone makes you or someone else feel unworthy, bullied, or denigrated. The status quo will remain if we don't act on injustice.—Jennifer Pramuk
- I think it is important to have a female support system but equally important to develop self-efficacy to overcome those times of perceived impasse because of someone or something. Women should unabashedly make

personal and career choices on their own timing and terms. It is important to cultivate creative thinking regarding how garnered skills can be transferred across situations and careers and to seize opportunities for growth and innovation in unexpected places.—Nicole Smolensky

- When I was a new Assistant Professor, I gave my female students the following advice in an attempt to make them fit into the mold of what we typically consider to be a successful herpetologist: try to speak with confidence (deep voice, be loud, avoid inflections at the end of sentences), don't show tattoos or dress in a revealing way, watch out about sharing your ideas but be outspoken in seeking collaboration. More recently, however, I have considered the fact that at least some of this advice is trying to mold them to be more masculine, which then continues to uphold the unfair and male-biased standards in our field. Now, I actively try to avoid messaging about these male-biased standards. If we ever want people to be treated fairly no matter their gender, then it is important that people feel that they can be professional with whatever types of gender expression, clothing, and other attributes they choose, so long as they behave professionally.—Emily Taylor
- Go for it! Expand your interests, sample broadly, be open to discussion, be well prepared and enthusiastic—all to find ways to satisfy and expand your interests.—Marvalee Wake
- Choose a group of organisms that interest you and become the world expert on it, learning all that you can about every aspect of their lives. That is the surest way to become regarded as an expert and it will get you invited to meetings where that group is of special interest—especially on the topic that you have adopted as your own. Observe them in a natural setting for a good amount of time even if you end up doing lab work. What you focus on should depend on what you have patience for. If you don't have patience for sitting and just watching animals behave, then don't work on behavior. If you like collecting, then focus on collecting specimens and studying their variation. If you have patience for getting lab apparatus to work, then do lab work instead. If you aren't passionate about something or for some reason motivated to tackle it as a problem, then don't do it.—Mary Jane West-Eberhard

We would be remiss not to give a booming shout-out to the many men who have long accepted and respected women field biologists as equals. Many interviewees emphasized the positive support and encouragement they received from their male mentors and role models, men who profoundly influenced the women's development as scientists and as field biologists. In the case of one of us (MLC), men offered amazing opportunities during a time when male field herpetologists far outnumbered female field herpetologists. Early opportunities included amphibian and reptile survey work at a remote site in eastern Ecuador in 1968; field assistant on a bird ecology study with conservation biologist Tom Lovejoy in Belém, Brazil, in 1969; and biodiversity survey work in a region to be established as the Yasuní National Park in Ecuador in 1977. Many of the more senior interviewees had few female role models in field biology. We have many men to thank for teaching, guiding, and supporting us, free of gender bias.

Section 3

Looking Toward the Future

9 Ongoing Challenges and Moving Forward

Imagine you are a female undergraduate science student. You are in your senior year, have research experience, and have been accepted to graduate school. When a position as a science laboratory manager opens up in your department, you apply. A male student, also a senior, with virtually the same credentials is offered the position instead of you. Should you be surprised?

Unfortunately, no, as reflected by a study published in 2012. One hundred and twenty-seven faculty members recruited from biology, chemistry, and physics departments at six research-intensive U.S. universities participated in a randomized, double-blind study.[1] Each was given the application materials of a fictitious student for a laboratory manager position. One set of application materials was used for all participants, the only difference being the applicant's name, randomly assigned either male (John; $n = 63$) or female (Jennifer; $n = 64$). To create enough ambiguity to allow for potentially biased responses, the student's application suggested that he/she was likely to succeed in science (had coauthored a publication after 2 years of research experience) but was not over-the-top impressive. Using validated scales, each of the 127 participants rated their applicant's competence and their own likelihood of hiring the student; selected an annual salary for the applicant; and indicated how much career mentoring they would provide for the student. Because all information on the application other than the student's name was held constant, any differences were attributable to the gender of the applicant.

The results are disturbing. The faculty participants rated the male student as significantly more competent and hirable than the female student. They selected a significantly higher salary for the male applicant ($30,238.10) as compared to the female applicant ($26,507.94), and they indicated they would offer significantly more career mentoring to the male applicant. Male and female faculty were equally likely to exhibit bias against the female applicant. This study was published 10 years prior to the time of our writing. It would be informative to repeat the study with the same design. With the greater awareness of, discussion about, and attempts to eliminate gender discrimination in academia, one would hope that the results would reveal less bias against women.

Women are less likely than men to enroll in STEM (science, technology, engineering, and math) majors, more likely to leave STEM fields, and less likely to be hired in STEM jobs.[2] Women who do acquire jobs in STEM fields are slower to get promoted than men. These facts have fueled the "leaky pipeline" metaphor, which describes the attrition of women from the STEM fields after they have begun train-

---

[1] Moss-Racusin et al. (2012).
[2] References cited in Cech and Blair-Loy (2019).

DOI: 10.1201/9781003311508-12

ing. The study of university faculty members referred to above reveals that women undergraduate science majors experience gender bias at early stages of their careers. Female undergraduates face gender bias not only from faculty but also from peers (see "prove-it-again" under "Gender Bias in Science" in the next section).

The general public still harbors conscious or unconscious biases and stereotypes against women scientists. Appearance of increasing femininity decreases the perception of being a scientist.[3] Children's science books feature males as scientists three times as frequently as females.[4] Researchers have been studying children's drawings of scientists for the past five decades.[5] In these "Draw-A-Scientist" studies, children are asked to draw a scientist and then the facilitator records the gender of the person in the picture. Between 1966 and 1977, less than 1% of the drawings made by children were of female scientists (28 of 5,000 drawings); all 28 were drawn by girls. Since that time, 78 additional experiments have been conducted, involving over 20,000 students from grades K-12. By 1985, 33% of the drawings made by girls were of female scientists. By 2016, that percentage had increased to 58%. That's the good news. The bad news is that nearly 90% of the boys included in the 78 experiments still draw pictures of male scientists. The other bad news is that both boys and girls are more likely to draw a male scientist as they get older. If scientists and society at large work to increase the visibility of female scientists beyond the handful that most people know about (e.g., Rachel Carson, Marie Curie, Dian Fossey, Rosalind Franklin, Jane Goodall, Margaret Mead, and Barbara McClintock), more children will depict women scientists in the next set of "Draw-A-Scientist" experiments.

Textbooks shape readers' perceptions of who scientists have been, who they are, and who they can be through the scientists' work they highlight. For this reason, books provide opportunities to show students the gender and racial diversity of scientists that actually exists. An examination of binary gender, race, and year of published work of 1,107 scientists highlighted in introductory biology textbooks revealed that we need to do much better at including diversity.[6] The scientists most commonly featured in the seven contemporary textbooks examined were White men—not unexpected, because for centuries White men dominated science. The percentage of women and scientists of color whose work was represented in the textbooks increased in more recent decades. However, through extrapolation, the authors estimated that if citation rates for various demographic groups increase at the current rate, it would take many years for representation to reflect the makeup of biology students using the textbooks: for women scientists = 18 years; Hispanic/Latinx scientists = 30 years; Asian scientists = 50 years; Black/African American scientists = nearly 500 years into the future. Clearly, the potential consequences of these estimates are disastrous; writers of textbooks must become more inclusive.

Although women have come a long way, they still face challenges in becoming accepted as scientists equal to men. Women still contend with the view that they don't belong in science, which makes them feel they are intruders in the "boys' club."

---

[3] Banchefsky et al. (2016).
[4] Caldwell and Wilbraham (2018).
[5] Miller et al. (2018), Terada (2019).
[6] Wood et al. (2020).

Women are told "You were only hired because we needed a woman." In some fields there is still a stigma against hiring a woman because "she's just going to get pregnant, take maternity leave, and eventually leave the job." Women who are already mothers often face a penalty once on the job market.[7]

Women scientists are targets of sexist jokes. Consider the infamous remarks made by British biochemist and molecular physiologist Sir Tim Hunt, Nobel Laureate and Fellow of the Royal Society, at the World Conference of Scientists in Seoul, South Korea in 2015: "Let me tell you about my trouble with girls ... Three things happen when they are in the lab: You fall in love with them, they fall in love with you, and when you criticize them, they cry."[8] Hunt was both describing and defending the sort of behavior Louis Leakey exhibited toward Jane Goodall (Chapter 5) and Dian Fossey (Chapter 6). Hunt went on to suggest that male and female scientists should work in separate labs because they distract each other. He later said that his remarks were meant to be lighthearted but that he stood by them. Comments like these are damaging to both men and women scientists. But, the good news is that times are changing and comments such as Hunt's are no longer tolerated by a large segment of society. Because of the strong backlash, voiced by men and women alike, Hunt resigned from University College London and was forced to step down from the European Research Council for his joke.

Before focusing on specific challenges that women in field biology face, let's look briefly at various aspects of gender bias, which affect all women scientists.

GENDER BIAS IN SCIENCE

We need to face the reality that, in general, male scientists are seen as having more potential than female scientists and thus afforded more opportunities than women. Women's scientific achievements often receive less recognition than those of their male counterparts.[9] Biases favor male scientists in hiring, salary, start-up funds, authorship of papers, highly supportive letters of recommendation, and invitations to give talks and speak on conference panels. Chapter 8 reflects the joy and satisfaction the interviewees experience as field biologists—they love what they do. But along with the rewards, the interviewees representing a range of age and experience, from younger to more senior individuals, shared frustrations and challenges they have experienced as women scientists. These are not isolated complaints. Rather, they reflect the published literature on gender bias that has repeatedly documented distinct patterns affecting women.[10]

PROVE-IT-AGAIN

Women feel they must provide more evidence of competence than men to be viewed as equally competent, stereotyping that reflects the perceived lack of fit between

---

[7] e.g., Correll et al. (2007).

[8] https://www.nytimes.com/2015/06/12/world/europe/tim-hunt-nobel-laureate-resigns-sexist-women-female-scientists.html; accessed 11 June 2020.

[9] Débarre et al. (2018).

[10] Reviewed in Williams et al. (2016).

being a woman and being a scientist.[11] Women scientists report a range of "prove-it-again" behaviors. Mistakes women make tend to be noticed more and remembered longer than those made by men.[12] A woman's success is often attributed to "luck;" a man's to "skill."[13] Women feel their successes often are discounted. Objectivity rules are applied more rigorously to women, who feel they are hyper-scrutinized when something goes wrong. Women feel they are under pressure to perform perfectly. Men dominate meetings; women often feel it is difficult to be heard and recognized. Women frequently mention that when they suggest an idea, for example, during a meeting, it falls on deaf ears. Later in the meeting when a man offers the same suggestion, it's "Wow, what a great idea." Female professors find they must prove their competence to students, more so than do male professors. Female graduate teaching assistants report that their male counterparts receive more respect from undergraduate students. Male scientists are more likely to be assumed to be competent; women must prove their worth.

This "prove-it-again" pattern of gender bias begins early and occurs in the social environment women experience in the science classroom. In a study of undergraduate students in a large introductory biology course at the University of Washington, participants were asked to anonymously list class peers who they felt were "strong in their understanding of classroom material."[14] The surveys were conducted after the first, second, and third exams for each of the three iterations of the course, each of which was majority female (55%–58%). Most students received very few nominations, and all of the three or four most nominated students for being knowledgeable about the course material in each of the six classes were males. Outspoken females with extremely high exam scores failed to reach the same "celebrity" status as their male classmates. Even when controlling for performance and outspokenness, the bias in nominations toward males was due to males over-nominating their male peers; females nominated males and females equally.

Imagine how this bias in the learning environment of a classroom affects a young woman's self-confidence and reinforces the assumption that men make better scientists than women. Belief in one's own abilities strongly influences motivation, achievement, and career trajectory.[15] When we realize that this type of bias exists during the early career stages for women, we can understand why women scientists feel the need to prove-it-again ... and again.

THE TIGHTROPE

Prescriptive gender bias results in women often feeling like they walk a tightrope between being seen as too feminine to be competent or too masculine to be likable.[16] Science often is seen as a masculine field, yet women are expected to

---

[11] e.g., Moss-Racusin et al. (2012).
[12] Grunspan et al. (2016).
[13] Swim and Sanna (1996).
[14] Grunspan et al. (2016).
[15] Reviewed in Usher and Pajares (2008).
[16] Fiske et al. (1999).

be feminine and sometimes even subordinate. If women come across as too feminine, however, they are viewed as lesser scientists. Women experience pressure to assume traditionally feminine roles such as "den mother" (providing emotional support and counseling to students who don't ask it from their male professors or advisors) or "dutiful daughter" (always willing to help). Women often are expected to perform more service work than men. A male scientist can dress any way he wants, including fitting the stereotype of a disheveled, absent-minded professor. Women struggle with how to present themselves. Many wear non-feminine clothes at work—dress like a man—to gain respect, because they have been or have seen others ridiculed for wearing flowery dresses. Many avoid revealing or tight clothes or large earrings that might distract, for fear of not being taken seriously. Women scientists walk a fine line between being viewed by their male colleagues as assertive (good) versus bitchy (bad). Studies have shown that when a man expresses anger, his perceived status is increased; when a woman expresses anger, her perceived status is decreased.[17]

And then there is the issue of praising oneself. Most women have watched and/or felt this gender difference. Self-promotion is seen as appropriate for men, but distasteful in women, who are stereotypically supposed to be modest.[18] As children, women are taught not to boast about their achievements. Rather, they are socialized to be nurturing and communally oriented. In contrast, men are socialized to be authoritative, dominant, and to speak well of themselves. The result is that men are more likely than women to emphasize the importance of their work, including using social media to draw attention to it. A 2019 study analyzed over 100,000 clinical research articles and more than six million published papers in the life sciences between 2002 and 2007 indexed in PubMed to examine for gender differences in the positive framing of research results in titles and abstracts.[19] The authors focused on 25 words that past research has identified as distinctly positive and that are used frequently in life sciences articles, words such as "novel," "excellent," "robust," "unprecedented," and "unique." Articles in which both the first and last authors were females were significantly less likely to use these positive words to describe research results as compared to papers in which the first and last authors were males; differences were greatest in the high-impact journals. This difference matters, as more positive presentations draw greater readership and are associated with more citations, a factor used in hiring, promotion, and research funding. The authors found no evidence that the male scientists' use of flashy descriptors stemmed from their science being more novel or innovative.

Self-promotion increases the perception of competence for women, but at the cost of social rejection for violating gender prescriptive behavior to be modest. On the other hand, women scientists pay a price for not self-promoting, in terms of respect and recognition of their accomplishments and ultimately career advancement. The dilemma is a classic Catch-22: damned if they do and damned if they don't.

---

[17] e.g., Brescoll and Uhlmann (2008).

[18] e.g., Rudman (1998).

[19] Lerchenmueller et al. (2019).

THE MATERNAL WALL

Motherhood triggers a form of gender bias. Men often assume that women become less engaged in their careers and less productive after they have children. In fact, the opposite may be true. Childless women often claim that their colleagues who are mothers seem more organized, focused, efficient, and productive than they are because every hour available for their career is precious.[20] Ironically, women who maintain their productivity after parenthood are viewed with suspicion—are they good mothers? Women often feel the need to prove they are high-achieving scientists *and* nurturing mothers. The maternal wall bias affects women without children in a different way. They may be viewed as somehow incomplete, without a personal life outside of work, and as such are expected to assume the brunt of entertaining, such as hosting socials and taking seminar speakers to dinner. In some professions, women without children are expected to work shifts that are not ideal for women with families, e.g., evenings, weekends, and holidays.

Parenthood often affects the careers of men and women differently, because many women still shoulder a disproportionate share of caregiving responsibilities. Many women feel they cannot be full-time scientists and mothers. A study of the career trajectories of full-time professionals in STEM fields after the birth or adoption of their first child revealed that parenthood is an important driver of gender imbalance.[21] In the study, which followed 841 professionals who had a first child and 3,365 who remained childless from 2003 to 2010, 43% of new mothers, compared to 23% of new fathers, left full-time STEM employment after their first child. These rates were significantly higher than the attrition rates for men and women without children. Women (and men) should have the choice of leaving their jobs or working part-time to attain preferred work-family balance. But too often women leave science careers not by their own choice after becoming mothers but because of discrimination from colleagues and administrators.

A classic example of the expectation that women should assume the childcare role came with the COVID-19 pandemic that began in 2020. The closing of daycare centers and schools created a disadvantage for many women scientists who ended up with the responsibility of childcare, including home schooling.[22] This was especially true for women whose partners were able to continue working outside of the home or who had structured jobs with scheduled online meetings. For many women scientist-mothers, research was the first activity to suffer when time was limited. Women found it difficult to continue collaborations, submit manuscripts to journals, and write grant proposals. Academics worry that the social repercussions of the coronavirus skewed a playing field that was not level in the first place and that women were more strongly disadvantaged than their male colleagues, especially those men for whom the coronavirus situation afforded them *more* time to write and submit manuscripts and grant proposals. If women took more of a professional hit than their male colleagues during the lockdown, how will this disparity affect tenure, promotion, and

---

[20] For example, Williams et al. (2016).
[21] Cech and Blair-Loy (2019).
[22] Fazackerley (2020), Kramer (2020), Malisch et al. (2020).

other forms of career advancement for women scientists over the coming decade? How will it affect the makeup of academic faculties in the near future?

It is relevant to point out that some women scientists are competing for recognition and salary raises with male colleagues who have wives who do not work outside the home and who perform most of the childcare duties, cooking, and housecleaning. Society fosters a double standard. Men are praised for spending time with their children ("That's fine to leave the meeting early. It's terrific that you pick up your kids from daycare."); it is expected of women ("I understand you need to leave the meeting early to pick up your kids from daycare."). Men generally pay less of a professional price than women for having children.

TUG-OF-WAR

As if women didn't have enough challenges proving themselves to men, women sometimes experience bias from other women in male-dominated fields, including science.[23] Studies have shown that women who have experienced gender discrimination during their careers commonly stereotype, criticize, and distance themselves from their female colleagues, especially in environments that exhibit gender bias.[24]

Various types of tug-of-war behaviors can occur. Women sometimes judge each other for failing to achieve the "proper" balance between presenting themselves as too feminine or too masculine. Rivalry and misunderstandings occur between mothers and non-mothers regarding the level of commitment required to be a successful scientist. Mothers feel judged for being mothers, and non-mothers feel judged for having chosen not to become mothers. Another problem results from tokenism, where women compete with each other for one "woman's spot." Some women who have struggled to get to where they are in a male-dominated environment expect other women to struggle also. Men are not alone in stereotyping women.

ISOLATION

A fifth pattern of gender bias, termed "isolation," has recently been identified by Williams and colleagues[25] through interviews with 60 women scientists of color (20 Black women, 20 Latinas, and 20 Asian-Americans; most of the interviewees were professors). The conversations often reflected a sense of "bleak isolation," because of negative racial stereotypes and discrimination. Many interviewees expressed a sense of exclusion, such as not being invited to social events on the assumption that because they would be the only minority present they might feel uncomfortable. Some women respondents said they do not socialize with colleagues after hours, particularly with men, afraid that doing so might negatively affect perceptions of their competence. Many women of color indicated they do not discuss their personal lives with work colleagues, out of fear of criticism and backlash.

---

[23] For example, Moss-Racusin et al. (2012).
[24] Reviewed in Williams et al. (2016).
[25] Williams et al. (2016).

Racial and ethnic minorities face discrimination, including verbal harassment, in the workplace.[26] Minority women face double jeopardy—discrimination both as women and as minorities. In a survey of 238 respondents (35 minority men, 88 minority women, 46 White men, and 69 White women) from five ethnically diverse organizations, 47% reported experiencing at least one episode of harassment within the previous 2 years; 23% had experienced at least one episode of ethnic harassment; and 38% had experienced at least one episode of sexual harassment.[27] Minority women experienced significantly more overall harassment (a combination of ethnic and sexual harassment) than did minority men, White women, or White men.

ADDITIONAL CHALLENGES FACED BY WOMEN FIELD BIOLOGISTS

Fieldwork comes with its own set of challenges for women. Consider hormones. Imagine being a reproductive-age woman living in primitive conditions at a remote tropical field site without plumbing or running water; the struggle to keep your birth control pills cool and dry (heat and humidity can change the molecular structure and make them less effective); the discovery that you are pregnant and unable to return to civilization for another 3 months. Imagine being a postmenopausal woman dealing with wildly fluctuating mood swings and hot flashes while working in triple-digit temperatures and 98% humidity at a remote site in lowland rain forest and you have run out of batteries for your laboratory and bedroom fans.

Motherhood can restrict fieldwork opportunities and productivity for diverse reasons, from feelings of guilt, concern, and loneliness associated with being away from the children while doing fieldwork without them, to the distractions and time commitments of motherhood while doing fieldwork with the children. Nonetheless, many field biologists who are mothers have been extraordinarily successful, as evidenced by many of the interviewees who have shared their stories in Chapters 7 and 8, women profiled in Section 1, and of course many other women field biologists not mentioned herein. Many factors can influence whether or not to take children into the field, such as age of the child(ren); type, duration, and location of the fieldwork; childcare options; and finances.

Isolation during fieldwork generates potential professional and personal challenges. Distance from mentors and colleagues, unavailability of telephone or email contact, and inability to access the literature can present logistical difficulties. Long periods of time away from a partner and children, friends, and extended family can strain relationships. Time in the field with male colleagues can be extremely rewarding, an opportunity to bond in friendship and support. If unrealistic expectations develop, however, these interactions can evolve into uncomfortable, awkward, or even dangerous situations.

The field as a working environment provides fertile ground for sexual harassment and even sexual violence.[28] Fieldwork often entails spending long periods of time in relative isolation with few other individuals. Personalities can clash, individuals

---

[26] For example, Schneider et al. (2000).

[27] Berdahl and Moore (2006).

[28] e.g., Joyce (2016), Wadman (2017).

can become controlling, and emotions can go haywire. Women scientists working in the field contend with sexual harassment at a much higher frequency than do male scientists. Sexual harassment encompasses a wide range of behaviors, from offensive comments and discrimination, to Louis Leakey-like unwelcome sexual advances or contact, to sexual assault including rape.

A 2020 documentary film entitled "Picture a Scientist," an Uprising Production, [29] highlights the personal experiences of three women scientists and shares current perspectives and efforts to make science more diverse, equitable, and welcoming to everyone. The film uses an iceberg analogy, with the tip of the iceberg being the traditional definition of sexual harassment. In the sciences, and undoubtedly in many other areas of life, most gender issues are less evident. Cumulatively, these experiences can have equally devastating effects on individuals.

A study published in 2014 surveyed 666 field scientists, 77.5% of whom identified as women, from across the life, physical, and social sciences.[30] The study revealed a systemic problem. Sixty-four percent of respondents reported they had personally experienced some form of sexual harassment, and over 20% reported they had personally experienced sexual assault. Women were 3.5 times more likely to report having experienced sexual harassment than men and 5.5 times more likely to have experienced sexual assault. Perpetrators of harassment and assault toward women were primarily their superiors.

The study revealed another problem. Fewer than half of the survey respondents recalled knowing about a Code of Conduct at their field sites. Of the respondents who said they had experienced sexual harassment, only 20% said they were aware of a mechanism to easily report harassment at the time it occurred; only 18% of those who said they had experienced sexual assault were aware of a mechanism to report the incident. For these reasons, few women reported the incidents they experienced at their field sites (67 of 361 who responded in the survey as having been harassed and 36 of 131 who responded in the survey as having been assaulted). Fewer than 20% of those who reported the harassment or assault incident at their field site were satisfied with the outcome. Recall that Jane Goodall and Dian Fossey dealt with Louis Leakey's advances but confronted him directly (Chapters 5 and 6). Women should expect to be able to engage in fieldwork without facing discrimination, unwanted verbal or physical sexual advances, or sexual assault.

A follow-up study with a subset of these respondents revealed that many women felt the "rules" of behavior (what constitutes professional conduct) and the consequences of misconduct at their field sites were ambiguous or absent.[31] Sometimes rules were present but not enforced. A hostile field environment leads to alienation and emotional distress for the victim and can negatively impact careers. Unless we can improve the social environment and reporting mechanisms, we will lose women in field biology. Policies that emphasize safety, inclusivity, and collegiality must be set in place. Everyone, from field personnel to leaders of the field crew, to principal investigators, must know and adhere to the designated rules. Procedures for

---

[29] https://www.pictureascientist.com; accessed 6 May 2021.
[30] Clancy et al. (2014).
[31] Nelson et al. (2017).

reporting violations must become common knowledge, and outcomes for victims must be improved.

Another example: Women in Ocean Science C.I.C. reported that sexual harassment is a widespread problem in marine science, a field in which men still hold most of the power.[32] A month-long online survey was launched in 2020, to which 980 self-identified women responded; all participants were currently working or studying, or had worked or studied, in marine science. An astounding 78% of respondents reported they had experienced sexual harassment in learning or workplace environments, most commonly in the university (45%) or during fieldwork (46%). Incidence of sexual harassment ranged from having a person expose oneself or perform sexual acts on oneself (13 respondents), physical acts of sexual assault or rape (48 respondents), and sexual coercion (50 respondents), to unwanted touching or other physical contact (346 respondents), lustful staring at the victim or the victim's body (386 respondents), and verbal remarks of a sexual nature, including jokes (669 respondents). Thirty-nine percent of respondents who had experienced sexual harassment did not report it; 31% did not know if their learning or working environments had sexual conduct policies in place; and 52% of respondents did not know how to report a sexual harassment incident.

The most common reasons given by respondents for not reporting sexual harassment (they could select multiple reasons) were: (1) management won't act, so there is no point (43%), (2) reporting might harm the witness/victim's career (52%), and (3) the harasser is in a position of power, which might influence the outcome (53%). The reasons are almost certainly typical of what women in other fields of science experience: fear of consequences and little faith in management taking complaints seriously.

Most recently during our writing of this book, a story broke on the sexual harassment and discrimination that has spanned more than a decade at the Smithsonian Tropical Research Institute (STRI) in Barro Colorado Island, Panama.[33] (See Chapter 4 for a discussion about the beginnings of the station.) The story reported in December 2021 interviewed 16 women scientists who described a pattern of sexual misconduct by men at the station—men who had the power to influence their careers. Repercussion from reporting misconduct included the perpetrator withholding data. Two of these women have left academia. STRI has long been considered a premier location for tropical research, but the revelation of pervasive sexual misconduct by powerful men at the station has tarnished its reputation and understandably will make women scientists leery of working there.

Other challenges relate to being the only woman on a field crew or expedition. One is getting permission in the first place. Recall that during the early 1900s Annie Alexander, philanthropist and paleontological collector, was unable to join a field expedition—even one she financed—as the only woman (Chapter 3). It simply wasn't proper. This requirement is no longer dogma, but in some circles being an only woman in the field with men is still frowned upon. If a woman is permitted to be the sole female in an otherwise all-male field crew, she may (even unintentionally)

[32] St. Clair (2021).
[33] Jha (2021).

be expected to assume stereotypically feminine roles such as organizing, planning, cooking, cleaning up camp, and offering emotional support. At the same time, she may feel pressure to avoid standing out as female. Superiors often advise women to wear baggy clothing in the field to avoid being a distraction. Unfortunately, the stigma of "wear tight or revealing clothes, and you're just asking for trouble" is still with us. Why is it a woman's problem to address rather than a male problem?

Field crews composed of all women face yet a different set of potential challenges. Women are more likely to be harassed by men from surrounding communities if the crew is all-female; for this reason, women often feel safer if the field crew is mixed-gender. All-female crews sometimes feel that the local communities do not respect them as scientists. This matters because, for example, all-women field crews feel they are not taken seriously when they react to field vandalism, including stolen equipment, removal of flagging tape, snapped traps, and disturbed quadrats and transects; men command more authority in these circumstances. All-female crews sometimes have trouble gaining access to field sites; add a male, even an undergraduate assistant, and access becomes more likely. Conscious or unconscious biases and stereotypes against women as scientists may be magnified for women in field biology, because for too long field biology belonged to men.

As expressed by many of the women we interviewed, another challenge is the vulnerability/safety issue of working alone in the field as a woman. Any time a woman camps alone, sleeps in her car alone, or runs a transect line through the forest alone, she takes a risk—a risk that men worry less about. Women should not feel ashamed to refuse to take this risk or feel the need to prove themselves.

ETHNIC AND RACIAL MINORITIES IN FIELD BIOLOGY

In addition to the ethical and moral reasons to make science more inclusive, research has documented that gender- and ethnic-diverse research teams and other working groups have greater problem-solving ability than do less diverse groups.[34] Increased diversity yields greater creativity and innovation. And yet after decades of attempts to make science more accessible to ethnic and racial minorities, minorities are still underrepresented in the scientific workforce. The effects of underrepresentation are compounded for people at the intersection of more than one minority identity, such as women of color. To begin to understand the reasons for this underrepresentation, we need to examine more broadly the lack of ethnic and racial diversity in science.

Just as we lose women in STEM fields as they climb the ladders of academic ranks and leadership roles, we lose PEERs (persons excluded due to ethnicity or race).[35] PEERs, including Blacks/African Americans, Latinx/Hispanics, and Indigenous peoples, are actually overrepresented among entering-college students planning to study one of the STEM fields. In 2017, PEERs made up about 30% of the U.S. population; students entering STEM fields were about 34% PEERs. But these students

---

[34] For example, Cooke and Kemeny (2017).

[35] David Asai (2020a, b) has noted that science has been an active participant in the exclusion of certain persons who participate, resulting in the underrepresentation of some ethnic and racial groups. These are PEERs—people excluded *because* of their ethnicity or race. Use of the term PEERs is meaningful because it addresses the *cause* of underrepresentation and not simply the outcome of that exclusion.

were twice as likely to leave STEM fields as their White and Asian American counterparts, typically leaving during the first year of college. By the bachelor's level,
PEERs represented 18% of individuals in STEM, and the percentage dropped to
11% at the PhD level; PEERs comprised only about 6% of tenure-track STEM faculty.[36] In general, these students are not dropping out from lack of interest or lack
of preparation.[37] David Asai, Senior Director for Science Education at the Howard
Hughes Medical Institute, has emphasized that although scientists have done a good
job recruiting PEER students and providing them with research experience and peer
advising, we need to do a better job of providing more inclusive learning environments. He has argued that we need to change the culture of science education so
that students feel they belong.[38] In addition, scientists need to remove the visible and
invisible barriers that prevent or exclude PEER students.

As indicated by many of the women field biologists interviewed (Chapter 8), a
sense of belonging is essential to one's happiness and success in science. The following study, which focused on ecology and evolutionary biology, fields that form
the framework for field biology, is further affirmation of this fact. Ethnic and racial
minorities are underrepresented in graduate programs, as postdoctoral researchers,
and as faculty members in ecology and evolutionary biology (EEB) departments in
the United States.[39] An online survey of over 1,900 monoracial college undergraduates (360 African American, 313 Latinx/Hispanic, 709 White, and 524 Asian/Asian
American) majoring in STEM disciplines attempted to identify factors that might
contribute to this underrepresentation.[40] Recognizing the importance of inclusivity,
the survey included questions concerning a sense of belonging. Compared to their
White counterparts, ethnic and racial minorities reported less prior exposure to the
field of ecology, less comfort in outdoor environments, and a lesser sense of belonging in the fields of EEB. Early life experiences with nature and outdoor activities add
to one's comfort level of being outside and interest in pursuing ecology and evolutionary biology—and thus field biology.

We must acknowledge that discrimination, whether due to explicit or implicit bias,
is another reason why people of color are underrepresented in EEB and therefore
field biology. A 2019 study examined how gender and racial stereotypes influence
perceptions of postdoctoral candidates.[41] Two hundred and fifty-one tenured and
tenure-track professors of biology and physics from eight public research universities
in the United States participated in a survey. Each was asked to read one of eight curricula vitae of a hypothetical doctoral graduate applying for a postdoctoral position
in their field and to rate the applicant for competence, hireability, and likeability. All
aspects of the curriculum vitae were held constant, except for the applicant's name,
which was used to manipulate perceived gender and race: White (Bradley Miller and

---

[36] Asai (2020a, b).
[37] Asai (2020a, b).
[38] Asai (2020a, b).
[39] O'Brien et al. (2020).
[40] O'Brien et al. (2020).
[41] Eaton et al. (2019). However, a major limitation of this study is that for statistical reasons, the authors
were unable to examine the effect of faculty participant race and gender on their ratings of the postdoctoral candidates. The paper does not indicate what percentage of the faculty participants were White.

Claire Miller), Asian (Zhang Wei [David] and Wang Li [Lily]), Black (Jamal Banks and Shanice Banks), and Latinx (José Rodriguez and Maria Rodriguez). A pretest of 20 faculty showed that all accurately indicated the intended race and gender of the name combinations. The biology subfield was evolutionary biology and included 157 faculty participants (65% self-identified as men). Ninety-four physics faculty members participated in the survey (90% self-identified as men).

The survey results are revealing. The physics faculty rated the male candidates as significantly more competent and more hirable than the females. They rated Asian and White candidates as more competent and hirable than Black and Latinx candidates. Although the biology faculty did not rate the male and female applicants significantly differently in competence or hireability, they rated Asian candidates as more competent and hirable than Black candidates and as more hirable than Latinx candidates. Across the departments, faculty rated the women applicants as significantly more likable than the males. Clearly, anti-bias interventions are needed to ensure that women and minorities are evaluated fairly in science. Decisions made at the postdoctoral level will affect gender and racial makeup of faculty. A more diverse faculty will not only increase the comfort level and sense of belonging experienced by undergraduate and graduate students, but it also will reduce obstacles that PEER students face.

All of these reasons help explain the underrepresentation of women of color in fieldwork. But there is more. Fieldwork in a predominantly White male environment can be emotionally and mentally draining for women of color and can create a feeling of isolation. Uncomfortable or dangerous situations encountered during fieldwork range from racial slurs and stalking, to having a gun pulled on them and sexual assault.[42] Being a person of color engaged in fieldwork, especially being a female alone in an unfamiliar area, requires constant awareness of actions to minimize risk and potential danger. Given the compounding difficulties that women of color face while conducting fieldwork, it is imperative to make field biology a safer and more welcoming profession.

This very real concern about safety for people of color while working in the field was illustrated for the world to witness by an incident that occurred on 25 May 2020 in Central Park, New York City. A White woman called the police on Christian Cooper, a Black man who was birdwatching. He asked her to leash her dog, which was running loose in an area where dogs are required to be on-leash. She fallaciously claimed an African American man was threatening her and her dog. The public outrage that ensued brought attention to the discomfort and potential danger resulting from discrimination and racial profiling that people of color sometimes experience when outside enjoying nature. Since then, in response, activities have showcased Black/African American biologists and nature enthusiasts (e.g., Black in Nature; Black Birders' Week, June 2020; Black in Animal Behavior Week, July 2020; Black in National Parks Week, August 2020; Black Mammalogists Week, September 2020; Black in Marine Science, November–December 2020; Black in Entomology Week, February 2021; Black in Natural History Museums, October 2021). Individuals have worked to create opportunities for people of color. For example, Carlee Jackson, a

---

[42] Demery and Pipkin (2021).

graduate student and shark enthusiast at Nova Southeastern University in Florida, co-founded the organization MISS—Minorities in Shark Science for girls and women of color interested in elasmobranch fishes.[43] They raised $25,000 in just a few weeks, with the goal of creating opportunities for minorities interested in sharks.

The incident in Central Park, as well as the aftermath of the killings of George Floyd, Ahmaud Arbery, Breonna Taylor, and others, precipitated some much-needed contemplation and action, highlighting identity-based harassment.[44] For example, the month following the Christian Cooper incident, two women biology graduate students at Cornell University began writing guidelines designed to protect researchers from identity-based harassment that might be experienced during fieldwork.[45] Evolutionary ecologist Michelle Tseng reached out to other people of color (including two women interviewed for this book, Diane Srivastava and Jessica Ware) to compile their strategies for navigating a sometimes unwelcoming academic environment in ecology and evolution.[46]

Although over the past two decades the number of women of color in field biology has increased, the numbers are far from reflecting the demographics of students planning to major in biology. Gender, racial, ethnic, and intersectional issues are being discussed in more open and honest ways than ever before. Now we need to effect real change.

MOVING FORWARD

There are scientifically established biological differences between men and women. For example, we know of no man who has given birth. *On average*, women have smaller bodies, and less muscle mass (therefore less strength) than men, which creates certain physical disparities.[47] There are certain situations, however, where women out-perform men physically. For example, women, *on average*, have more endurance,[48] one trait among many that makes them valued as Type I Hotshots and Smoke Jumpers fighting wildfire.[49] Women, *on average*, excel in social situations and in circumstances demanding empathy.

Neuroanatomical studies examining gender differences in brain regions confirm a neuronal basis for observed contrasts in physical and behavioral tendencies,[50] but these described differences are not so great as to suggest the existence of gender-specific brain types.[51] Studies examining the use of language (a neurologic function) also demonstrate gender differences. In conversation, men *on average* tend to be more assertive and try to dominate (reflecting a desire to be independent) while women *on*

---

[43] Luxor (2020).

[44] Viglione (2020).

[45] Demery and Pipkin (2021).

[46] Tseng et al. (2020).

[47] Miller et al. (1993).

[48] Hunter (2016).

[49] https://www.outsideonline.com/2413566/first-women-hotshots-wildland-firefighters; accessed 23 August 2018; Hollerbach et al. (2017).

[50] Ingalhalikar et al. (2013).

[51] Rippon (2019).

average tend to be more expressive, tentative, and polite (reflecting their desire to be social and cooperative).[52]

These gender differences in communication styles might make women appear subordinate to men. Spurious connections between these conversational styles/goals and social status may have influenced Aristotle's conclusions concerning gender differences (in addition to his belief that women are flawed physically). Recall that he considered women to be immature, imperfect, and deficient and that a woman's place is in the home, controlled by her husband (Chapter 1). He further believed that men's strength lies in "commanding" while women's is in "obeying." Despite these measured differences in what male and female brains (and therefore behaviors) emphasize, equally rigorous studies show no differences between males and females in overall intelligence.[53] There is a large variation in these averages and therefore substantial overlap in the abilities of men and women. In fact, variation from the mean is the hallmark of the women we highlight here, none of whom can be considered average by just about any metric you choose to use.

Despite the ongoing challenges for women in field biology, men and women alike are more attuned to discriminatory issues than ever before. This greater awareness is resulting in efforts to address and resolve problems. For example, we are increasingly recognizing that women scientists' reluctance to promote themselves and their own research has profound consequences on career development and ultimately on personal and professional satisfaction. Universities and other institutions are addressing this issue by providing training courses to help women scientists build self-confidence and promote themselves and their work.

Women are still underrepresented in senior leadership roles in corporations, politics, across academic disciplines, and in many other professions. Academic societies are being recognized as effective avenues for supporting and promoting leadership by women scientists.[54] Governing board positions, including president, treasurer, and other officer positions, are elected. Because the terms are generally no more than a few years, the positions have high turnover, offering abundant opportunities for leadership. Participation in society leadership roles can be an effective way to climb the professional career ladder of visibility and influence.

Sexual harassment in science is being addressed. For example, France Córdova, director of NSF (National Science Foundation), dealt head-on with the issue of NSF-supported male scientists harassing their female students, staff, and colleagues. As of 2018, institutions that accept an NSF grant must notify the foundation of any sexual harassment reports by leading scientists on the grant. Funds can be lost as a penalty for harassment. Individuals can also report incidences of harassment directly to NSF, which then might conduct its own investigation. Again, if warranted, the grant funds will be suspended. The announcement in *The New York Times*[55] of this new requirement included a profound statement: "That the N.S.F.'s new sexual harassment policy

---

[52] Citations in Merchant (2012).

[53] Halpern (2012).

[54] Potvin et al. (2018).

[55] Harmon (2018).

was put in place by a woman who controls a $5 billion research budget captures the bittersweet nature of the #MeToo moment for many scientists."

Leaders of field crews are increasingly being trained in the interpersonal skills of conflict management, negotiation, and resolution to minimize bullying, alienation, and harassment behaviors before they can escalate. If negative behaviors do escalate, field crew leaders then have mechanisms in place to address and defuse the situations. These efforts are critical to ensure that students and other early career scientists do not face the embarrassment, humiliation, and alienation from harassment that can make them leave science.

Women are particularly vulnerable to harassment at scientific meetings. In response, many scientific societies have written a Code of Conduct, to which all participants must comply. For example, the JMIH (Joint Meeting of Ichthyologists and Herpetologists) Code of Conduct, established in 2019, forbids the following categories of conduct at its annual meeting: (1) discrimination on the basis of a protected trait (e.g., age, race, ethnicity, gender, religion, and disability), (2) harassment on the basis of a protected trait, (3) sexual harassment, (4) retaliation, (5) bullying, and (6) other forms of unacceptable behavior (e.g., verbal abuse of a speaker, unwelcome physical contact, and disrespectful or inappropriate disruption of talks or poster sessions). Procedures are in place for reporting an incident and for investigating and resolving alleged prohibited conduct as well as appeals and disciplinary action. The goal is to make scientific meetings more welcoming to everyone. Codes of conduct are also being adopted by institutions and departments.

One of the most critical mechanisms to improve the chances that a woman succeeds in science is the support of at least one mentor or role model who wholeheartedly believes in her.[56] If one of those mentors or role models is a woman, so much the better. Support and guidance from such a person who can encourage an aspiring scientist when she questions her own abilities will instill self-confidence and a sense of belonging. Formal mentoring programs are now widespread, from in-person mentoring experiences to virtual platforms (e.g., Entering Mentoring; Million Women Mentors; MUSE; National Research Mentoring Network; PEER Women in Science Mentoring Program; Science Network Mentor Program; and Women in Science).

Look how far women have come! Twenty-three hundred years ago Aristotle wrote that women were intellectually inferior to men. By the Age of Reason/Age of Enlightenment, from the late seventeenth through the eighteenth centuries, reasoning, the scientific method, and experimentation were promoted to understand the world. But science belonged to men. Rousseau, one of the more famous of the Enlightenment philosophers, believed that women should be subservient to men, and that their primary roles are as wife and mother. According to Rousseau, "participation in science required a certain strength that women simply lack." Women were considered such a distraction to men and so dangerous to academic life that until the late 1800s, male faculty at the University of Oxford and the University of Cambridge were required to be celibate.

---

[56] e.g., Zeldin and Pajares (2000), Usher and Pajares (2008), Young et al. (2013).

In the early years of the United States, private religious-supported schools existed for girls and young women. But not everyone had access to private schools, could afford them, or believed that women should be educated. Until the 1820s in much of the country, women's formal education stopped after elementary school. The first public high schools for girls were opened in 1826 (Boston and New York). Think about it—it has been less than 200 years since girls in the United States routinely received an education past primary school. Once it became accepted that women should be educated, women could finally go to a few selected colleges (e.g., Oberlin College and Wesleyan College) and take science courses.

By the late 1800s in the United States, once women could be educated in science they were "allowed" to work in science, but with conditions. For many, their only role was as a research associate. Women often worked without pay. Many worked in a corner of their husband's lab and received no credit for their work. For many decades, women in the United States accepted these peripheral positions just to be able to do science. By the early 1900s, women were accepted as faculty members and in other professional positions, but tenure, promotion, and equitable pay were often out of reach. There were rare exceptions—women scientists who successfully worked as paid professionals and advanced in their careers in what was still largely a man's world.

Women's acceptance in field biology was slow in coming. During the 1800s and early 1900s, North American field biology involved mostly collecting and describing species of plants and animals to document the continent's biodiversity. The expeditions, many of which were government-sponsored, employed mostly men. A woman's place was in the home or classroom, so most women naturalists during this time were writers, artists, or educators. Not all would be kept out of field biology, however. Some were passionate about travel, illustrating and studying nature, and collecting specimens for scientific study. Katherine Brandegee, Alice Eastwood, Ynés Mexía, Annie Alexander, and others defied gender expectations and showed that women are equal or superior to men when outdoor work calls for physical endurance and emotional perseverance. These pioneers proved to men, and inspired women who would follow in their footsteps, that women field biologists are as capable as men at observing, studying, appreciating, understanding, and communicating nature. They demonstrated that Aristotle and Rousseau were just plain wrong.

REFERENCES

Asai, D. 2020a. Excluded. Inclusive science: Editorial. *Journal of Microbiology & Biology Education* 21:1–2.

Asai, D. 2020b. Race matters. Commentary. *Cell* 181:754–757.

Banchefsky, S., J. Westfall, B. Park, and C. M. Judd. 2016. But you don't look like a scientist!: women scientists with feminine appearance are deemed less likely to be scientists. *Sex Roles* 75:95–109.

Berdahl, J. L., and C. Moore. 2006. Workplace harassment: double jeopardy for minority women. *The Journal of Applied Psychology* 91:426–436. doi: 10.1037/0021-9010.91.2.426; accessed 17 November 2021.

Brescoll, V. L., and E. L. Uhlmann. 2008. Can an angry woman get ahead? *Psychological Science* 19:268–275.

Caldwell, E. F., and S. J. Wilbraham. 2018. Hairdressing in space: depiction of gender in science books for children. *Journal of Science & Popular Culture* 1:101–118. doi: 10.1386/jspc.1.2.101_1.

Cech, E. A., and M. Blair-Loy. 2019. The changing career trajectories of new parents in STEM. *Proceedings of the National Academy of Sciences* 116:4182–4187.

Clancy, K. B. H., R. G. Nelson, J. N. Rutherford, and K. Hinde. 2014. Survey of academic field experiences (SAFE): trainees report harassment and assault. *PLoS One* 9:e102172. doi: 10.1371/journal.pone.0102172.

Cooke, A., and T. Kemeny. 2017. Cities, immigrant diversity, and complex problem solving. *Research Policy* 46:1175–1185. doi: 10.1016/j.respol.2017.05.003; accessed 17 November 2021.

Correll, S. J., S. Benard, and I. Paik. 2007. Getting a job: is there a motherhood penalty? *American Journal of Sociology* 112:1297–1338.

Débarre, F., N. O. Rode, and L. V. Ugelvig. 2018. Gender equity at scientific events. *Evolution Letters* 2018. doi: 10.1002/evl3.49; accessed 17 November 2021.

Demery, A.-J. C., and M. A. Pipkin. 2021. Safe fieldwork strategies for at-risk individuals, their supervisors and institutions. *Nature Ecology & Evolution* 5:5–9. doi: 10.1038/s41559-020-01328-5.

Eaton, A. A., J. F. Saunders, R. K. Jacobson, and K. West. 2019. How gender and race stereotypes impact the advancement of scholars in STEM: professors' biased evaluations of physics and biology post-doctoral candidates. *Sex Roles* 82:127–141. doi: 10.1007/s11199-019-01052-w.

Fazackerley, A. 2020. Women's research plummets during lockdown—but articles from men increase. *The Guardian,* 12 May 2020. https://www.theguardian.com/education/2020/may/12/womens-research-plummets-during-lockdown-but-articles-from-men-increase; accessed 17 November 2021.

Fiske, S. T., A. J. Cuddy, and J. Xu. 1999. (Dis)respecting versus (Dis)liking: status and interdependence predict ambivalent stereotypes of competence and warmth. *Journal of Social Issues* 55:473–489.

Grunspan, D. Z., S. L. Eddy, S. E. Brownell, B. L. Wiggins, A. J. Crowe, and S. M. Goodreau. 2016. Males under-estimate academic performance of their female peers in undergraduate biology classrooms. *PLoS One* 11:e0148405. doi: 10.1371,journal.pone.0148405.

Halpern, D. F. 2012. *Sex differences in cognitive abilities.* 4th edition. New York: Taylor & Francis, Psychology Press.

Harmon, A. 2018. 'Enough is enough': science, too, has a problem with harassment. *The New York Times*, November 19, 2018. https://www.nytimes.com/2018/11/19/science/gender-harassment-science-universities.html; accessed 17 November 2021.

Hollerbach, B., K. Heinrich, W. S. C. Poston, C. K. Haddock, A. Kehler, and S. A. Jahnke. 2017. Current female firefighters perceptions, attitudes, and experiences with injury. *International Fire Service Journal of Leadership and Management* 11:41–48.

Hunter, S. K. 2016. The relevance of sex differences in performance fatigability. *Medicine & Science in Sports & Exercise* 48:2247–2256. doi: 10.1249/MSS.0000000000000928.

Ingalhalikar, M., A. Smith, D. Parker, T. D. Satterthwaite, M. A. Elliott, K. Ruparel, H. Hakonarson, R. E. Gur, R. C. Gur, and R. Verma. 2013. Sex differences in the structural connectome of the human brain. *Proceedings of the National Academy of Sciences of the United States of America* 111:823–828.

Jha, N. 2021. Women scientists described a culture of sexual misconduct at the Smithsonian's Tropical Research Institute. *BuzzFeed News.* Posted December 9, 2021. https://www.buzzfeednews.com/article/nishitajha/smithsonian-tropical-research-institute-metoo; accessed 9 December 2021.

Joyce, K. 2016. Out here, no one can hear you scream. Huffpost, March 17, 2016. https://highline.huffingtonpost.com/articles/en/park-rangers; accessed 2 July 2018.

Kramer, J. 2020. The virus moved female faculty to the brink. Will universities help? *The New York Times*, October 6, 2020. https://nyti.ms/3lhH5im; accessed 17 November 2021.

Lerchenmueller, M. J., O. Sorenson, and A. B. Jena. 2019. Gender differences in how scientists present the importance of their research: observational study. *British Medical Journal*. doi:_10.1136/bmj.l6573.

Luxor, S. 2020. Young shark scientist working to bring more women of color into marine biology. *SunSentinel*, August 13, 2020. https://www.sun-sentinel.com/community/riverside-times/fl-cn-davie-carlee-jackson-shark-science-20200813-ugmjblursvbqnjm-33kiawmm43i-story.html; accessed 17 November 2021.

Malisch, J. L., B. N. Harris, S. M. Sherrer, K. A. Lewis, S. L. Shepherd, et al. 2020. In the wake of COVID-19, academia needs new solutions to ensure gender equity. *Proceedings of the National Academy of Sciences USA* 117:15378–15381. www.pnas.org/cgi/doi/10.1073/pnas.2010636117.

Merchant, K. 2012. How men and women differ: gender differences in communication styles, influence tactics, and leadership styles. CMC Senior Theses, Paper 513. Claremont, CA: Claremont College. Available at http://scholarship.claremont.edu/cmc_theses/513; accessed 2 January 2021.

Miller, A. E., J. D. MacDougall, M. A. Tarnopolsky, and D. G. Sale. 1993. Gender differences in strength and muscle fiber characteristics. *European Journal of Applied Physiology and Occupational Physiology* 66:254–262.

Miller, D. I., K. M. Nolla, A. H. Eagly, and D. H. Uttal. 2018. The development of children's gender-science stereotypes: a meta-analysis of 5 decades of U. S. Draw-A-Scientist studies. *Child Development* 89; accessed 20 March 2018.

Moss-Racusin, C. A., J. F. Dovidio, V. L. Brescoll, M. J. Graham, and J. Handelsman. 2012. Science faculty's subtle gender biases favor male students. *Proceedings of the National Academy of Sciences* 109:16474–16479. www.pnas.org/cgi/doi/10.1073/pnas.1211286109.

Nelson, R. G., J. N. Rutherford, K. Hinde, and K. B. H. Clancy. 2017. Signaling safety: characterizing fieldwork experiences and their implications for career trajectories. *American Anthropologist* 119:710–722. doi: 10.1111/aman.12929.

O'Brien, L. T., H. L. Bart, and D. M. Garcia. 2020. Why are there so few ethnic minorities in ecology and evolutionary biology? Challenges to inclusion and the role of sense of belonging. *Social Psychology of Education* 23:449–477. doi: 10.1007/s11218-019-09538-x.

Potvin, D. A., E. Burdfield-Steel, J. M. Potvin, and S. M. Heap. 2018. Diversity begets diversity: a global perspective on gender equality in scientific society leadership. *PLoS One* 13:e0197280. doi: 10.1371/journal.pone.0197280.

Rippon, G. 2019. *The gendered brain: the new neuroscience that shatters the myth of the female brain*. London: The Bodley Head Limited.

Rudman, L. A. 1998. Self-promotion as a risk factor for women: the costs and benefits of counterstereotypical impression management. *Journal of Personality and Social Psychology* 74:629–645.

Schneider, K. T., R. T. Hitlan, and P. Radhakrishnan. 2000. An examination of the nature and correlates of ethnic harassment experiences in multiple contexts. *Journal of Applied Psychology* 85:3–12.

St. Clair, M. 2021. Sexual harassment in marine science. Women in Ocean Science C.I.C. https://www.womeninoceanscience.com/sexual-harassment; accessed 17 November 2021.

Swim, J. K., and L. J. Sanna. 1996. He's skilled, she's lucky: a meta-analysis of observer's attributions for women's and men's successes and failures. *Personality and Social Psychological Bulletin* 22:507–508.

Terada, Y. 2019. 50 years of children drawing scientists. *edutopia*, May 22, 2019. https://www.edutopia.org/article/50-years-children-drawing-scientists.

Tseng, M., R. W. El-Sabaawi, M. B. Kantar, J. H. Pantel, D. S. Srivastava, and J. L. Ware. 2020. Strategies and support for Black, Indigenous, and people of colour in ecology and evolutionary biology. *Nature Ecology & Evolution* 4:1288–1290. doi: 10.1038/s41559-020-1252-0.

Usher, E. L., and F. Pajares. 2008. Sources of self-efficacy in school: critical review of the literature and future directions. *Review of Educational Research* 78:751–796.

Viglione, G. 2020. Scientists speak up about harassment in field research. *Nature* 585:3. https://media.nature.com/original/magazine-assets/d41586-020-02328-y/d41586-020-02328-y.pdf.

Wadman, M. 2017. Disturbing allegations of sexual harassment in Antarctica leveled at noted scientist. *Science*, AAAS, October 6, 2017. https://www.sciencemag.org/news/2017/10/disturbing-allegations-sexual-harassment-antarctica-leveled-noted-scientist.

Williams, J. C., K. W. Phillips, and E. V. Hall. 2016. Tools for change: boosting the retention of women in the STEM pipeline. *Journal of Research in Gender Studies* 6:11–75.

Wood, S., J. A. Henning, L. Chen, T. McKibben, M. L. Smith, M. Weber, A. Zemenick, and C. J. Ballen. 2020. A scientist like me: demographic analysis of biology textbooks reveals both progress and long-term lags. *Proceedings of the Royal Society B* 287:20200877. doi: 10.1098/rspb.2020.0877.

Young, D. M., L. A. Rudman, H. M. Buettner, and M. C. McLean. 2013. The influence of female role models on women's implicit science cognitions. *Psychology of Women Quarterly* 37:283–292.

Zeldin, A. L., and F. Pajares. 2000. Against the odds: self-efficacy beliefs of women in mathematical, scientific, and technological careers. *American Educational Research Journal* 37:215–246.

Index